AF477133

Sensors Fault Diagnosis Trends and Applications

Sensors Fault Diagnosis Trends and Applications

Editor

Piotr Witczak

MDPI • Basel • Beijing • Wuhan • Barcelona • Belgrade • Manchester • Tokyo • Cluj • Tianjin

Editor
Piotr Witczak
University of Zielona Góra
Poland

Editorial Office
MDPI
St. Alban-Anlage 66
4052 Basel, Switzerland

This is a reprint of articles from the Special Issue published online in the open access journal *Sensors* (ISSN 1424-8220) (available at: https://www.mdpi.com/journal/sensors/special_issues/sensor_fault_diagnosis).

For citation purposes, cite each article independently as indicated on the article page online and as indicated below:

LastName, A.A.; LastName, B.B.; LastName, C.C. Article Title. *Journal Name* **Year**, *Volume Number*, Page Range.

ISBN 978-3-0365-1048-4 (Hbk)
ISBN 978-3-0365-1049-1 (PDF)

Contents

About the Editor

Piotr Witczak (PhD) received his M.Sc. degree in computer science from University of Zielona Gora, Poland in 2012. In 2018, at the same University he received his PhD in the field of robust and fault-tolerant control. His main research interests focus on model predictive control, model based fault diagnosis, neural modelling and control of aerodynamic systemsand embedded systems. Recently he pursues also tropical geometry implementations in computer science and control engineering. Piotr Witczak is longtime practitioner and assistant professor at his Alma Mater.

Editorial

Trends in Sensors Fault Diagnosis

Piotr Witczak

Faculty of Computer, Electrical and Control Engineering, Institute of Control and Computation Engineering, University of Zielona Gora, 65-417 Zielona Gora, Poland; p.witczak@issi.uz.zgora.pl

Citation: Witczak, P. Trends in Sensors Fault Diagnosis. *Sensors* **2021**, *21*, 2224. https://doi.org/10.3390/s21062224

Received: 11 March 2021
Accepted: 17 March 2021
Published: 23 March 2021

Publisher's Note: MDPI stays neutral with regard to jurisdictional claims in published maps and institutional affiliations.

Recently, the automation of processes has been widely demanded. Many industries seek optimization in production time, costs or scrap. One may consider here only a few highly automated branches, such as silicone chips production, waste and water management, and fluid transportation, but due to the recent pandemic, also other benefits in automatic control. However, to fully utilize its potential, reliable and fast fault detection is necessary (either for predictive maintenance or fault-tolerant control). This Special Issue of *Sensors* tries to depict current trends in fault diagnosis in industrial applications. Wide coverage of branches, as well as plurality of recent developments, were the main reasons to compose it. Starting at fault detection for pneumatic valves, through sensors networks, autonomous vehicles (either ground or aerial), water pipelines to much more common gearboxes, roll bearings or brakes. All of those exciting topics are covered within this Special Issue, in all cases bringing new perspective to fault diagnosis applications.

The authors of [1] incorporate the autoregressive neural network with exogenous inputs for behavioral prediction in fault detection and diagnosis of pneumatic valves. IT enables the creation of the signature matrix and the decision tree. Using soft computing methods brings more and more satisfying results as the quality of methods increases in parallel with computational power required in the process. The Kinsman method is also used in [2] for fault detection and fault-tolerant control of magnetic brake system. In this case, however, a neural network is used to tune the state-space model of a system, increasing the quality of more conventional residual-based fault detection mechanism operating with feedback controller. Additionally, [3] incorporates a shallow artificial neural network to scan-chains (using failure feature vectors). The mentioned papers utilize shallow networks, whereas in [4], the authors implement a deep network in form of a stacked autoencoder for the fault diagnosis of roll bearings. Similarly, in [5], the authors present convolutional a deep neural network to classify the potential faults. Some analytical model-based approaches are also in this area of interest. Closer examination of [6], as well as [7], may bring the reader to a conclusion that especially linear models are worth studying. In the case of [6], this is a simple, but powerful linearization around an operating point, whereas in [7], the authors propose a Takagi–Sugeno fuzzy model composed of a number of linear sub-models. While similar in modeling approaches, both papers deal with fault diagnosis in a slightly different way. The authors of the first utilize the well-known Kalman filters and the authors of the latter implement a form of modified Luenberger observer. The above-mentioned approaches share some similarities and thus advantages and disadvantages, but also differ in many ways. One must carefully study both to pick a suitable solution or idea and adapt it for their own advantages.

In constrast to the above mentioned, the authors of [8] focus their efforts in enhancing the quality of frequency-based methods in fault diagnosis. They propose data enhancement for subsequent wavelet transform. The fault is then detected by an SVM classifier. The main idea for fault detection and performance degradation prediction in [9] is also based on the Support Vector Machine. However, the authors proposed a heavily modified algorithm in the form of kernel function hybrid, optimized by krill herd. Here, as in [8], the methods are applied in the frequency domain. The aforementioned SVM is also a part of the fault diagnosis method proposed in [10], where the one-against-one multiclass support vector machine is fed by a signal with noise reduced by using the adaptive noise filtering technique. The noise reduction method is based on the LMS filter with Gaussian reference noise generator. This very interesting technique is applied in the frequency domain.

Finally, [11] compares a magnitude of approaches on fault detection, isolation, identification and even recovery for automotive sensors, such as LIDAR. The authors compare 120 papers from the last 15 years and give a comprehensive insight into the undertaken topic. The authors explicitly classify faults of perception sensors into seven categories, provide exemplary methods to deal with them and review them. Additionally, some sub-classes are considered. One of the interesting outcomes of this survey is a conclusion that, recently, most researchers are focused on environmental-related faults, such as rain, fog or snow. This seems valid and reasonable for elucidating whether weather conditions are most the common source of faults and noises, as they cannot be controlled.

Comparing the proposed approaches, it becomes clear that there is no universal solution for all fault-diagnosis problems; rather, a solution needs to be fitted to the particular case. All mentioned papers clearly show that this is the best approach. Considering either analytical or soft-computing methods, one must first deeply understand an underlying system and pick a most suitable method. The editor has a deep hope that this short Special Issue on sensors fault diagnosis will help many readers choose their path and discover solutions fitted to their needs.

Funding: This research received no external funding.

Acknowledgments: The guest editor of this Special Issue would like to thank all authors who have submitted their manuscripts, the reviewers for their hard work during the review process, referees who have provided very useful and thoughtful feedback and the editors of *Sensors* for their kind help and support. I hope that the readers enjoy reading the articles within this Special Issue and that the published works reflects current trends and will be inspirational for the further development of Fault Diagnosis and its applications.

Conflicts of Interest: The author declares no conflict of interest.

References

1. Andrade, A.; Lopes, K.; Lima, B.; Maitelli, A. Development of a methodology using artificial neural network in the detection and diagnosis of faults for pneumatic control valves. *Sensors* **2021**, *21*, 853. [CrossRef] [PubMed]
2. Patan, K.; Patan, M.; Klimkowicz, K. Sensor fault-tolerant control design for magnetic brake system. *Sensors* **2020**, *20*, 4598. [CrossRef] [PubMed]
3. Lim, H.; Cheong, M.; Kang, S. Scan-chain-fault diagnosis using regressions in cryptographic chips for wireless sensor networks. *Sensors* **2020**, *20*, 4771. [CrossRef] [PubMed]
4. He, J.; Ouyang, M.; Yong, C.; Chen, D.; Guo, J.; Zhou, Y. A Novel Intelligent Fault Diagnosis Method for Rolling Bearing Based on Integrated Weight Strategy Features Learning. *Sensors* **2020**, *20*, 1774. [CrossRef] [PubMed]
5. Yao, Y.; Zhang, S.; Yang, S.; Gui, G. Learning attention representation with a multi-scale cnn for gear fault diagnosis under different working conditions. *Sensors* **2020**, *20*, 1233. [CrossRef] [PubMed]
6. Geng, K.; Chulin, N.A.; Wang, Z. Fault-tolerant model predictive control algorithm for path tracking of autonomous vehicle. *Sensors* **2020**, *20*, 4245. [CrossRef] [PubMed]
7. Pazera, M.; Witczak, M.; Kukurowski, N.; Buciakowski, M. Towards simultaneous actuator and sensor faults estimation for a class of takagi-sugeno fuzzy systems: A twin-rotor system application. *Sensors* **2020**, *20*, 3486. [CrossRef] [PubMed]
8. Luong, T.T.; Kim, J.M. The enhancement of leak detection performance for water pipelines through the renovation of training data. *Sensors* **2020**, *20*, 2542. [CrossRef] [PubMed]
9. Liu, F.; Li, L.; Liu, Y.; Cao, Z.; Yang, H.; Lu, S. HKF-SVR optimized by krill herd algorithm for coaxial bearings performance degradation prediction. *Sensors* **2020**, *20*, 660. [CrossRef] [PubMed]
10. Nguyen, C.D.; Prosvirin, A.; Kim, J.M. A reliable fault diagnosis method for a gearbox system with varying rotational speeds. *Sensors* **2020**, *20*, 3105. [CrossRef] [PubMed]
11. Goelles, T.; Schlager, B.; Muckenhuber, S. Fault detection, isolation, identification and recovery (Fdiir) methods for automotive perception sensors including a detailed literature survey for lidar. *Sensors* **2020**, *20*, 3662. [CrossRef] [PubMed]

Article

Development of a Methodology Using Artificial Neural Network in the Detection and Diagnosis of Faults for Pneumatic Control Valves

Ana Andrade, Kennedy Lopes *, Bernardo Lima and André Maitelli

Laboratory of Petroleum Automation—LAUT, Federal University of Rio Grande do Norte-UFRN, Natal 59078-970, Brazil; acca.engpet@gmail.com (A.A.); bernardo1411@hotmail.com (B.L.); maitelli@dca.ufrn.br (A.M.)
* Correspondence: kenreurison@dca.ufrn.br; Tel.: +55-84-9916-8998

Abstract: To satisfy the market, competition in the industrial sector aims for productivity and safety in industrial plant control systems. The appearance of a fault can compromise the system's proper functioning process. Therefore, Fault Detection and Diagnosis (FDD) methods contribute to avoiding any undesired events, as there are techniques and methods that study the detection, isolation, identification and, consequently, fault diagnosis. In this work, a new methodology that uses faults emulation to obtain parameters similar to the Development and Application of Methods for Diagnosis of Actuators in Industrial Control Systems (DAMADICS) benchmark model will be developed. This methodology uses previous information from tests on sensors with and without faults to detect and classify the situation of the plant and, in the presence of faults, perform the diagnosis through a process of elimination in a hierarchical manner. In this way, the definition of residue signature is used as well as the creation of a decision tree. The whole process is carried out incorporating FDD techniques, through the Non-Linear Auto-Regressive Neural Network Model With Exogenous Inputs (NARX), in the diagnosis of the behavioral prediction of the signals to generate the residual values. Then, it is applied to the construction of the decision tree based on the most significant residue of a certain signal, enabling the process of acquisition and formation of the signature matrix. With the procedures in this article, it is possible to demonstrate a practical and systematic method of how to emulate faults for control valves and the possibility of carrying out an analysis of the data to acquire signatures of the fault behavior. Finally, simulations resulting from the most sensitized variables for the production of residuals that is generated by neural networks are presented, which are used to obtain signatures and isolate the flaws. The process proves to be efficient in computational time and makes it easy to present a fault diagnosis strategy that can be reproduced in other processes.

Keywords: fault detection and diagnosis; artificial neural network; NARX; control valve; decision tree; signature matrix

Citation: Andrade, A.; Lopes, K.; Lima, B.; Maitelli, A. Development of a Methodology Using Artificial Neural Network in the Detection and Diagnosis of Faults for Pneumatic Control Valves. *Sensors* **2021**, *21*, 853. https://doi.org/10.3390/s21030853

Received: 20 October 2020
Accepted: 3 January 2021
Published: 27 January 2021

1. Introduction

Currently, with industries presenting an increasingly competitive profile, companies in the oil, petrochemical and natural gas sector, among others, are beginning to demonstrate a need for improvement in productivity, reliability in system stabilization and safety of their industrial plants.

Although there is a continuous search for increasingly efficient productivity, safely and with quality, the search for improvement in processes can be threatened due to the need for the accelerated use of the equipment that make up an industrial plant, making them increasingly likely to show signs of degradation such as: wear due to incorrect operation or repetitive movement, the principle of corrosion, the appearance of erosion, the accumulation of debris (sedimentation), the discovery of an (obstruction), among others. The appearance of these symptoms can be an indication of the presence of abnormalities in the system and qualify failures that can lead to a permanent interruption in the ability to perform a certain function during operation [1].

Although the controllers are able to meet several types of disturbances, there are some changes, during the process, that they are unable to deal with correctly. These unexpected situations can characterize a fault, which can be defined as being an unallowed deviation from at least a certain characteristic property or process variable of the system [2,3].

In this respect, faults in the process, production and power generation industries, among others, can be classified according to their type as faults of actuators, sensors and plant components or parameters. These possible failures interrupt the action in the control process and they can produce substantial measurement errors or even change the dynamic input/output properties of the system, causing an increase in operating costs and even a degradation in the performance of the industrial plant [4–6].

In the industrial processes, to solve problems, early diagnosis can allow the performance of necessary procedures such as prevention. Thus, fault detection becomes a fundamental measure in the system control process, consequently, the importance of developing reliable methodologies for the diagnosis of the system becomes essential [7]. Therefore, FDD techniques can be used to monitor the functioning of systems, detect anomalies and in the occurrence of faults, be able to classify them. The FDD has been acting in research for many decades with different applications, aiming in the most diverse areas to solve problems through the control of industrial processes which, as presented in [8], the search for well-designed controls results in increased system reliability, to avoid the occurrence of faults. Other applications of FDD techniques can be related to the use of technologies in building constructions as seen in [9], whose work mentions a situation contrary to the physical redundancy process, in which virtual redundancy does not increase the cost and complexity, thus bringing, benefits such as the application of automatic FDD methods which, in turn, allows the comparison of duplicate signals, making it possible to detect divergences from each other. Then, fault-tolerant control can be achieved by duplicating a physical sensor with a virtual one so that the system can continue to function even if it fails. Still, with a wide range of applications, a review study is presented in [10], based on automated methods of detecting and diagnosing faults in heating, ventilation and air conditioning systems [11]. As previously introduced, we know the importance of a stabilized control loop. Then, in [12] the author proposed a method capable of detecting faults in the pneumatic control valve to obtain better mesh stability through a technique that determines the fault in a valve through the analysis of the output vibration data of the valve. We can also include the use of FDD techniques in the approach of neural networks, in [13] in the process of detecting and isolating faults in the control valve, a conventional detector that aimed to correct errors in the fault threshold regions, received as complement a neural network, whose proposed detector that included the network design, resulted in a better performance in the faults detection and isolation stage.

In the field of engineering, a certain relevance was observed in articles that use neural networks to solve Fault Detection and Isolation (FDI) and FDD problems, although the development of research with neural network applications is not something so recent [14–18]. Its use in fault detection has been extensively studied, showing wide applicability for nonlinear systems in the context of recurrent networks, which is why researchers explore its use with purposes for several general predictive finalities, especially when referring to process control [19–23]. Thus, several scholars have noticed neural networks as something promising that would represent the knowledge of the faults and their classification [24–28]. Therefore, as a trend in the faults classification stages, data fusion has been one of the main areas of focus in data analysis, used by some authors for the development of research. Among the several works that approach data fusion as a proposed methodology, some have as main objective the use of a data fusion of multiple sensors applied to the method of detection and classification of faults in distributed physical processes for onboard auxiliary systems, resulting in the generation of heterogeneous data by the dynamics of the interactive process [29]. With the same principle in the use of the data fusion technique, this other work proposes an adaptive multisensor fusion method based on Deep Convolutional Neural Networks (DCNN) for fault diagnosis, the results reported better diagnostic accuracy in relation to the comparative methods in the experiment [30]. In [31], based on detailed data fusion research, a Hybrid Deep Belief Network (HDBN) learning model was developed, capable of integrating data in various ways for the intelligent diagnosis of faults applied to a vehicle drive system. Then, as a proposal, three data fusion methods were developed: data union, data join and data hybrid, in order to improve the overall performance of the proposed model without collecting more data.

Finally, in [32] the author presented a method of fault diagnosis based on Improved Detrended Fluctuation Analysis (IDFA) and multisensor data fusion, with applicability for rolling element bearings, proving the effectiveness of the proposed method through the validation of experimental data.

With an overview of the works already mentioned related to data fusion, it was possible to verify that when combining and analyzing measurements together, an appropriate approach to detect failures of complex systems was observed specifically in order to obtain safer and more efficient processes.

As an example of application of the presented methodology, an initial study of the faults that are presented in the DAMADICS benchmark was used to evaluate different types of faults through tests that do not necessarily need to have defective valves. The procedure is carried out through the emulation of signals, which intend to incorporate in a controlled system, situations that could occur in the wear of the equipment. The signals resulting from these emulations determined a database used in the selection of the most sensitized variables for the production of residual patterns generated by neural networks. Once the plant signals of faults situations are captured, neural networks are trained to learn how the output behaves. Thus obtaining different neural networks trained and expert for each behavior. Then, techniques of the neural network NARX were used to predict the behavior of the signals. The residues are generated by comparing the signal predicted by the NARX network with the real values of the plant, in order to present this information for the FDD methods.

We decided to use this methodology to make the system for detection and diagnosis as autonomous as possible. The main objective was to define a way to analyze the faults without having to adjust several parameters. When we work with artificial neural networks we have the advantage of offering generalizations to the problems that, once trained and understood by the networks, the settings are regulated directly by the synaptic weights and not at the discretion of the system specialists.

For the process of applying the FDD methods, a combination models method is performed. The data-oriented model with the use of the neural network for the generation of residues, fundamental for the faults diagnosis stage, and the other model is based on rules in the patterns classification, that is, the specialist's knowledge, in the case of network, which must be developed to determine a set of rules describing the behavior of the system. Thus, some faults will be described by rules, allowing better identification and precision in the diagnosis. Then, for the FDD process, the decision tree will be built to assist in the diagnosis of the most significant residue for a given signal. A decision is made at each new level of the tree by signing the standards that provide the fault isolation.

The article presents as novelty the application in an iterative way in the search of the faults diagnosis. The procedure is based on tracing the faults through a decision tree that eliminates the possible causes of the faults. In this sense, an innovative method of building this decision tree was introduced, composed of a set of neural networks to detect and isolate the flaws learned from the signals tests. The rest of the article is organized as follows: in Section 2, the proposed method is described in detail. Section 3 presents the case study, Section 4 results and analysis and finally, the conclusions are given in Section 5.

2. Materials and Methods

In this section, we describe the development of the methodology for the detection and diagnosis of failures using neural network techniques.

As a necessary step to substantiate the work, the plant must maintain the same conditions for each type of fault, such as level stability, flow maintenance, pressure control, among others. As a result, the variables characterized as defective may sensitize only some signs of interest. In this way, the signals of interest will be defined as those with the greatest differential of means and variances for each specific fault. The most suitable situation for identifying the behavior of a fault is characterized by the ability that each fault has to produce different behavior in the plant. Thus, an intelligent system can learn to diagnose the fault.

In order to make the understanding of the FDD method more evident, the high-level architecture can be separated into two stages, as described below according to the flowchart of Figure 1.

At first, a study is carried out on the effect of each fault with certain intensities. This preparation is carried out based on the work on DAMADICS and adapting the situations of faults in the emulation of AUTHOMATHIKA. With that in mind, the controllers are adjusted to try to stabilize the tank level (TK-1001) even if faults are occurring. The robustness of the controller guarantees an adequate level for the plant, but for this reason, some sensors perceive an unusual activity when compared to normal situations. These perceptions are affected by the intensity of the fault and its characteristics. Then, all relevant signals from the plant are captured and stored for future analysis. The analyses are performed according to which will be the best neural network capable of producing the best residue (with the least approximation error). This residue is defined as the difference in the time signals from the neural network, which simulates the plant's behavior, with its value stored in times of fault.

After analyzing and extracting the best networks for each fault situation, the development of the decision tree is put into action through the fault isolation procedure. This second process develops the elimination of faults until the moment that the candidates can no longer be eliminated, thus finalizing the diagnosis. The conclusion will not always be unique and the results will have to be analyzed by specialists for a refinement, whose wear is causing this behavior. However, in this methodology, we present an approach on how to reduce the investigation of possible causes of the present behavior.

As seen in Figure 1, the necessary condition to start the whole process was the control of the already static level, which allows the simulation of all faults, from F0 that characterizes the situation without the occurrence of the fault until Fault F19. Thus, the data were acquired and analyzed for the selection of certain variables, such as, for example, inlet and outlet flow rate, actual valve opening, desired valve opening, among others. In this way, assisting the diagnosis, it was possible to analyze all the faults for the process of the next step referring to the construction of the decision tree for the generation of residue, which was analyzed according to the performance of the faults for the construction of the matrix of signatures, based on the evaluation of the residual signal.

The signature matrix was developed according to Figure 2, whose trained neural network must be chosen specifically to determine which network best classified a given fault.

As shown in Figure 2, the signature matrix was developed to relate each neural network as a specialist in observing each type and intensity of fault. For example, a given fault has the most appropriate diagnosis according to the network, which will keep the residue within the defined thresholds. Situations like these demonstrate that the residue from this NARX gets an indication of zero in the signature. Values of $+1$ or -1 occur when the residue is above or below these limits, respectively.

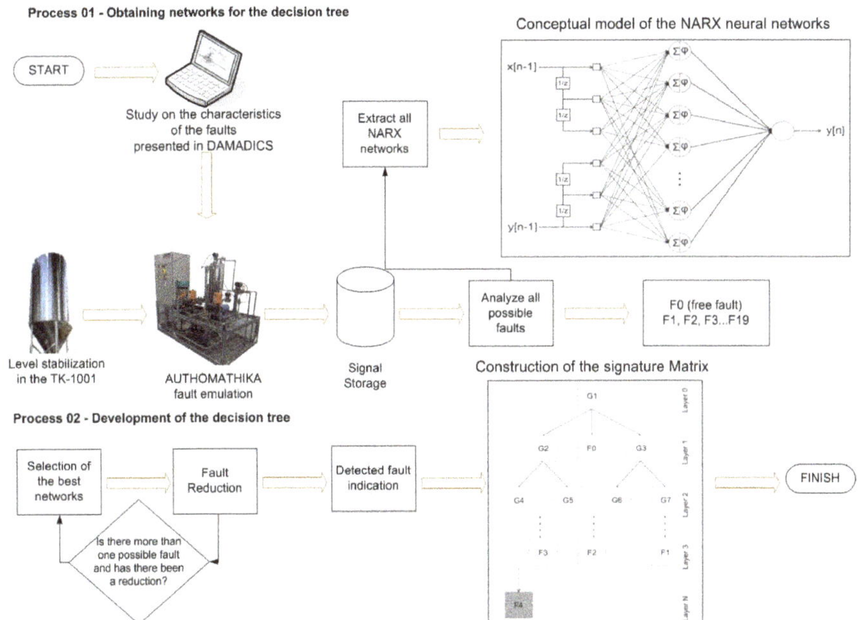

Figure 1. General flowchart of the methodology developed, applied to the study of faults for pneumatic control valve.

Networks are trained to become experts in detecting each type of fault. In the previous analysis of the sensors, different network configurations are trained with all fault situations that may occur in the plant. At the end of this procedure, the neural networks are organized in order of the best and worst network in the classification for each type of failure. In the example shown in Figure 2, 13 neural networks specialized in detecting 13 plant situations are presented: the first indicates the possibility of nonoccurrence of the faults and the rest showing that the faults of (F2 ... F13) may be active.

The presentation of a signature represents a step in the construction of the decision tree. Diagnostics are performed at classification levels by re-evaluating the data. The data will be evaluated until the result indicates only an active neural network or that the active neural networks can no longer be separated to distinguish the fault. In this way, a new structure similar to Figure 2 is used with the same data presented in the previous layer but using the second best diagnostic networks according to the categorization developed in the training process. The process continues until there are no more discarded networks, which are the networks in which the activations obtain a response of 0.

From the interest in the detection and diagnosis of faults, neural network techniques can be applied; in this case, the NARX was used as a viable tool in the training process for analysis in the detection of faults in order to guarantee their time series forecasting effectiveness and consistency.

The NARX networks were chosen because they can incorporate the dynamics of process signals. To make this possible, it is important to use exogenous inputs that are the feedback of your outputs. In this sense, networks were trained to incorporate the dynamics of the signal residues: the difference between the signal presented and an approximation of the specialist NARX network for each fault.

Thus, different configurations of NARX neural networks were designed, shown in Figure 3. As the training data were available, two types of network structures were tested, one in which the inputs y[n−1] are fed back by the output itself to produce the y[n] and another input x[n−1] where the output is produced with the values y[n] approximated by the values made available offline of the emulations of the faults.

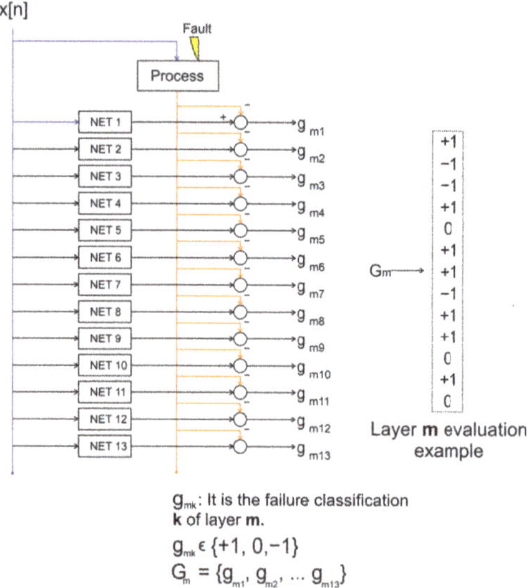

g_{mk}: It is the failure classification **k** of layer **m**.

$g_{mk} \in \{+1, 0, -1\}$

$G_m = \{g_{m1}, g_{m2}, \cdots g_{m13}\}$

Figure 2. Obtaining the fault signature by analyzing the thresholds present in the residues.

The proposal to use the NARX neural network is defined to simulate the behavior of the faults from the perspective of each one. With this, when the process is being carried out, the neural network, together with the actual outputs of the plant, will cause the residue necessary for the definition of the signatures.

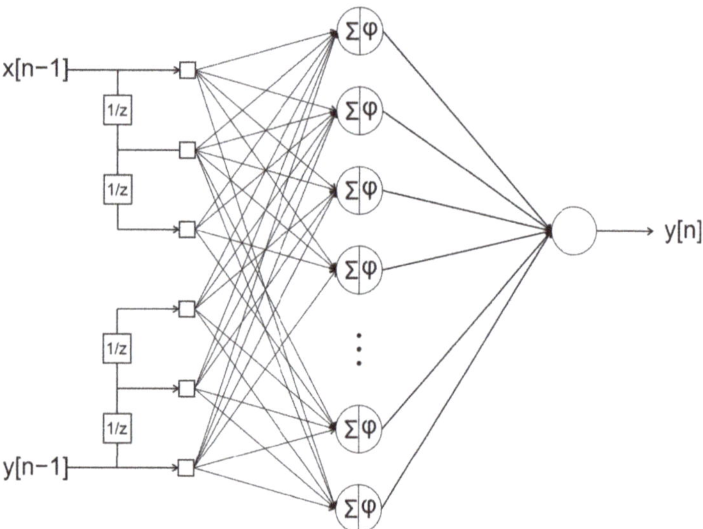

Figure 3. Conceptual model of the NARX neural network.

As part of the proposed methodology, a code was developed that makes it possible to systematically train neural networks for different elaborated structures. Tests and analyses were applied to compare their performances, so that the choice of the best neural network topology to be used for each specific fault with certain predicted intensities, presents some relevant aspects: delays in the feedback output, different amounts of neurons in the single hidden layer and definition of the input-output pattern. Thus, the stage of searching always for the best network in relation to a certain condition allows for the presentation of the following structure:

- Delays at the exit;
- With or without feedback;
- Single hidden layer with varying amounts of neurons;
- Definition of the input-output standard.

For training with all standards that have or do not have faults, applied to each number of neural networks, the best networks should be presented as results for the training, validation and test phases.

In order to ensure that the results of the neural networks are consistent with the defined topology, it is recommended that the training procedure be repeated a few times. The methodology idealized in the search for the best network will be finalized when the fault detection occurs efficiently, deciding for only one fault or when the system is no longer able to distinguish between possible faults.

In the presence of a chain of the best networks to determine each situation, groupings with different patterns can be created that best suit a given network belonging to the first layer. Thus, Artificial Neural Network (ANN) that is not suitable for that circumstance will then be discarded. This process is repeated with the other auxiliary layers using tiebreakers until the moment the fault is isolated.

The best topologies are those whose signal simulation with the neural network has managed to keep the output within the pre-established limits of the appropriate situation, that is, when it is previously established which are the best neural networks, which implies that they are the networks that manage to keep the signal forecast within the limits of what would happen if a fault of that presented pattern occurred.

For the construction of the decision tree in the residual assessment process, the detection step will be essential for the analysis of the magnitude of the residues. In a first analysis, not all residue will have a zero value, nor will a single network indicate this value. However, this pattern composed of all networks, previously known through training, indicates what is happening in the plant.

Even so, there may be doubts about what is happening. As all networks are being analyzed simultaneously, other situations for diagnosis are evaluated. This procedure is presented by an analysis scheme of other layers of clusters. This is like the scheme shown in Figure 4.

We trained 48 neural networks and classified them as to their performance in diagnosing each of the 12 faults as well as the normal situation. Thus, there is a priority on what are the best settings for each type of pattern. If there is any doubt that the network has not adequately isolated a single type, these hypotheses will be analyzed by another layer that will examine another group of NARX networks. It is not necessary to present all the data again for analysis, they must immediately go through all networks and only the sets that indicate the groups are analyzed.

The decision tree has a hierarchical structure made up of branches in order to simplify the process. Signatures contribute to the detection of faults, and their isolation is aided by taking advantage of the tree structure.

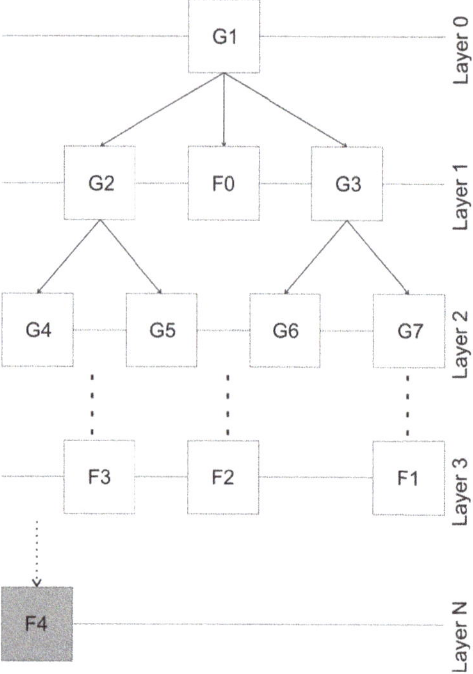

Figure 4. Logical representation of fault signature layers.

3. Case Study

With the development and progress of science and technology, the research and application of methods using FDD attract more and more attention and, thus, many results are obtained. Thus, different approaches have been used when applying FDD techniques in order to solve problems [33].

An example of these applicabilities is the use of techniques to describe different perspectives related to hybrid methods in the detection, identification, behavior and diagnosis of faults from residue, including in the presence of noise as well as in the process of building decision trees [34].

Another categorization regarding its applicability provides a synthesized classification of the FDD methods and is represented in the diagram in Figure 5. However, we can extrapolate the field of applications in other processes and or systems [9].

According to this classification of the FDD methods presented, the present work brought together two of these techniques to obtain a practical procedure for discovering valve faults. While we have an automated data collection process for each fault, using a data-oriented methodology, the classification of these data follows a system of rules characterized by the generation of signatures, more precisely the known patterns of faults that are diagnosed through a vector of characteristics produced as a comparison of the plant with NARX neural networks.

To apply FDD techniques and use ANN for the development of the methodology, it made it possible to make the system autonomous for the detection and diagnosis process, with the main objective being to determine a way to analyze the faults without having to adjust several parameters. Thus, when we work with ANN we have the advantage of offering generalizations to the problems that, once trained and understood by the networks, the settings are regulated directly by the synaptic weights and not at the discretion of the system specialists.

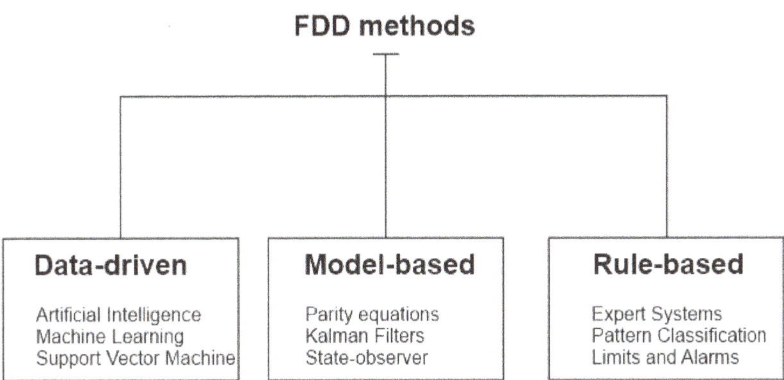

Figure 5. Diagram adapted from the classification of FDD methods [9].

Then, in this section the case study will be presented, which was based on the DAMADICS actuator system, well known in the literature for being a benchmark project, enabling studies related to the fault detection in valves, mainly in the area of engineering in process control. In this work, the set of faults of the DAMADICS benchmark was used as a base model to be emulated and tested in the didactic plant AUTHOMATHIKA PDH-1002, thus making it possible to validate the proposed methodology for the detection and diagnosis of faults using neural networks and, finally, the description of the analysis of the results obtained.

3.1. DAMADICS

The DAMADICS simulator integrates the benchmark model that was developed according to descriptions from a real industrial actuator. This simulator aims to control the flow of water from a boiler that is part of an evaporation station. Its physical composition consists essentially of three elements, as seen in Figure 6: the control valve (V), the pneumatic servomotor (S) and the positioner (P). These three main components are basically composed of a set of physical measurement values: flow sensor measurement (F), pressure sensor at the valve inlet (P1), pressure sensor at the valve outlet (P2), liquid temperature (T1), stem displacement (X) and process control external signal (CV). As for the (V1), (V2) and (V3) valves, they can be activated manually in case of any eventuality, thus, the alarm on the actuator must trigger for the closing of the (V1) and (V2) valves, as well as the manual control by the (V3) valve [35,36].

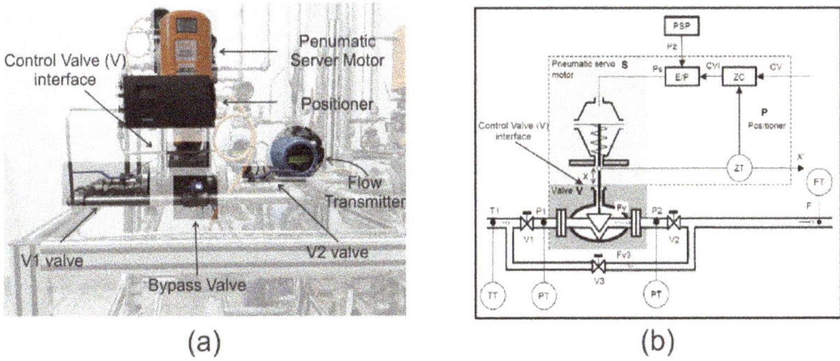

Figure 6. Similarity scheme of actuators: (**a**) AUTHOMATHIKA actuator based on (**b**) actuator model DAMADICS [35].

For the case study, the context of the applicability of the fault simulator used DAMADI CS and the didactic plant AUTHOMATHIKA is not related to their differences but to the conceptual use of the faults of DAMADICS, which were emulated in AUTHOMATHIKA. This process resembles the real situation due to the analysis of signals, since AUTHOMATHI KA is a physical plant of laboratory size, for educational purposes, emphasizing the supervision and instrumentation of systems, as well as the process of identification and control of faults.

In the detection and identification of faults, the composition of the benchmark model makes it possible to simulate a set of nineteen faults (F1, F2, ... F19) of the actuator, classified into four distinct groups: control valve failures (F1 ... F7) ; pneumatic servomotor faults (F8 ... F11), positioner faults (F12 ... F14) and general faults/external faults (F15 ... F19). Thus, the faults previously chosen by the DAMADICS actuator to act in the didactic plant were: obstruction fault (F1), sedimentation fault (F2), and erosion fault (F3) [37,38].

All of these selected faults belong to the fault class in the control valve and are related to the valve internals. They were studied and analyzed as to their type and mode of action, classified as incipient, occurring gradually or abruptly, faster. However, some problems can still result in failures of the type: grabbing, external and internal leaks present in the control valve, as well as faults in the pneumatic actuators, occurred due to certain complications in the connection or connector, springs and even, diaphragm or piston. The valve model shown in Figure 6 is known as a globe valve. The types of faults on that device are already cataloged and once the fault is discovered, it is possible to correct it. As an example, it may happen that the stem moves with difficulty or even remains stationary. In this situation, the probable cause may be insufficient air supply, the positioner is malfunctioning or even the force of the actuator is insufficient. As a corrective action: check if there is an air leak in the actuator or an instrument signal; the positioner must be consulted in its installation, operation and maintenance manual; check the pressure of the actuator supply. A description of another type of fault that can occur in this valve is excessive leakage through the valve seat. This may be due to incorrect shutter adjustment, the shutter may be worn or damaged or the seat is worn or damaged. As a corrective action: reassemble the valve for the correct adjustment of the shutter; disassemble the valve to replace the shutter; disassemble the valve to change the seat.

3.2. AUTHOMATHIKA PDH-1002

The AUTHOMATHIKA plant identified by Figure 7 was developed for use in research projects at the Petroleum Automation Laboratory (LAUT-UFRN). The aim is to facilitate teaching in control and identification of industrial systems, offering various resources and operational facilities. The plant also has several predefined and illustrated meshes in the Processview Supervisory System, assisting the control process.

Analyzing through Figure 8, as an essential step for the development of the work, it was necessary to establish the tank level control (TK-1001) with the setpoint value already set at 30 cm and compare it with the real level value. Through the manipulation of the inlet and outlet flows, the $G_2(s)$ controller acts on the top of the tank, determining through the difference between the intended level and the desired setpoint what should be the flow required to eliminate the level error. The inlet flow of the tank (desirable) must be compared with the outlet flow (intended) obtained by the plant sensors so that it is presented to the $G_1(s)$ controller. The same controller uses the difference of the desired and current flow rates, to determine the appropriate frequency to be modified by the frequency inverter itself, in order to keep the pump level (B-1002A) stabilized.

The most external control, on the other hand, allows for the changing of the level through the $G_3(s)$ controller. This is by means of signals belonging to the levels emitted by the sensors, to the level controller as desired and thus, to describe what will be the intended output flow to manipulate the desired valve opening (LCV-1001) so that the error between the setpoint and the intended level remains close to zero. This whole procedure enabled the didactic plant to be able to work on a static level of the tank, with different failures analyzed under the same conditions. As the outlet flow is dependent on the level and given that the dimensions of the tank are large enough for the outlet flow to be significant, good control was necessary to be robust in order to ensure that the level does not differ from the operating point.

Figure 7. Didactic plant AUTHOMATHIKA PDH-1002.

In the process of modeling the faults for the detection and identification steps, it was necessary to acquire the database of the didactic plant to validate the procedure of the dynamic system that operates presenting situations in a normal (without the occurrence of faults) and abnormal (in the faults). As an initialization step, some signals were evaluated for the selection of certain variables, namely: inlet flow (desirable) from the tank; outflow (intended) from the tank; effective (real) opening of the valve; the desired opening of the valve, and finally, the variable level of the tank which must always operate stably due to the control of the flow system level of the didactic plant, thus, there is no possible destabilization that will interfere in the identification of the detected faults.

In the simulation, to present the normal behavior of the valve, what was emitted in the actuation necessarily occurs in the effective opening. As the valve itself has an internal control and consequently, its dynamics between the actuation signal and the one returned by the positioner, the differences may not always constitute an actuation fault. Then, the fault assessment process (F1, F2 and F3), already presented in the DAMADICS section, presented an alternating reference signal that was sent to the valve to cover the entire opening and closing path. If any defect is present, the valve's behavior should not obey the reference signal but the failed signal. For all faults, the situations were analyzed with different intensity levels: obstruction (10%, 20%, 30% and 40%), sedimentation (90%, 85%, 80% and 75%) and erosion (10%, 15%, 20% and 25%).

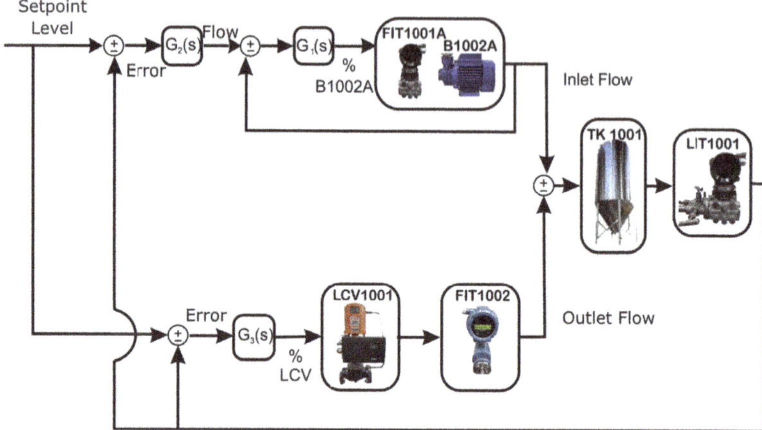

Figure 8. Control grid of the didactic plant.

The F1 Fault emulation was performed by sending the setpoint signal to the closed 0% level in a stochastic manner. Thus, different intensities of 10%, 20%, 30% and 40% were defined as the probability for the occurrence of valve obstruction at each interval of change of reference signal. In situations where the plant has a simulated obstruction, the tank level tends to increase because the control is not designed for that situation. If the level rises indefinitely during the presence of F1, the defective valve's behavior cannot be captured in the same situation when the fault ends.

In the F2 Fault, which is characterized by sedimentation, the open valve maintenance time is increased by the controller to compensate for the deficiency of not being able to open completely. The simulation of this fault occurred by limiting the valve travel superiorly, so that the simulation prevents the valve from reaching any value above the determined limit. Something similar happened in Fault F3, in which erosion prevents the valve from reaching a lower value than stipulated.

In the use of the NARX network for the detection and diagnosis of faults, at the stage of the process of training the networks for different structures, the search for the topology of the best network that fits the given fault and specific intensity was developed with the following structure:

- Several exit delay variables, which in this work were used from 1 to 6;
- With or without feedback;
- A hidden layer with different numbers of neurons was applied: (5, 10, 15 and 20);
- Definition of the input-output standard (three inputs for one output).

The proposal to use the NARX neural network is defined to simulate the behavior of the faults from the perspective of each one of them. Thus, when the process is being carried out, the neural network together with the actual outputs of the plant cause the residue necessary for the definition of the signatures.

4. Results and Analysis

In the presented methodology, the residues that were within the thresholds were used to filter the faults. The data are evaluated by layers, from neural networks, for the production of residues that are labeled with +1, −1 and 0, representing the logical situations for comparing the behavior of the NARX standard with the process output.

The representation of each network can be found by the diagram in Figure 9. The networks are identified by indexes, in which 48 different configurations are defined, trained for all faults and normal situation. As shown in this figure, each neural network was named Training Rule (TR01...TR26), in which the classification of networks specialized in detecting 13 situations present in the plant are exposed.

Indices of the best networks ordered in descending order													
Fault/ Intensity	TR01	TR02	TR03	TR04	TR05	TR06	TR07	TR08	TR09	TR10	TR11	TR12	TR13
No fault	45	46	41	43	40	42	39	29	48	17	44	38	25
Type 01 — 1	40	34	35	42	36	31	48	38	44	28	39	46	30
Type 01 — 2	48	44	32	36	47	39	37	40	28	46	31	35	43
Type 01 — 3	40	42	39	38	27	32	47	33	46	28	26	35	48
Type 01 — 4	35	40	36	43	27	39	28	38	45	47	44	46	30
Type 02 — 5	47	32	48	44	28	40	13	39	35	38	36	31	46
Type 02 — 6	17	10	1	28	31	48	42	44	36	46	38	32	40
Type 02 — 7	5	21	42	40	34	48	35	44	43	38	47	46	28
Type 02 — 8	39	40	47	35	31	44	37	36	38	41	46	45	43
Type 03 — 9	40	48	32	35	47	36	43	46	42	44	30	38	45
Type 03 — 10	35	44	32	28	43	31	39	34	38	47	48	29	42
Type 03 — 11	36	31	44	28	40	39	33	30	45	42	43	35	34
Type 03 — 12	31	38	44	26	46	40	48	36	27	43	39	32	42

Indices of the best networks ordered in descending order													
Fault/ Intensity	TR14	TR15	TR16	TR17	TR18	TR19	TR20	TR21	TR22	TR23	TR24	TR25	TR26
No fault	35	9	15	36	37	31	34	11	27	5	1	32	30
Type 01 — 1	47	43	26	33	29	45	41	27	32	37	25	22	17
Type 01 — 2	27	41	29	42	34	26	30	33	38	45	25	6	5
Type 01 — 3	43	34	45	41	30	31	29	25	37	36	44	5	17
Type 01 — 4	31	34	32	48	42	37	41	33	25	29	26	6	17
Type 02 — 5	43	37	26	34	27	29	45	42	41	1	33	30	2
Type 02 — 6	35	47	43	33	39	34	30	37	27	26	25	41	29
Type 02 — 7	39	1	32	27	36	37	41	45	26	33	29	31	25
Type 02 — 8	42	48	32	27	28	34	33	26	30	3	29	25	21
Type 03 — 9	27	33	41	39	28	37	26	34	29	25	31	4	9
Type 03 — 10	46	27	40	30	45	36	41	25	26	33	37	2	6
Type 03 — 11	38	47	48	46	27	29	32	37	26	41	25	22	17
Type 03 — 12	25	47	33	30	41	35	37	29	45	34	28	5	13

Figure 9. The 26 best networks with the best training performances, in descending order.

The indexes in Figure 9 must be selected from a set of Cartesian Product, on the chosen properties of the neural network architecture. Each number represents an index that names the neural network itself to detect each type of fault with its due intensity. The existing possibilities and the indexes that identify them are shown in Figure 10. Thus, it is still possible to perform other TR from the total possible configurations of NARX networks.

#	Open/Closed Loop	Output delays	Hidden Neurons	#	Open/Closed Loop	Output delays	Hidden Neurons	#	Open/Closed Loop	Output delays	Hidden Neurons
1	Closed	1	5	17	Closed	5	5	33	Open	3	5
2	Closed	1	10	18	Closed	5	10	34	Open	3	10
3	Closed	1	15	19	Closed	5	15	35	Open	3	15
4	Closed	1	20	20	Closed	5	20	36	Open	3	20
5	Closed	2	5	21	Closed	6	5	37	Open	4	5
6	Closed	2	10	22	Closed	6	10	38	Open	4	10
7	Closed	2	15	23	Closed	6	15	39	Open	4	15
8	Closed	2	20	24	Closed	6	20	40	Open	4	20
9	Closed	3	5	25	Open	1	5	41	Open	5	5
10	Closed	3	10	26	Open	1	10	42	Open	5	10
11	Closed	3	15	27	Open	1	15	43	Open	5	15
12	Closed	3	20	28	Open	1	20	44	Open	5	20
13	Closed	4	5	29	Open	2	5	45	Open	6	5
14	Closed	4	10	30	Open	2	10	46	Open	6	10
15	Closed	4	15	31	Open	2	15	47	Open	6	15
16	Closed	4	20	32	Open	2	20	48	Open	6	20

Figure 10. Identification of the networks in Figure 9, in relation to the number of neurons, the hidden layer, delays in the outputs used and whether the network is open or closed loop.

Example of a Failed Signal F1—Multiple Network Layers

The example shown in Figure 11 indicates the production of residue when the valve system is under the condition of the first fault. The black signals represent the average value of the residue in the presented range and, the dotted lines, the acceptance thresholds for that type of fault. The first graph defines the limits of a signal considered normal for the first layer determined by the average of the signal predicted by the NARX network with half a standard deviation upwards and downwards. The other graphs perform the same procedure with the difference that neural networks are specialists in detecting faults. In this sense, the neural network that offers the least residue, already trained for this purpose, is the first neural network of faults.

The signature is acquired by the standard, in which the mean of the residues are located above $+1$, below -1 or between the thresholds 0, which depend on the size of the defined deviation.

The choice of high deviations can cause many false negatives since the faults are of low intensity and would not be captured by such a high variation. Thus, we chose to choose values less than a deviation, even if it detects false positives, as they would be confirmed or denied by the various networks. Thus, a signal can be characterized by more than one type of fault or even a normal situation. Concentrating on the example of Fault F1, for the first layer of networks shown in Figures 11–13, it is observed that a group of networks indicates what the proper analysis of the fault may be. In this way, the indices presented refer to the order in which each network is able to classify each fault. For example: $(\mathrm{NARX}_0[1])$ represents the Layer 0 (zero) NARX which is the specialist in determining Fault 1. That is, if the data captured is from Fault 1 (Type 1, Intensity 1).

Figures 11–13 show the residual values when the same data from the sensors with the same fault are presented to different NARX networks. Each NARX specializes in a different type of fault. Figure 11 shows the residues for specialists in the signal without fault (F0) and those responsible for detecting Faults 1 (F1), 2 (F2) and 3 (F3). Figure 12 for Faults F4, F5, F6 and F7; Figure 13 for F8, F9, F10, F11 and F12. Through the more detailed analysis of the representative graphs in Figures 11–13, we can obtain the signatures by positioning the black continuous line, which represents the average of the residue and the dotted lines that determine the variation around the average value. Depending on the location, the signature indicates the activation of each NARX network. If they have values of 0, they will be marked as a possible diagnosis. Those that are different are immediately discarded. All of this methodology is dependent on the viewing window and they need to always choose the same observation interval to make the diagnosis.

For the second layer of the tree, the analysis is only for residues that had their average values within the limits of acceptance of a possible diagnosis. In this sense, Figure 14 analyzes F0, F1, F3 and F4 and Figure 15 analyzes the possibilities of Faults F6, F9, F10, F11 and F12.

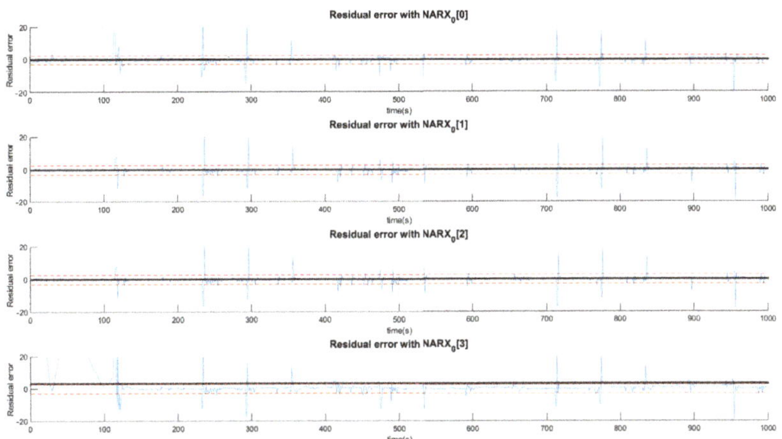

Figure 11. Production of plant residues for the networks expert to detect no presence of faults ($NARX_0[0]$) and for NARX expert to Faults F1, F2 and F3. All of these are produced in the first layer (Layer 0) of the decision tree.

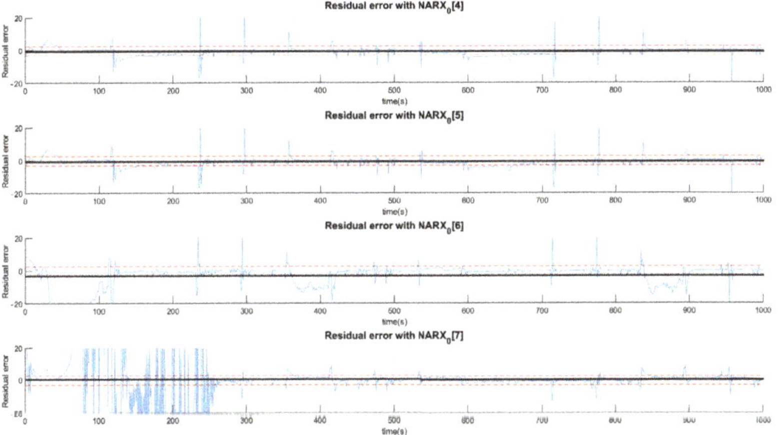

Figure 12. Production of plant residues when NARX networks expert to Faults F4, F5, F6 and F7 are used. All of these are produced in the first layer (Layer 0) of the decision tree.

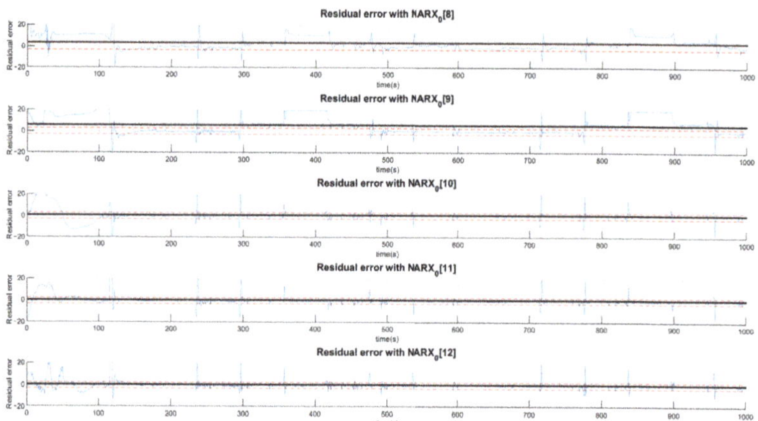

Figure 13. Production of plant residues when NARX networks expert to Faults F8, F9, F10, F11 and F12 are used. All of these are produced in the first layer (Layer 0) of the decision tree.

The G_1 fault set, which contains the faults diagnosed by this pattern is, Equation (1):

$$G_1 = \{0, 1, 3, 4, 6, 9, 10, 11, 12\} \tag{1}$$

What defines the experts for each fault is only the specialized NARX network architecture for each fault. In this situation, the correspondence of the neural network to the first layer can be verified through the first column of the indices shown in Figure 9, with its characteristics shown in Figure 10 and summarized in Table 1. As an example, note that the $(NARX_0[0])$ (specialist in detecting signal without fault) represents the neural network number 45 for the first layer. In turn, network 45 defines an open-loop neural network with six delays at the output and five hidden neurons.

Table 1. Correspondence between the type of failure including the intensity values with the best response for the production of residue.

Fault Type/Intensity	0	1	3	4	5	6	10	11	12
Network	45	48	40	35	17	40	35	36	31

It is not yet possible to adequately isolate the fault using this analysis. Although the F1 Fault belongs to this set, other specialists produce a similar residue that prevents them from being distinguished.

The distinction procedure is continued through a second layer. In this situation, the possibilities of defining other faults that do not belong to G_1, are eliminated, determining that the subscriptions of the networks belonging to the group with the second best Train Rule (TR02) that were in G_1 are analyzed on a second level. The results for this second layer are specified in Figures 14 and 15.

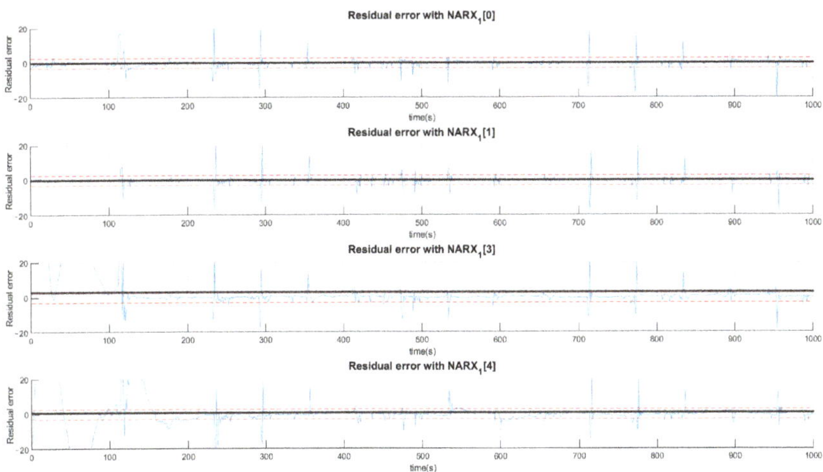

Figure 14. Residual analysis with the second layer of the network excluding those that did not have limits close to zero in the first layer. In this excerpt, residual values are presented for the situations: no faults (NARX$_1$[0]), Faults F1, F3 and F4.

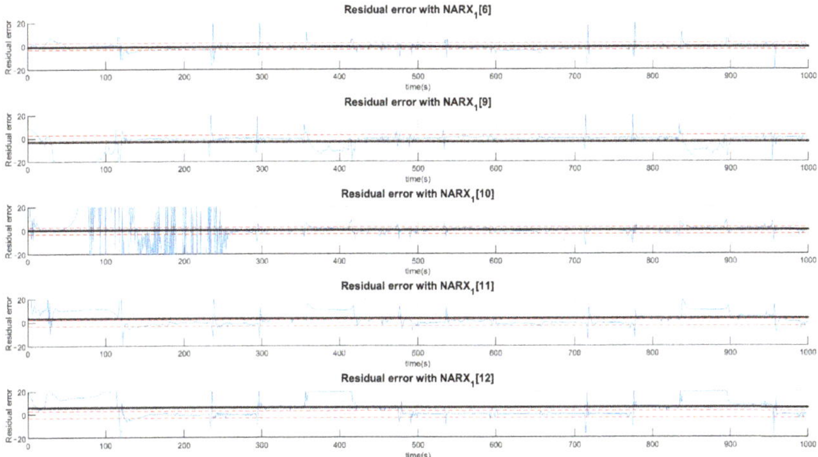

Figure 15. Residual analysis with the second layer of the network excluding those that did not have limits close to zero in the first layer. In this excerpt the residual values for the situations are presented: Faults F6, F9, F10, F11 and F12.

With layer analysis, there is no problem with minor or major residual errors. The important thing is that the residues that do not represent the standard are discarded until it is no longer possible to isolate a fault.

The entire fault isolation process can be described as Figure 16. Initially, the input signals previously obtained from the process are presented to all networks of all layers. The number of layers is a parameter to be chosen. The greater the number of defined layers, the greater the insulating capacity. After this stage, the signature is continuously updated until the moment when it is no longer possible to find a signature with values of -1 or $+1$ (isolated faults) or all faults situations are discarded (inconclusive).

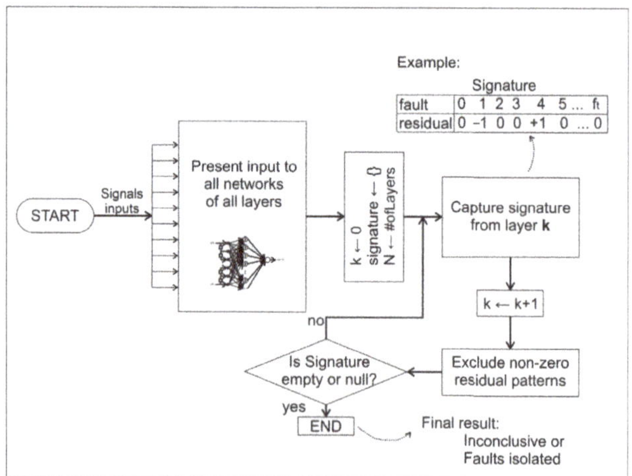

Figure 16. Iterative process of obtaining signatures and fault isolation.

After performing the analysis for Layer 2, only 5 residue patterns fit the model shown in Figure 17. Fault options are limited to patterns 0 (normal situation), 1 (F1 Fault, with Intensity 1), 4 (F1 Fault, with Intensity 4) and 10 (F3 Fault, with Intensity 2). The number 6 (F2 Fault, with Intensity 2) will be discarded for a new layer since it is not within the limits that classify it.

This methodology presents the steps to be developed in the isolation of faults. In the graphs presented, a signal with the F1 Fault was selected, with a length of 1000 s. In this situation, the intensity of the fault is small enough to be confused with the normal situation itself; however, until the analysis of the third layer, the situation of this fault has not been ruled out.

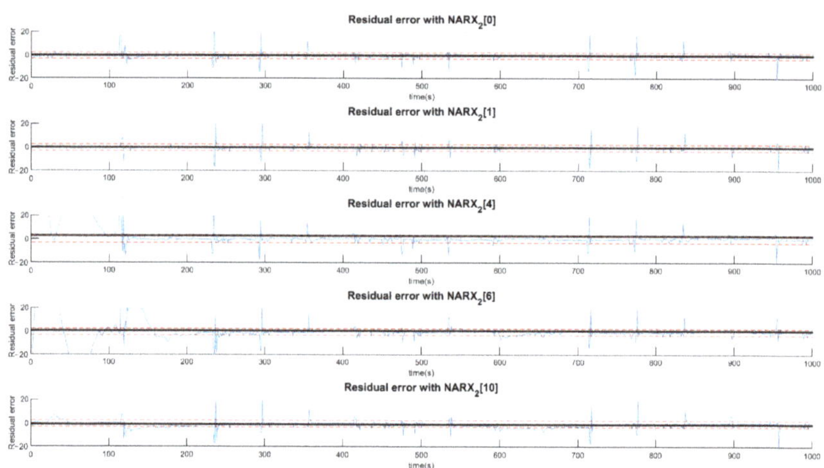

Figure 17. Residual analysis with the third network layer (Layer 2), excluding those that did not have limits close to zero in the second layer.

A complete six-layer analysis of this fault is analyzed for that same interval. The reading of these results can be viewed in the shape of a tree, in which the process of eliminating the patterns is aided by the association of signatures, Figure 18.

In Layer 0 the universe of all situations that may occur from previously emulated faults are present. From this set, below, is the result of the residues of all types represented by a complete signature. It is observed that in 2, 5, 7 and 8, there are behaviors other than zero, which proposes the disposal of these situations. In each layer, the same analysis is performed until there is nothing more to be discarded.

In Layer 5 there is the intended outcome intended in this methodology, which is the greatest possible fault isolation. In the example presented, the fault is difficult to distinguish because it is an example with a low intensity that can be confused both with the normal situation and with all other faults of small intensities in the complete set studied. However, all other situations could be ruled out and, in an intervention to correct the device, the diagnoses considered may be the ones isolated by this methodology.

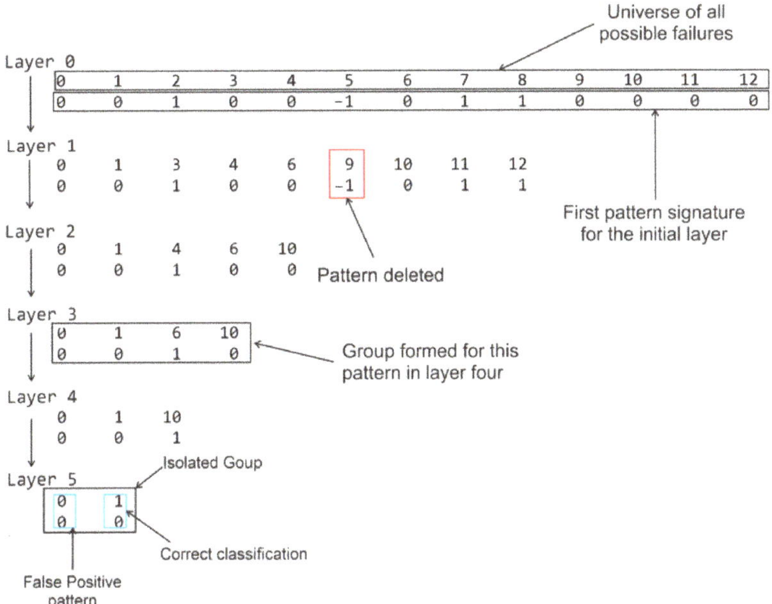

Figure 18. Fault isolation process with the construction of a decision tree for the standard with Fault F1 and Intensity 1.

Regarding the computational effort, the time required to perform this analysis is constant depending on the number of layers, since all data can be presented to all layers in a single step. In addition, when a fault is discarded, the upper layer neural networks that reproduce the residue are turned off and only the cases being evaluated are still under analysis.

5. Conclusions

The present work presented a methodology on how to diagnose fault through the isolation using residual patterns and the construction of a decision tree. This proposal takes advantage of the knowledge acquired in a previous analysis of emulated fault patterns so that in an analysis step they can be considered or discarded until the separation of the standards is irreducible. The practical process of selecting specialists with various neural network architectures that could be trained and classified as to the best diagnosis on the behavior of a device in the industrial process, which in this case was a pneumatic valve, was also exposed. Given the above, the methodology proved to be efficient in achieving, in a simplified way, the restriction of the diagnosis of possible faults that may occur. The methodology has as a limitation a previous study, in offline mode, of device tests with the presence of faults. This procedure is not always allowed for control systems as it is necessary for the plant to interrupt current processes. Future work can explore a set of faults with different natures in processes involving other devices and the equipment operating history to make the methodology more comprehensive. In these works, a performance analysis can be verified using different time windows to obtain signatures and confirmation that the fault isolation is still correct.

Author Contributions: Conceptualization, A.A., K.L. and B.L.; Methodology, A.A., K.L. and B.L.; Software, K.L.; Validation, K.L.; Formal Analysis, A.A., K.L. and B.L.; Investigation, A.A., K.L. and B.L.; Resources, A.A., K.L. and B.L.; Data Curation, A.A., K.L. and B.L.; Writing—Original Draft Preparation, A.A., K.L. and B.L.; Writing—Review and Editing, A.A., K.L. and B.L.; Visualization, A.A., K.L. and B.L.; Supervision, A.A., K.L., B.L. and A.M.; Project Administration, A.A., K.L.; Funding Acquisition, A.M. All authors have read and agreed to the published version of the manuscript.

Funding: This study was financed in part by the Coordenação de Aperfeiçoamento de Pessoal de Nível Superior—Brazil (CAPES)—Finance Code [001].

Institutional Review Board Statement: Not applicable.

Informed Consent Statement: Not applicable.

Acknowledgments: The authors thank the Coordenação de Aperfeiçoamento de Pessoal de Nível Superior (CAPES), the financial support for the research and the support of the Laboratory of Automation in Petroleum (LAUT/UFRN).

Conflicts of Interest: The authors declare no conflict of interest.

References

1. Bezerra, C.G.; Costa, B.S.J.; Guedes, L.A.; Angelov, P.P. An evolving approach to unsupervised and real-time fault detection in industrial processes. *Expert Syst. Appl.* **2016**, *16*, 134–144. [CrossRef]
2. Isermann, R.K.; Ballé, P. Trends in the application of model-based fault detection and diagnosis of technical processes. *Control Eng. Pract.* **1997**, *5*, 709–719. [CrossRef]
3. Van Schrick, D. Remarks on terminology in the field of supervision, fault detection and diagnosis. *IFAC Proc. Vol.* **1997**, *30*, 959–964. [CrossRef]
4. Lunze, J.; Schröder, J. Sensor and actuator fault diagnosis of systems with discrete inputs and outputs. *IEEE Trans. Syst. Man Cybern. Part Cybern.* **2004**, *34*, 1096–1107. [CrossRef] [PubMed]
5. Venkatasubramaniam, V. Process fault detection and diagnosis: Past, present and future. *IFAC Proc. Vol.* **2001**, *34*, 1–13. [CrossRef]
6. Mehranbod, N.; Soroush, M.; Panjapornpon, C. A method of sensor fault detection and identification. *J. Process. Control* **2005**, *15*, 321–339. [CrossRef]
7. Kourd, Y.; Lefebvre, D.; Guersi, N. Neural networks with decision trees for diagnosis issues. *Comput. Sci. Inf. Technol. CS IT* **2013**, *3*, 29–39.
8. de Campos, M.C.M.M.; Teixeira, H.C.G. *Controles tíPicos de Equipamentos e Processos Industriais*; Blucher: São Paulo-SP, Brasil, 2010.
9. Mattera, C.G.; Quevedo, J.; Escobet, T.; Shaker, H.R.; Jradi, M. A method for fault detection and diagnostics in ventilation units using virtual sensors. *Sensors* **2018**, *18*, 3931. [CrossRef]
10. Kim, W.; Katipamula, S. A review of fault detection and diagnostics methods for building systems. *Sci. Technol. Built Environ.* **2018**, *24*, 3–21. [CrossRef]
11. Yu, Y.; Woradechjumroen, D.; Yu, D. A review of fault detection and diagnosis methodologies on air-handling units. *Energy Build.* **2014**, *82*, 550–562. [CrossRef]
12. Venkata, S.K.; Rao, S. Fault detection of a flow control valve using vibration analysis and support vector machine. *Electronics* **2019**, *8*, 1062. [CrossRef]
13. Sánchez, A.J.; Bustos, G.S.; Reyes, B.A.; Guillermo, R.V. Improvements in failure detection of DAMADICS control valve using neural networks. In Proceedings of the IEEE Symposium on Evolving and Autonomous Learning Systems (ETCM), Salinas, Ecuador, 16–20 October 2017; pp. 1–5.
14. Chen, J.; Patton, R.J. *Robust Model-Based Fault Diagnosis for Dynamic Systems*; Kluwer Academic Publishers: Boston, MA, USA, 1999; p. 354, ISBN 0-7923-8411-3.
15. Gertler, J.J. *Fault Detection and Diagnosis in Engineering Systems*; Marcel Dekker: New York, NY, USA, 1998; p. 504, ISBN 0-8247-9427-3.
16. Venkatasubramaniam, V.; Rengaswamy, R.; Yin, K.; Kavuri, S.N. A review of process fault detection and diagnosis: Part I: Quantitative model-based methods. *Comput. Chem. Eng.* **2003**, *27*, 293–311. [CrossRef]

17. Venkatasubramaniam, V.; Rengaswamy, R.; Yin, K.; Kavuri, S.N. A review of process fault detection and diagnosis: Part II: Qualitative models and search strategies. *Comput. Chem. Eng.* **2003**, *27*, 313–326. [CrossRef]
18. Venkatasubramaniam, V.; Rengaswamy, R.; Yin, K.; Kavuri, S.N. A review of process fault detection and diagnosis: Part III: Process history based methods. *Comput. Chem. Eng.* **2003**, *27*, 327–346. [CrossRef]
19. Dreiseitl, S.; Ohno-Machado, L. Logistic regression and artificial neural network classification models: A methodology review. *J. Biomed. Inform.* **2002**, *35*, 352–359. [CrossRef]
20. Kazyssztof, P. *Artificial Neural Networks for the Modelling and Fault Diagnosis of Technical Processes*; Springer: Berlin, Germany, 2008; p. 228.
21. Schmitz, G.P.J.; Aldrich, C.; Gouws, F.S. ANN-DT: An algorithm for extraction of decision trees from artificial neural networks. *IEEE Trans. Neural Netw.* **1999**, *10*, 1392–1401. [CrossRef]
22. Calado, J.M.F.; Korbicz, J.; Patan, K.; Patton, R.J.; da Costa, J.S. Soft computing approaches to fault diagnosis for dynamic systems. *Eur. J. Control.* **2001**, *7*, 248–286. [CrossRef]
23. Korbicz, J.; Kościelny, J.; Kowalczuk, Z.; Cholewa, W. *Robust Model-Based Fault Diagnosis for Dynamic Systems*; Springer: Berlin/Heidelberg, Germany, 2004; ISBN 978-3-642-62199-4.
24. Sivaprakasam, R.; Venkumar, P.; Devaraj, D.; Kavuri, S.P.R. Artificial neural network approach for fault detection in rotary system. *Appl. Soft Comput.* **2008**, *8*, 740–748.
25. Kourd, Y.; Lefebvre, D.; Guersi, N. New technique for online faults diagnosis based on faulty models design: Application to DAMADICS actuator. In Proceedings of the 2012 20th Mediterranean Conference on Control & Automation (MED), Barcelona, Spain, 3–6 July 2012; pp. 297–302.
26. Taqvi, S.A.; Tufa, L.D.; Zabiri, H.; Maulud, A.S.; Uddin, F. Fault detection in distillation column using NARX neural network. *Neural Comput. Appl.* **2020**, *32*, 3503–3519. [CrossRef]
27. Amiruddin, A.A.A.M.; Zabiri, H.; Taqvi, S.A.; Tufa, L.D. Neural network applications in fault diagnosis and detection: An overview of implementations in engineering-related systems. *Neural Comput. Appl.* **2020**, *32*, 447–472. [CrossRef]
28. Jiang, W.; Hu, W.; Xie, C. A new engine fault diagnosis method based on multi-sensor data fusion. *Appl. Sci.* **2017**, *7*, 280. [CrossRef]
29. Sarkar, S.; Sarkar, S.; Virani, N.; Ray, A.; Yasar, M. Sensor fusion for fault detection and classification in distributed physical processes. *Front. Robot.* **2014**, *1*, 16.
30. Jing, L.; Wang, T.; Zhao, M.; Wang, P. An adaptive multi-sensor data fusion method based on deep convolutional neural networks for fault diagnosis of planetary gearbox. *Sensors* **2017**, *17*, 414. [CrossRef] [PubMed]
31. Zhang, T.; Li, Z.; Deng, Z.; Hu, B. Hybrid data fusion DBN for intelligent fault diagnosis of vehicle reducers. *Sensors* **2019**, *19*, 2504. [CrossRef]
32. Song, Q.; Zhao, S.; Wang, M. On the accuracy of fault diagnosis for rolling element bearings using improved DFA and multi-sensor data fusion method. *Sensors* **2020**, *20*, 6465. [CrossRef]
33. Katipamula, S.; Brambley, M.R. Review article: Methods for fault detection, diagnostics, and prognostics for building systems—A review, Part I. *HVAC R Res.* **2005**, *11*, 3–25. [CrossRef]
34. [CrossRef] Gertler, J. Analytical redundancy methods in fault detection and isolation—Survey and synthesis . *IFAC Proc. Vol.* **1991**, *24*, 9–21. [CrossRef]
35. Sundarmahesh, R.; Kannapiran, B. Fault diagnosis of pneumatic valve with DAMADICS simulator using ANN based classifier approach. In Proceedings of the International Conference on Innovations In Intelligent Instrumentation, Optimization And Signal Processing (ICIIIOSP), Coimbatore, Tamil Nadu, India, 1–2 March 2013; pp. 11–17.
36. Kalisch, M.; Przystalka, P.; Timofiejczuk. Application of selected classification schemes for fault diagnosis of actuator systems. In Proceedings of the 2014 Federated Conference on Computer Science and Information Systems, Warsaw, Poland, 7–10 September 2014; pp. 1381–1390.
37. Bartyś, M.; Patton, R.; Syfert, M.; de Las Heras, S.; Quevedo, J. Introduction to the DAMADICS actuator FDI benchmark study. *Control Eng. Pract.* **2006**, *14*, 577–596. [CrossRef]
38. Katunin, A.; Amarowicz, M.; Chrzanowski, P. Faults diagnosis using self-organizing maps: A case study on the DAMADICS benchmark problem. In Proceedings of the Federated Conference on Computer Science and Information Systems (FedCSIS), Lodz, Poland, 13–16 September 2015; pp. 1673–1681.

Article

Fault-Tolerant Model Predictive Control Algorithm for Path Tracking of Autonomous Vehicle

Keke Geng [1,*], **Nikolai Alexandrovich Chulin** [2] **and Ziwei Wang** [1]

[1] School of Mechanical Engineering, Southeast University, Nanjing 211189, China; 213162730@seu.edu.cn
[2] School of Automation Systems, Moscow Bauman State Technical University, Moscow 109807, Russia; nchulin@yandex.ru
* Correspondence: jsgengke@seu.edu.cn; Tel.: +86-18851663852

Received: 29 June 2020; Accepted: 28 July 2020; Published: 30 July 2020

Abstract: The fault detection and isolation are very important for the driving safety of autonomous vehicles. At present, scholars have conducted extensive research on model-based fault detection and isolation algorithms in vehicle systems, but few of them have been applied for path tracking control. This paper determines the conditions for model establishment of a single-track 3-DOF vehicle dynamics model and then performs Taylor expansion for modeling linearization. On the basis of that, a novel fault-tolerant model predictive control algorithm (FTMPC) is proposed for robust path tracking control of autonomous vehicle. First, the linear time-varying model predictive control algorithm for lateral motion control of vehicle is designed by constructing the objective function and considering the front wheel declination and dynamic constraint of tire cornering. Then, the motion state information obtained by multi-sensory perception systems of vision, GPS, and LIDAR is fused by using an improved weighted fusion algorithm based on the output error variance. A novel fault signal detection algorithm based on Kalman filtering and Chi-square detector is also designed in our work. The output of the fault signal detector is a fault detection matrix. Finally, the fault signals are isolated by multiplication of signal matrix, fault detection matrix, and weight matrix in the process of data fusion. The effectiveness of the proposed method is validated with simulation experiment of lane changing path tracking control. The comparative analysis of simulation results shows that the proposed method can achieve the expected fault-tolerant performance and much better path tracking control performance in case of sensor failure.

Keywords: autonomous vehicle; model predictive control; path tracking control; fault detection and isolation

---

1. Introduction

Fault signal detection and isolation, as well as fault-tolerant control systems, are important contents in the research field of autonomous vehicle and prerequisites for ensuring the driving safety in complex traffic scenarios. The failure of autonomous vehicles mainly occurs in the process of sensing information acquisition and motion state transmission. Fault detection and isolation algorithms are widely used in various unmanned systems, such as ground autonomous vehicles [1], underwater robots [2], and autonomous helicopters [3]. For autonomous vehicle systems, fault signal detection and isolation algorithms play an important role in autonomous environment perception, decision making, and motion control. In order to effectively detect and isolate fault signals and perform stable motion control, many schemes and technologies have been proposed, which can be divided into: the nonlinear algorithms [4] and the linear algorithms [5,6]. With the continuous development of autonomous vehicle technology, the types and number of on-board sensors also continue to increase, and the fault sensor signals have become the main reason for vehicle failure [7–9].

Sensor failure detection and isolation based on model information is a commonly used method [10]. Marzat J. and Avram R.C. et al. proposed sensor fault detection and isolation algorithms using control model information [11] and sliding mode observer [12], respectively. The evaluation and performance of model-based fault detection and isolation algorithms always depend on the accuracy of the system model used. Because the vehicle is a highly coupled and complex nonlinear system, it is impossible to obtain some vehicle parameters accurately, and the vehicle system model established always has uncertainty. In order to better detect faults and isolate fault signals at the site, some methods and strategies that are not model-based are proposed. Methods such as fuzzy logic [13,14], neural network [15,16], and Kalman filter [17] are used to estimate uncertain parameters in nonlinear systems. The performance of model-based fault detection and isolation methods rely on accurate linear system modeling. For the nonlinear systems [18], satisfactory results cannot be obtained by using these methods. However, the model-based methods that require less calculation and real-time performance are also better. Therefore, those model-based methods are still widely used in solving real engineering problems.

Recently, model-based fault detection and isolation algorithms for vehicle systems have been extensively studied. In particular, Chamseddine [19] used a sliding mode observer based on a quarter car model to detect sensor failures in vehicle systems. Although fault detection algorithms are robust to interferences, additional sensors such as displacement sensors are still used to replace the commonly used sensor configurations in the commercial vehicles [20]. In References [21,22], the parity check space method, which is typical model-based method, is proposed for fault detection. The applicability and stability of the model-based fault detection and isolation algorithms are usually poor because of the model uncertainty, interference, and sensor noise. In addition, few scholars have studied fault signal detection and isolation algorithms for robust path tracking control autonomous vehicles.

In order to overcome these limitations, this paper proposes a robust fault-tolerant model predictive control algorithm for path tracking of autonomous vehicle. The nonlinear single-track dynamic vehicle model is established as the research object and the linearization is carried out by using Taylor expansion. The model predictive control algorithm is designed for lateral path tracking control of autonomous vehicle. A fault signal detection and isolation algorithm is proposed and its implementation process can be described as follows: first, based on the output error covariance and weighted data fusion method, the optimal motion state information of the autonomous vehicle is obtained. Then, we designed a fault signal detector, composed of a main Kalman filter, three sub-Kalman filters, two state Chi-square detectors, and residuals Chi-square detectors, with fault signal detection matrix as output. Finally, the fault signal matrix is constructed as a diagonal matrix, which will be multiplied by the sensor signal matrix and the weight coefficient matrix to realize fault signal isolation. The single lane changing path tracking control simulation results confirmed the effectiveness of the proposed method in this paper.

The flowchart of the fault-tolerant model predictive control algorithm designed in this paper is shown in Figure 1. Vehicle motion status information can usually be obtained by GPS combined navigation system, visual odometer, and LIDAR SLAM. These sensors and algorithms together form an on-board perception system.

The rest of this paper is organized as follows. In Section 2, the vehicle dynamic model is established and model linearization process is described. In Section 3, the constraints and objective functions are constructed, and the path tracking control algorithm based on model predictive control is designed. In Section 3, the multi-sensor information fusion algorithm, fault signal detection algorithm, and isolation algorithm are described. Section 4 presents experimental verification using lane changing path tracking scenario. Conclusions are given in Section 5.

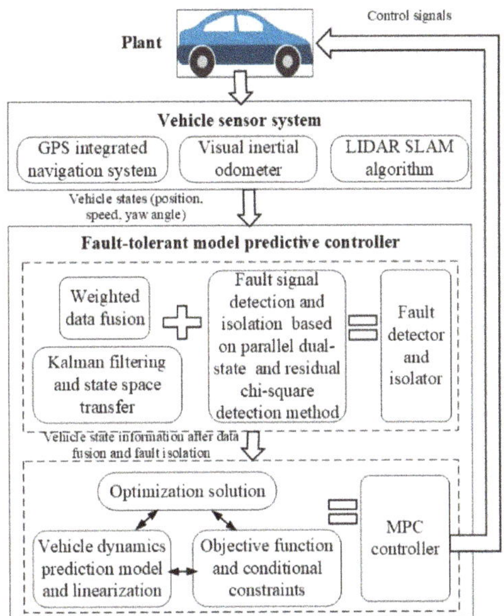

Figure 1. The flowchart of the fault-tolerant model predictive control algorithm.

2. Modeling and Problem Linearization

The impacts of the vehicle suspension characteristics are relatively small in relation to the research content of vehicle motion control. In this work, the vehicle-tire model is selected, which means no in-depth research on the characteristics of vehicle suspension. At the same time, the dynamic model established in this paper is mainly used to design the predictive model in the model predictive controller. It is required to simplify as much as possible on the basis of more accurately describing the vehicle's dynamic characteristics and reducing the amount of calculation. The following idealized assumptions are first proposed when performing dynamic modeling: (1) Ignoring road fluctuations and assuming that the vehicle is always driving on a flat road without vertical motion; (2) Ignoring suspension motion and the effect of the suspension structure on the coupling relationship; (3) The load movement of the front and rear axles is not considered, and the left and right transfer of the load is ignored; (4) Only the tire cornering characteristics are considered, and the vertical and horizontal coupling relationships are ignored; (5) Mechanical effects of steering system are also ignored. In this paper, a 3-degree-of-freedom single-track vehicle dynamics model is constructed, including longitudinal motion, lateral motion, and yaw (see Figure 2).

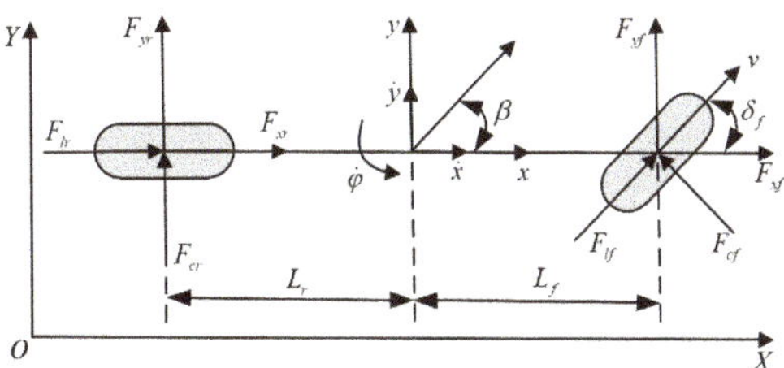

Figure 2. Schematic of the single-track vehicle model.

The longitudinal force, lateral force, and yaw motions of the vehicle can be written as:

$$\begin{cases} m(\ddot{x} - \dot{y}\dot{\varphi}) = \sum F_x = 2F_{lf}\cos\delta_f - 2F_{cf}\sin\delta_f + 2F_{lr} \\ m(\ddot{y} + \dot{x}\dot{\varphi}) = \sum F_y = 2F_{lf}\sin\delta_f - 2F_{cf}\cos\delta_f + 2F_{cr} \\ I_z\ddot{\varphi} = \sum M_z = 2L_f(F_{lf}\sin\delta_f + F_{cf}\cos\delta_f) - 2L_rF_{cr} \end{cases} \tag{1}$$

where m is the vehicle mass; φ is the yaw angle; x and y are the longitudinal and lateral position, respectively; δ_f is the front wheel rotation angle; I_z is the z-axis moment of inertia; F_x is the total longitudinal force on the vehicle; F_y is total lateral force on the vehicle; M_z is the total yaw moment on the vehicle; F_{cf}, F_{cr} are the lateral forces on the front and rear tires of the vehicle, and are related to the corner stiffness and corner angle of vehicle tires; F_{lf}, F_{lr} are longitudinal forces on the front and rear tires of the vehicle, which are related to the longitudinal stiffness and slip rate of the tire; F_{xf}, F_{xr} are the forces on the front and rear tires in the x direction; F_{yf}, F_{yr} are the forces on the front and rear tires in the y direction; L_f and L_r are the distances from the front and rear axis to the center of mass.

According to Equation (1), the vehicle dynamics model involves vehicle tire forces. The longitudinal and lateral forces are related to vertical load, road friction coefficient, slip rate, and tire corner angle:

$$\begin{cases} F_l = f_l(F_z, s, \mu, \alpha) \\ F_c = f_c(F_z, s, \mu, \alpha) \\ s = \begin{cases} rw_t/v - 1, & (v > rw_t, \; v \neq 0) \\ 1 - rw_t/v, & (v < rw_t, \; w_t \neq 0) \end{cases} \\ \alpha = \tan^{-1}(v_c/v_l) \end{cases} \tag{2}$$

where F_z is the vertical load; s is the slip rate; μ is the road surface adhesion coefficient; α is the tire cornering angle; w_t is the wheel speed; r is the wheel radius; v_l is the longitudinal speed; and v_c is the lateral speed, which can be expressed by v_x and v_y:

$$\begin{cases} v_l = v_y\sin\delta_y + v_x\cos\delta_y \\ v_c = v_y\cos\delta_y - v_x\sin\delta_y \end{cases} \tag{3}$$

Generally, the tire speed of a vehicle is difficult to obtain directly, which can be obtained by calculating the vehicle speed:

$$\begin{cases} v_{yf} = \dot{y} + L_f\dot{\varphi} \\ v_{yr} = \dot{y} - L_r\dot{\varphi} \\ v_{xf} = \dot{x} \\ v_{xr} = \dot{x} \end{cases} \tag{4}$$

When constructing the vehicle dynamics model, the front and rear axle load movements have been ignored. Therefore, the vertical load on the front and rear wheels of the vehicle can be calculated as:

$$\begin{cases} F_{zf} = L_rmg/(2L_f + 2L_r) \\ F_{zr} = L_fmg/(2L_f + 2L_r) \end{cases} \tag{5}$$

Generally, a vehicle in a stable driving state has a small variation angle and a slip rate. According to the Semi-Empirical Tire-Model [23], we know that the tire dynamics, including the tire longitudinal force, the tire lateral force, and the tire aligning torque, have obvious nonlinear characteristics, but the simulation results in References [24,25] show that the tire forces can be approximated by a linear equation when the longitudinal slip rate and the tire variation angle change in a small range. In addition, there are a large number of trigonometric functions in the vehicle dynamics model. Since each angle involved in the dynamics model is in a small angle interval, each trigonometric function can satisfy the following approximate conditions: $cos\theta \approx 1$, $sin\theta = 0$, $tan\theta = \theta$. After introducing the corner stiffness, corner angle, longitudinal stiffness, and slip rate, the tire force of the vehicle can be expressed as:

$$
\begin{cases}
F_{lf} = C_{lf}s_f \\
F_{lr} = C_{lr}s_r \\
F_{cf} = C_{cf}\left[\delta_f - \left(\dot{y} + L_f\dot{\varphi}\right)/\dot{x}\right] \\
F_{cr} = C_{cr}\left(L_r\dot{\varphi} - \dot{y}\right)/\dot{x}
\end{cases}
\tag{6}
$$

where C_{cf}, C_{cr} are the lateral stiffness of the front and rear tires; C_{lf}, C_{lr} are the longitudinal stiffness of the front and rear tires; S_f, S_r are the slip ratio of the front and rear tires.

Nonlinear vehicle dynamics model can be written as:

$$
\begin{cases}
\ddot{x} = \dot{y}\dot{\varphi} + (2/m)\left\{C_{lf}s_f - C_{cf}\left[\delta_f - \left(\dot{y} + L_f\dot{\varphi}\right)/\dot{x}\right]\delta_f + C_{lr}s_r\right\} \\
\ddot{y} = -\dot{x}\dot{\varphi} + (2/m)\left\{C_{lf}s_f\delta_f + C_{cf}\left[\delta_f - \left(\dot{y} + L_f\dot{\varphi}\right)/\dot{x}\right] - C_{cr}\left(\dot{y} - L_r\dot{\varphi}\right)/\dot{x}\right\} \\
\dot{\varphi} = \dot{\varphi} \\
\ddot{\varphi} = \left(2L_f/I_z\right)\left\{C_{lf}s_f\delta_f + C_{cf}\left[\delta_f - \left(\dot{y} + L_f\dot{\varphi}\right)/\dot{x}\right]\right\} + \left(2L_r/I_z\right)C_{cr}\left(\dot{y} - L_r\dot{\varphi}\right)/\dot{x} \\
\dot{Y} = \dot{x}\sin\varphi + \dot{y}\cos\varphi \\
\dot{X} = \dot{x}\cos\varphi - \dot{y}\sin\varphi
\end{cases}
\tag{7}
$$

For the convenience, $\xi_{dyn} = \left[\dot{x}, \dot{y}, \varphi, \dot{\varphi}, X, Y\right]^T$ are system state quantities and $u_{dyn} = \delta_f$ is the system control quantity.

Linearization of Vehicle Dynamics Model

For autonomous vehicle, the lateral motion control is to control the front wheel rotation angle, and then realize path tracking. Therefore, this paper selects path tracking as the ultimate goal of autonomous vehicle lateral control, and the tracking accuracy as the main indicator to measure the performance of the control system. Model predictive control can be divided into linear time-varying model predictive control (LMPC) [26] and nonlinear model predictive control (NMPC) [27]. Compared with NMPC, the LMPC uses the linear predictive model and has better real-time performance, which is a very important character for the motion control of autonomous vehicles. Thus, the LMPC is used in this work.

The vehicle model established in this work is a nonlinear model, which needs to be linearized. The state quantity and control quantity of the system satisfy the following relationship:

$$
\dot{\xi}_r = f(\xi_r, u_r)
\tag{8}
$$

Perform Taylor expansion at (ξ_r, u_r), retain the first-order terms, and ignore the higher-order terms, we get:

$$
\dot{\xi} - f(\xi_r, u_r) + \left.\frac{\partial f}{\partial f\xi}\right|_{\substack{\xi = \xi_r \\ u = u_r}} (\xi - \xi_r) \; | \; \left.\frac{\partial f}{\partial u}\right|_{\substack{\xi = \xi_r \\ u = u_r}} (u - u_r)
\tag{9}
$$

The formula can be transformed into:

$$
\dot{\xi} = f(\xi_r, u_r) + J_f(\xi)(\xi - \xi_r) + J_f(u)(u - u_r)
\tag{10}
$$

where $J_f(\xi)$ and $J_f(u)$ are the Jacobian matrixes of f relative to ξ and u, respectively.

Subtract the formula to get:

$$
\dot{\tilde{\xi}} = A(t)\tilde{\xi} + B(t)\tilde{u}
\tag{11}
$$

The linearized system equation can be written as:

$$\dot{\xi} = A(t)\xi(t) + B(t)u(t)$$
$$y = C\xi(t) \tag{12}$$

where $A(t) = \partial f/\partial \xi$, $B(t) = \partial f/\partial u$, $C = (0,0,0,0,1,0)^T$, and:

$$A(t) = \begin{bmatrix} -\dfrac{2C_{cf}\delta_{f,t-1}\left(\dot{y}_t+L_f\dot{\varphi}_t\right)}{m\dot{x}_t^2} & \dot{\varphi}_t + \dfrac{2C_{cf}\delta_{f,t-1}}{m\dot{x}_t} & 0 & \dot{y}_t + \dfrac{2C_{cf}\delta_{f,t-1}L_f}{m\dot{x}_t} & 0 & 0 \\[2mm] \dfrac{2\left[C_{cf}\left(\dot{y}_t+L_f\dot{\varphi}_t\right)-C_{cr}\left(L_r\dot{\varphi}_t-\dot{y}_t\right)\right]}{m\dot{x}_t^2} - \dot{\varphi}_t & -\dfrac{2\left(C_{cf}+C_{cr}\right)}{m\dot{x}_t} & 0 & -\dot{x}_t + \dfrac{2\left(C_{cr}L_r-C_{cf}L_f\right)}{m\dot{x}_t} & 0 & 0 \\[2mm] 0 & 0 & 0 & 1 & 0 & 0 \\[2mm] \dfrac{2\left[C_{cf}L_f\left(\dot{y}_t+L_f\dot{\varphi}_t\right)+C_{cr}L_r\left(L_r\dot{\varphi}_t-\dot{y}_t\right)\right]}{I_z\dot{x}_t^2} & \dfrac{2\left(C_{cr}L_r-C_{cf}L_f\right)}{I_z\dot{x}_t} & 0 & \dfrac{-2\left(C_{cf}L_f^2+C_{cr}L_r^2\right)}{I_z\dot{x}_t} & 0 & 0 \\[2mm] \cos(\varphi_t) & -\sin(\varphi_t) & -\dot{x}_t\sin(\varphi_t)-\dot{y}_t\cos(\varphi_t) & 0 & 0 & 0 \\[2mm] \sin(\varphi_t) & \cos(\varphi_t) & \dot{x}_t\cos(\varphi_t)-\dot{y}_t\sin(\varphi_t) & 0 & 0 & 0 \end{bmatrix} \tag{13}$$

$$B(t) = \begin{bmatrix} -\dfrac{2C_{cf}}{m}\left(2\delta_{f,t-1} - \dfrac{\dot{y}_t+L_f\dot{\varphi}_t}{\dot{x}_t}\right) & \dfrac{2\left(C_{lf}s_f+C_{cf}\right)}{m} & 0 & \dfrac{2L_f\left(C_{lf}s_f+C_{cf}\right)}{m} & 0 & 0 \end{bmatrix} \tag{14}$$

Discrete the above formula using the first-order difference quotient method to obtain the discrete state space equation:

$$\begin{cases} \xi(k+1) = A(k)\xi(k) + B(k)u(k) \\ \zeta(k) = C\xi(k) \end{cases} \tag{15}$$

where $A(k) = I + TA(t)$, $B(k) = I + TB(t)$, and T is sampling time.

After introducing the incremental model, the state-space equation can be written as:

$$\begin{cases} \Delta\xi(k+1) = A(k)\Delta\xi(k) + B(k)\Delta u(k) \\ \zeta(k) = C\Delta\xi(k) \end{cases} \tag{16}$$

3. Model Predictive Controller for Vehicle Lateral Motion Control

3.1. Construct the Objective Function

This paper uses the following objective function:

$$J(k) = \sum_{i=1}^{N_p} \|\Delta\eta(k+i|k)\|_Q^2 + \sum_{i=1}^{N_c-1} \|\Delta u(k+i|k)\|_R^2 + \rho\varepsilon^2 \tag{17}$$

where ρ is the weight coefficient; ε is the relaxation factor; N_p is the prediction time domain and N_c is the control time domain; Q is the state weighting matrix; R is the control weighting matrix; $\Delta\eta(k+i|k)$ is output deviation; and $\Delta u(k+i|k)$ is control deviation.

Since the vehicle dynamic model is used and the number of constraints is increased, in order to avoid the occurrence of no optimal solution, a relaxation factor ε is added to the objective function.

3.2. Construct the Constraints

The most prominent feature of the model predictive control is that it can easily handle the multi-constraint problem. In order to ensure that the reference path can be smoothly tracked, this paper uses the front wheel declination constraint, the front wheel declination incremental constraint, and the tire lateral angle dynamic constraints.

3.2.1. Front Wheel Declination and Its Incremental Constraints

Restrictions on the front wheel deflection angle and front wheel deflection angle of the vehicle can be set according to the actual physical parameters of the vehicle. The control quantity constraint expression is:

$$u_{\min}(k+t) < u(k+t) < u_{\max}(k+t),\ k = 0,1,\cdots,N-1 \tag{18}$$

The expression of the incremental constraints is:

$$\Delta u_{\min}(k+t) < \Delta u(k+t) < \Delta u_{\max}(k+t),\ k = 0,1,\cdots,N-1 \tag{19}$$

In the objective function and constraints, the optimized variable is the control quantity increment in the control time domain. Therefore, the control variable must first be converted into the matrix form of Δu.

The relationship between the control increment and the control quantity can be obtained:

$$\begin{cases} u(t+1) = \Delta u(t) + u(t) \\ u(t+2) = \Delta u(t+1) + \Delta u(t) + u(t) \\ \cdots \\ u(t+N) = \Delta u(t+N-1) + \Delta u(t+N-2) + \cdots + \Delta u(t) + u(t) \end{cases} \tag{20}$$

Convert Equation (18) to $U_{\min} \leq A\Delta U + U_t \leq U_{\max}$, where:

$$A = \begin{bmatrix} 1 & 0 & \cdots & 0 \\ 1 & 1 & \cdots & 0 \\ \vdots & \vdots & & \vdots \\ 1 & 1 & \cdots & 1 \end{bmatrix},\ \Delta U = \begin{bmatrix} \Delta u_{\min}(t+1) \\ \Delta u_{\min}(t+2) \\ \vdots \\ \Delta u_{\min}(t+N) \end{bmatrix}_{N\times 1},\ U_t = \begin{bmatrix} u(t) \\ u(t) \\ \vdots \\ u(t) \end{bmatrix}_{N\times 1} \tag{21}$$

The constraints of control quantity and control increment ensure that the control output generated by the model prediction controller is physically achievable, but for driving safety and comfort, the dynamic constraints of the vehicle also need to be introduced.

3.2.2. Dynamic Constraint of Tire Cornering

The vehicle sideslip due to wet or slippery roads may cause various accidents. Therefore, it is particularly important to increase vehicle dynamics constraints and reduce the possibility of vehicle sideslip.

The sideslip of the vehicle is closely related to the tire slip angle. When the vehicle runs straight on a horizontal road, the tire slip angle $\alpha = 0$; when the tire is elastically deformed by lateral force without lateral slip, $\alpha \leq \alpha_{max}$; when the tire is subjected to excessive lateral force, the vehicle slips, $\alpha > \alpha_{max}$. It can be concluded that the slip angle of the vehicle tire directly reflects whether the vehicle is slipping, and limiting the tire slip angle limits the occurrence of sideslip.

Since the established vehicle dynamics state-space equation does not take the tire slip angle as a state quantity and cannot directly constrain the tire slip angle, this paper needs to find the relationship between the tire slip angle α and the state quantity $\xi(k,t)$. The relationship is to restrain the tire slip angle by imposing a specific relationship constraint on the state quantity.

By Equations (2)–(4), the available tire front and rear wheel angles are:

$$\begin{cases} \alpha_f = \left(\dot{y} + L_f\dot{\varphi}\right)/\dot{x} - \delta_f \\ \alpha_r = \left(\dot{y} - L_r\dot{\varphi}\right)/\dot{x} \end{cases} \tag{22}$$

Using ξ_{dyn} as the state quantity and u_{dyn} as the control quantity, linearize the above formula to obtain:

$$\begin{cases} \alpha = E\xi(k,t) + Fu_{dyn}(k,t) \\ E = \begin{bmatrix} \frac{1}{\dot{x}} & -\frac{\dot{y}+L_f\dot{\varphi}}{\dot{x}^2} & 0 & \frac{L_f}{\dot{x}} & 0 & 0 \\ \frac{1}{\dot{x}} & -\frac{\dot{y}-L_r\dot{\varphi}}{\dot{x}^2} & 0 & -\frac{L_r}{\dot{x}} & 0 & 0 \end{bmatrix} \end{cases} \tag{23}$$

where $\alpha = \left[\alpha_f, \alpha_r\right]^T$ is the tire corner angle matrix, $F = [-1,0]^T$ is the direct transfer matrix, and E is the output matrix.

Based on the above objective function and constraints, the optimization problem of the controller can be described as:

$$\begin{cases} \min_{\Delta U, \varepsilon} \sum\limits_{i=1}^{N_p} \|\Delta\eta(t+i|t)\|_Q^2 + \sum\limits_{i=1}^{N_c-1} \|\Delta u(t+i|t)\|_R^2 + \rho\varepsilon^2 \\ \Delta U_{\min} \leq \Delta U \leq \Delta U_{\max} \\ \Delta U_{\min} \leq A\Delta U + U_f \leq \Delta U_{\max} \\ \alpha_{\min} \leq \Delta\alpha \leq \Delta\alpha_{\max} \\ \varepsilon \geq 0 \end{cases} \tag{24}$$

Solving the above formula can get the incremental sequence of control input in each control time domain:

$$\Delta U^*(k) = [\Delta u^*(k/k), \Delta u^*(k+1/k), \cdots, \Delta u^*(k+N-1/k)] \tag{25}$$

Apply the first element of the incremental sequence to the controller as the actual input increment:

$$u(k/k) = u(k-1) + \Delta u^*(k/k) \tag{26}$$

Repeating the above process, the optimal control input to the front wheel angle can be obtained.

3.2.3. Multi-Sensor Information Data Fusion and Fault Signal Isolation

If multi-sensors are used to measure one vehicle motion parameter, we can fuse the outputs of all sensor system using the weight assignment method [28], which can be written as follows:

$$O = WMI = [w_1, w_2, \cdots, w_n] diag[m_1, m_2, \cdots, m_n][i_1, i_2, \cdots, i_n]^T \tag{27}$$

where O is the result of data fusion; $W = [w_1, w_2, \cdots, w_n]$ is the weight matrix; $I = [i_1, i_2, \cdots, i_n]^T$ is outputs of each sensor; n is the number of sensors; and M is fault detection matrix.

The principle of the method can be written as follows:

$$\begin{cases} w_j = 1/\left[\sigma_j^2 \sum\limits_{i=1}^{n} \left(1/\sigma_i^2\right)\right] \\ \sum\limits_{i=1}^{n} w_j = 1 \end{cases} \tag{28}$$

where σ_i and σ_j are the output error dispersion of the i-th and j-th sensors; $i, j = 1, 2, \ldots n$.

The true values of vehicle motion states cannot be obtained, and the traditional methods determine the average value of the different sensors outputs as the true value. However, these methods are not suitable in our situation, since different sensors in different conditions can give a deliberately low accuracy or even failure; due to this reason, the average value may have a large difference with the true value, which is extremely harmful to vehicle safety.

If at moment k, the sensor j gives the measured value $T_j(k)$, then:

$$\begin{cases} \Delta T_j(k) = T_j(K) - \hat{T}_j(k) \\ \Delta\overline{T}_j = \frac{1}{N}\sum_{k=1}^{N} \Delta T_j(k) \\ \sigma_j(k) = \frac{1}{N}\sum_{k=1}^{N}\left[\Delta T_j(k) - \Delta\overline{T}_j\right]^2 \end{cases} \qquad k = 1, 2, \cdots, N \tag{29}$$

where $\Delta T_j(k)$ is the measurement error of the j-th sensor at time k; $\Delta\overline{T}_j$ is the average value of the j-th sensor at moment k; $\sigma_j(k)$ is the variance of the output error of the j-th sensor at time k; $\hat{T}_j(k)$ is prognostic assessment obtained using the Kalman filter; N is the number measurements from each sensor.

Since the following filtering process includes the isolation of unreliable data sources and error correction, we can approximately consider the estimated information as the true value.

3.3. Fault Signal Detector Design

Figure 3 shows the block diagram of proposed faults signals detector.

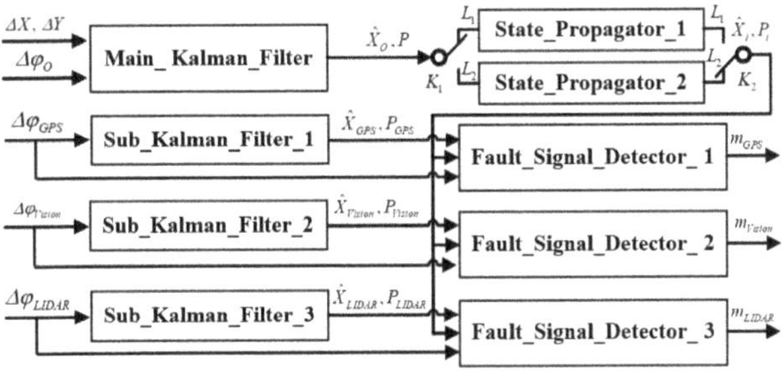

Figure 3. The block diagram of proposed faults signals detector.

Each filter in this detector is standard, we take the Sub_Kalman_Filter_1 for GPS signal channel as an example. The state vector and measurement vector can be written as:

$$\begin{cases} X_{GPS} = \left[\Delta\dot{x}, \Delta\dot{y}, \Delta\varphi, \Delta\dot{\varphi}, \Delta X, \Delta Y\right] \\ Y_{GPS} = \Delta\varphi \end{cases} \tag{30}$$

The state-space and measurement-space equations of Kalman Filter from moment $k-1$ to moment k can be written as:

$$\begin{cases} X_{GPS,k} - F_{GPS,k}X_{GPS,k-1} + w_{GPS,k} \\ Y_{GPS,k} = H_{GPS,k}X_{GPS,k} + v_{GPS,k} \end{cases} \tag{31}$$

where $F_{GPS,k}$ is the state transition matrix, $W_{GPS,k}$ is the process noise, $H_{GPS,k}$ is the measurement transition matrix, and $v_{GPS,k}$ is the measurement noise.

The equations of Kalman Filter can be written as:

$$\begin{cases} \hat{X}_{GPS,k|k-1} = F_{GPS,k}\hat{X}_{GPS,k-1|k-1} \\ P_{GPS,k|k-1} = F_{GPS,k}P_{GPS,k-1|k-1}F_{GPS,k}^{T} + Q_{GPS,k} \\ K_{GPS,k} = P_{GPS,k|k-1}H_{GPS,k}^{T}\left(H_{GPS,k}P_{GPS,k|k-1} + R_{GPS,k}^{T}\right)^{-1} \\ \hat{X}_{GPS,k|k} = \hat{X}_{GPS,k|k-1} + K_{GPS,k}\left(Y_{GPS,k} - H_{GPS,k}\hat{X}_{GPS,k|k-1}\right) \\ P_{GPS,k|k} = \left(I - K_{GPS,k}H_{GPS,k}\right)P_{GPS,k|k-1} \end{cases} \tag{32}$$

where $P_{GPS,k|k}$ is covariance matrix, $K_{GPS,k}$ is the gain matrix, $Q_{GPS,k}$ is the process noise covariance matrix, and $R_{GPS,k}$ is the measurement noise covariance matrix.

The equation of state propagator can be written as:

$$\begin{cases} \hat{X}_{i,k|k-1} = F_{i,k}\hat{X}_{i,k-1|k-1} \\ P_{i,k|k-1} = F_{i,k}P_{i,k-1|k-1}F_{i,k}^T + Q_{i,k} \end{cases}, \quad i = 1,2 \tag{33}$$

The χ^2 test method is widely used to detect faults in stochastic dynamic systems based on correspondence between the observed and reference signals [29]. This method can be divided into three types: χ^2 test for residual error; χ^2 test for state with a single state propagator; χ^2 test for state with double state propagators. These methods have their own advantages and disadvantages:

(1) If the test statistics are calculated using the residual error, it is almost impossible to detect the fault in the state transfer process, although the fault of the sensors can be easily detected;

(2) If the test statistics are calculated using the state vector, it is possible to detect the fault in the state transfer process and to evaluate the fault of the sensor indirectly. If only one state propagator is used for correction of the state prediction error, the error accumulates and may diverge with time increases;

(3) If two state propagators are used and alternately reset the outputs, then accumulation of errors can be avoided. However, for this method, if M sub-filters are used, then $2M$ state propagators should be used, which not only complicates the structure of the algorithm but also affects the speed of calculations.

It can be seen that these methods have their advantages and disadvantages. In this work, a novel robust fault detector is proposed with a structure that simultaneously implements a χ^2 test for residual error and a χ^2 test for state. Double state propagators are used only for the main Kalman filter (see Figure 3), which allows to simplify the structure of the algorithm and increase the speed of calculations. As an example, Figure 4 shows the diagram of the Fault_Dignal_Detector_1.

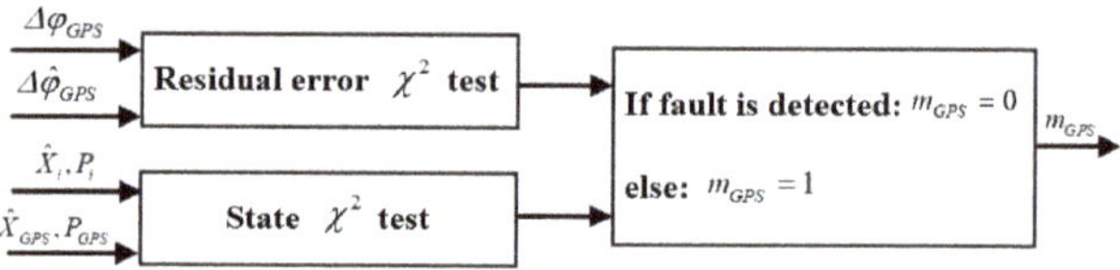

Figure 4. The diagram of the Fault_Signal_Detector_1.

As long as one test channel detected a fault, it can be considered as fault signal. In addition, since the state vectors of all filters are the same, we use only two state propagators. Derive the state χ^2 test with two state propagators and define error state vectors as:

$$\begin{cases} e_{i,k} = X_k - \hat{X}_{i,k} \\ e_{GPS,k} = X_k - \hat{X}_{GPS,k} \end{cases} \tag{34}$$

where X_k is true state vector, $\hat{X}_{GPS,k}$ is estimation from Sub_Kalman_Filter_1 for GPS signal channel, and $\hat{X}_{i,k}$ is estimation error from state propagator i. In this paper, we consider the following variable to detect fault signal from Sub_Kalman_Filter_1 for GPS signal channel:

$$\beta_{GPS,k} = e_{i,k} - e_{GPS,k} = \hat{X}_{GPS,k} - \hat{X}_{i,k} \tag{35}$$

The variance of this variable can be written as:

$$T_{GPS,k} = E\left\{\beta_{GPS,k}\beta_{GPS,k}^T\right\} = E\left\{e_{GPS,k}e_{GPS,k}^T - e_{GPS,k}e_{i,k}^T - e_{i,k}e_{GPS,k}^T + e_{i,k}e_{i,k}^T\right\} = P_{GPS,k} - P_{GPS\_i,k} - P_{i\_GPS,k} + P_{i,k} \tag{36}$$

where $P_{GPS\_i,k}$ and $P_{i\_GPS,k}$ is the cross-covariance.

We set the same initial conditions for the Sub_Kalman_Filter_1 for GPS signal channel and the state propagator i, then we can obtain $P_{GPS\_i,k} = P_{i\_GPS,k} = P_{i,k}$, therefore, the variance can be written as:

$$T_{GPS,k} = P_{GPS,k} - P_{i,k} \tag{37}$$

Define the fault detection function:

$$\lambda_{GPS,k} = \beta_{GPS}^T T_{GPS,k}^{-1} \beta_{GPS} \tag{38}$$

The fault decision criteria can be written as:

$$\begin{cases} \lambda_{GPS,k} \geq \varepsilon_\beta, \ fault \\ \lambda_{GPS,k} < \varepsilon_\beta, \ no \ faul \end{cases} \tag{39}$$

where the threshold ε_β is determined by the function of false alarm rate based on statistical results.

In this paper, we use the state χ^2 test with two state propagators. The working principle can be described as follows: during period t_k a fault occurs and the switch K_1 is located at position " L_1", switch K_2 is located at position " L_2", the output of State_Propagator_1 is uncorrected due to this fault, but the output of State_Propagator_2 is obtained using the previous correct state, which can be used for fault correction. During period t_{k+1}, errors in State_Propagator_1 are corrected using the outputs of Kalman filter. After a time period Δt, the K_1 switch is located at position "L_2", the switch K_2 is located at position "L_1", and the State_Propagator_2 is used to correct the fault.

Test χ^2 residual error of Sub_Kalman_Filter_1 can be written as:

$$d_{GPS,k} = \Delta\varphi_{GPS,k} - \Delta\hat{\varphi}_{i,k} \tag{40}$$

Covariance residual error:

$$S_{GPS,k} = H_{GPS,k}P_{GPS,k}H_{GPS,k}^T + R_{GPS,k} \tag{41}$$

Define the fault detection function:

$$\gamma_{GPS,k} = d_{GPS,k}^T S_{GPS,k}^{-1} d_{GPS,k} \tag{42}$$

Fault decision criteria can be written as:

$$\begin{cases} \gamma_{GPS,k} \geq \varepsilon_d, \ fault \\ \gamma_{GPS,k} < \varepsilon_d, \ no \ fault \end{cases} \tag{43}$$

where the threshold ε_d is determined by the function of false alarm rate based on statistical results.

If the state χ^2 test or the residual error χ^2 test detected a fault, then $m_g = 0$, otherwise $m_g = 1$.

4. Simulation Experiment Verification and Discussion

4.1. Working Conditions Description

In order to verify the feasibility and effectiveness of the proposed method, the Driving Scenario Designer was used to build a simulated driving environment with two straight lanes, and the reference path and yaw angle were collected, as shown in Figure 5a. The system model was built in the Matlab/Simulink environment.

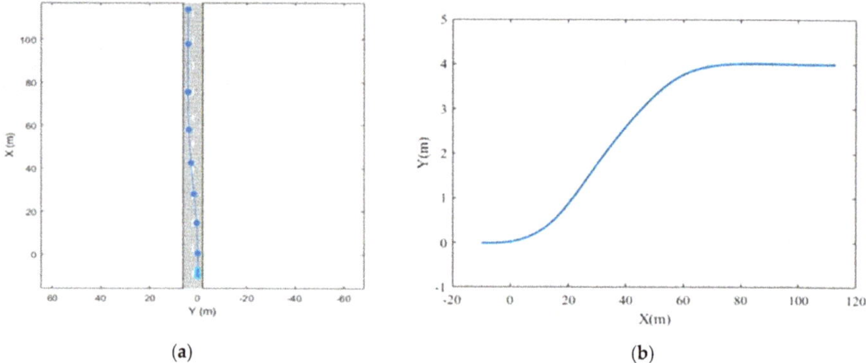

(a)   (b)

Figure 5. Driving scenario and reference path: (**a**) driving scenario; (**b**) reference path.

In Figure 5a, the simulation vehicle starts from the right lane and then changes lanes to the left. From Figure 5b, we can observe that the simulation vehicle is initially located at $x = -10, y = 0$, and then changes lanes at $x = 0, y = 0$. The lane changing process is completed at $x = 80$ m, and the position of the center of mass on the Y-axis reaches $y = 4$ m.

Figure 6a shows the values of vehicle yaw angle obtained by different sensors in the simulation process. Figure 6b shows the reference yaw angle of the vehicle and the yaw angle after data fusion.

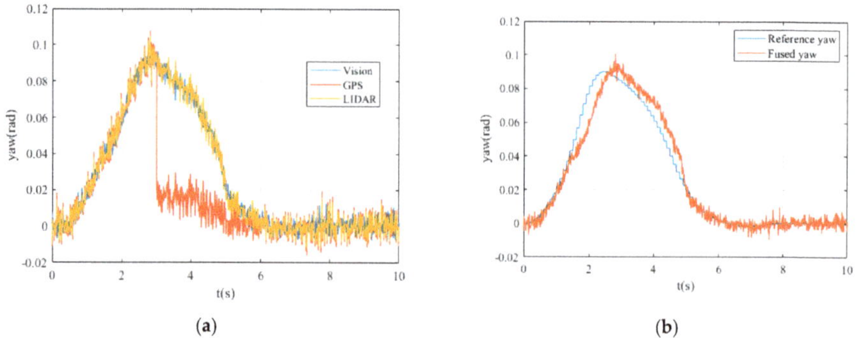

(a)   (b)

Figure 6. The yaw angle: (**a**) yaw angle detected using different sensors; (**b**) the reference and fused yaw angle.

The horizontal axis represents time, and the vertical axis represents the values of yaw angle. It can be seen that from moment 0s to 3s, the measured values of Vision, GPS, and LIDAR systems are similar. In the simulation process, we assume that the GPS signal has interfered from 3 to 6 s, and the obtained yaw angle value has a large deviation. After 6s, the GPS signal returns to normal.

It can be seen that the fused yaw angle is always less than the value of the reference yaw angle before 2.3s and the fused yaw angle is greater than the value of the reference yaw angle between 2.3s to 5s. According to the trend of the two curves in the image, we know that the path tracking control performance of straight line is significantly better than the curve line during the process of lane change.

The type of simulated autonomous vehicle used in our paper is passenger car. The simulation environment and initial simulation condition settings are shown in Table 1:

Table 1. The simulation parameters.

Parameters	Value	Dimension
Number of lanes:	2	
Lane width:	4	m
Lane length	127	m
Mass of car:	1575	kg
Moment of inertia	2875	kg.m^2
Length of car	4.7	m
Width of car	1.8	m
Height of car	1.4	m
Front overhang	0.9	
Rear overhang	1	
Longitudinal speed	15	km/h
Sampling time interval	0.01	s
The process noise covariance matrix Q	diag [0.001,0.001,0.001,.0.002]	
The measurement noise R	0.001	

4.2. Effectiveness of the Proposed Method

The yaw angle errors are shown in Figure 7.

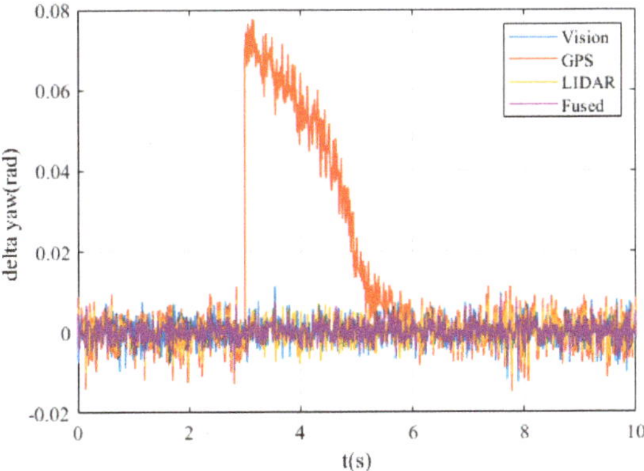

Figure 7. The yaw angle errors.

It can be seen that except GPS, the deviation between other measurement data and reference value is extremely small, and their absolute values are not more than 0.01 rad. However, when it comes to GPS data, the deviation is about 0.7rad at 3s, which is unacceptable considering the maxim yaw angle about 0.09 rad.

Figure 8 shows the value of simultaneous interpreting of three different sensors based on Chi-square test.

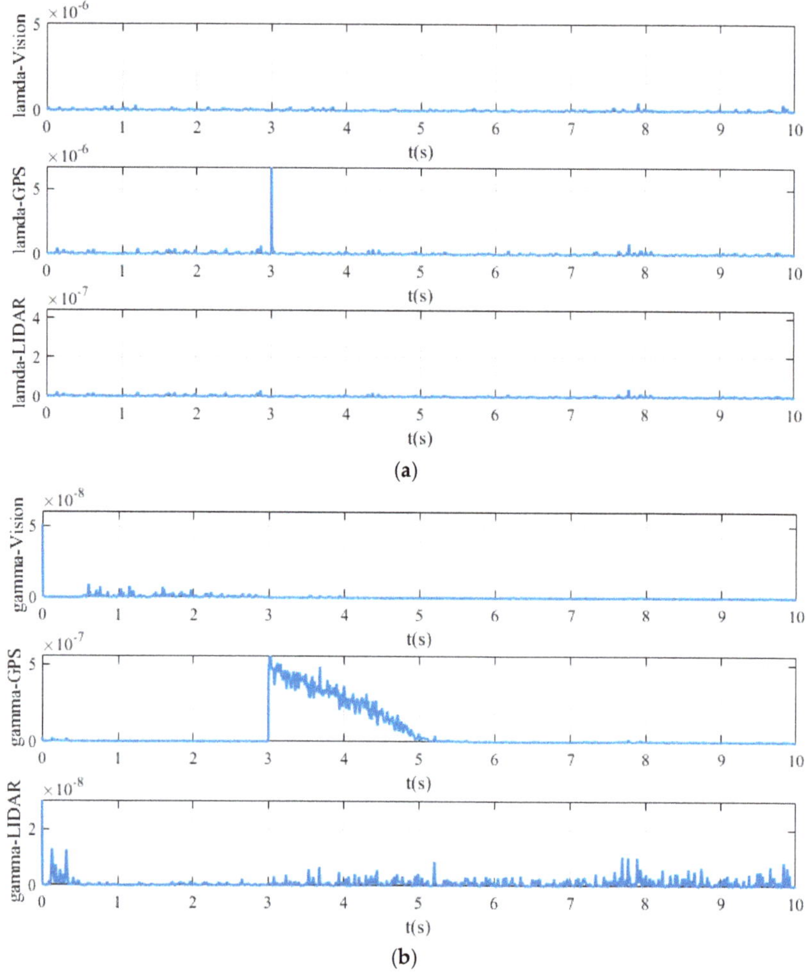

Figure 8. Yaw test statistic of: (**a**) state Chi-square test; (**b**) residuals Chi-square test.

Figure 8a is the value obtained by state chi-square test, and Figure 8b is the value obtained by residuals error chi-square test. In Figure 8a, the calculated values λ of state chi-square test of Vision, GPS, and LIDAR channels are very small, which are always below the order of magnitude 10^{-7}, when all sensor systems work without fault. In contrast, when the fault of GPS signal occurs, the calculated values of state chi-square test of GPS channel sharply increase at 3s, reaching more than the order of magnitude 5×10^{-6}. In Figure 8b, the calculated values γ of state residuals test of Vision, GPS, and LIDAR channels are generally not exceed 2×10^{-8}, when all sensor systems work without fault. However, in the period of 3s-5s, when the GPS signal has fault, the calculated values γ of residuals error chi-square test of GPS signal channel is much higher, and even reaches more than 5×10^{-7} at 3 s.

Figure 9a shows the lateral position changes of the vehicle obtained by Vision, GPS, and LIDAR sensor systems. Figure 9b shows the reference lateral position of the vehicle and the lateral position after data fusion.

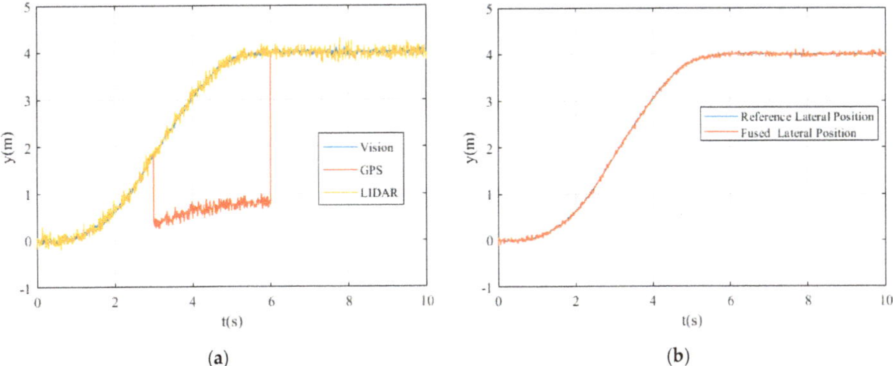

(a) (b)

Figure 9. The lateral position: (**a**) lateral position detected using different sensors; (**b**) the reference and fused lateral position.

It can be seen from Figure 9 that the lateral position obtained by GPS signal has obvious deviation in the time period of 3-6s, and the lateral position obtained by multi-sensor data fusion method is basically consistent with the reference lateral position. However, a small deviation between the reference and fused lateral position still exists due to the performance of path tracking controller.

Figure 10 shows the deviation of the measured lateral position by different sensor systems.

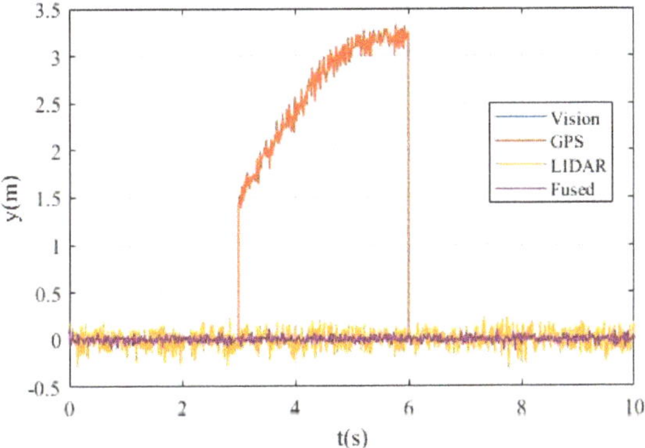

Figure 10. The lateral position error.

It can be seen from Figure 10 that the measurement deviations of vision and LIDAR have remained at a very small range, while the measurement error of GPS data gradually increases from about 1.5m to more than 3m in the period of 3-6s due to sensor failure. Considering that the lane width is only 4m, the GPS sensor fault is unacceptable and needs to be isolated.

Figure 11 shows the value of the fault detection function calculated by chi-square test for different sensor systems. Figure 11a shows the value obtained by state chi-square test and Figure 11b shows the value obtained by residuals chi-square test.

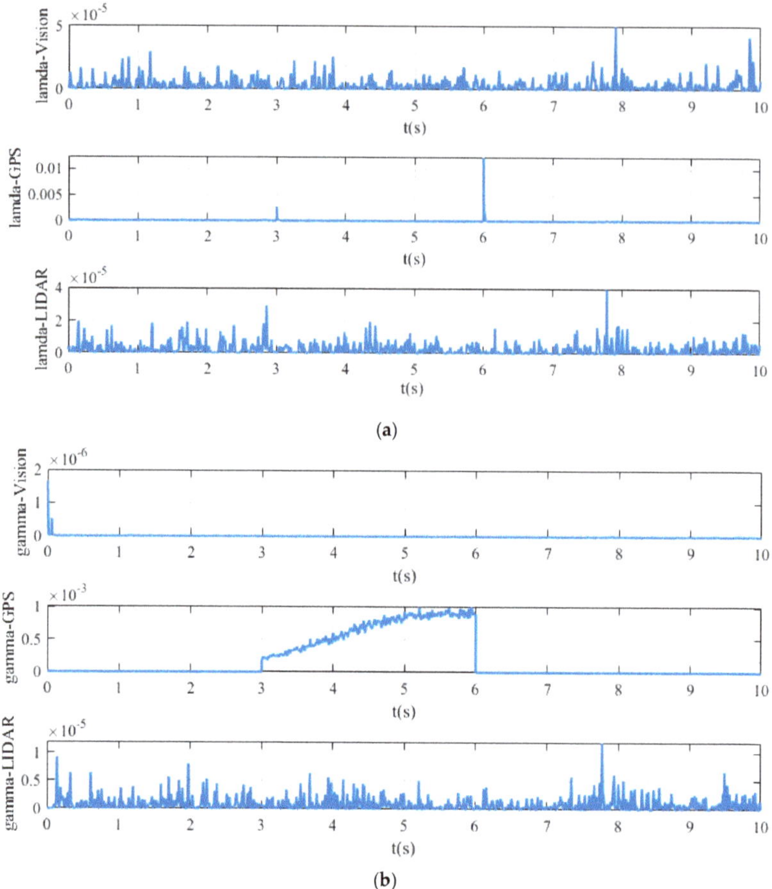

Figure 11. Lateral position test statistic of: (**a**) state Chi-square test; (**b**) residuals Chi-square test.

In Figure 11a, it can be seen that the calculated values of λ state chi-square test of vision and LIDAR channels are very small, basically below 2×10^{-5}, occasionally exceeding the order of magnitude 10^{-5}, but they are still extremely small. Comparatively, the data of GPS channel reaches the order magnitude of 10^{-3} at 3 s, or even more than 10^{-2} at 6 s, which is much larger than that of vision and LIDAR channels. In Figure 11b, the calculated value γ of vision channel reaches about 1×10^{-6} at the initial stage but always below the order magnitude of 10^{-6} in the following time. The calculated value γ of GPS data is extremely large from 3 s to 6 s, reaching the order magnitude of 10^{-3} due to the GPS sensor failure. The calculated values γ of LIDAR data fluctuate greatly, but they are all below 1×10^{-5}, which is also a reasonable interference situation.

From the above description and analysis of Figure 7 to Figure 11, it can be seen that the proposed method can detect the fault signal robustly when the sensor failure occurs.

Figure 12 shows the simulation results of path tracking control with and without fault isolation. The three curves in Figure 12a represent the reference lateral position, after fault isolation and before fault isolation. Figure 12b shows the lateral position error before and after fault isolation in the simulation process. From Figure 12a,b, it can be seen that without fault isolation, it is almost impossible to realize the path tracking control due to sensor faults. After fault isolation, the deviation between the actual lateral position and the reference lateral position is maintained within a very small range, which validates the effectiveness of the proposed fault-tolerant MPC algorithm for path tracking of autonomous vehicle.

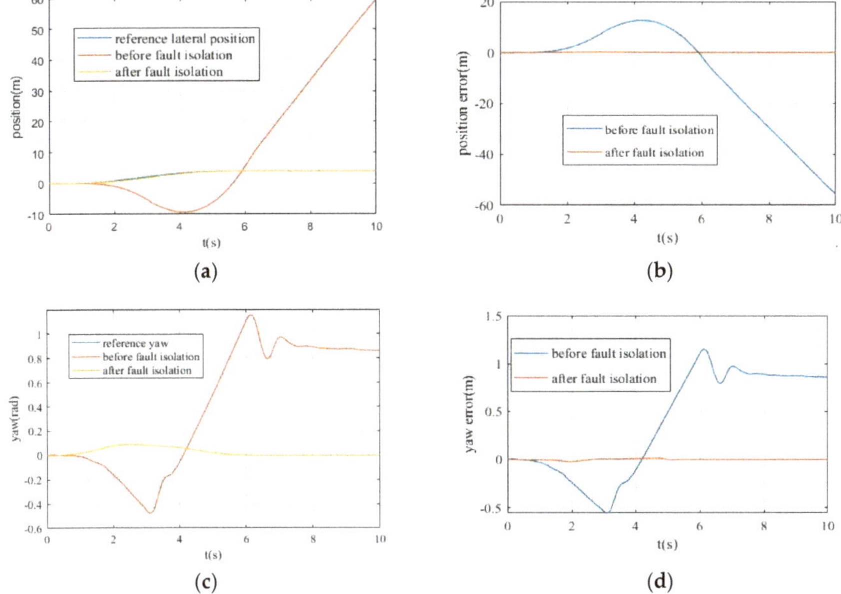

Figure 12. The path tracking control simulation results: (**a**) the lateral position changes; (**b**) the lateral position error before and after fault isolation; (**c**) the yaw angle changes; (**d**) the yaw angle error before and after fault isolation.

Figure 12c,d show the yaw angle obtained before and after fault isolation and the yaw angle error with respect to the reference values. From Figure 12c,d, it can be seen that with fault isolation, actual yaw angle of vehicle almost coincides with the reference yaw angle, while without fault isolation the yaw angle error is extremely large after 1s. From Figure 12d, the yaw angle error without fault isolation can reach—0.5 rad to 1.1 rad, while the reference value of yaw angle is always below 0.1rad, which means yaw angle error without fault isolation may reach 5–10 times more than the reference, and vehicle cannot track the reference path under this condition, which may lead to serious consequences. On the contrary, after fault isolation processing, the yaw angle error is always below 0.02 rad, which is extremely small for path tracking control.

4.3. Discussion of the Background and Outcomes of Our Work

With the continuous increasing requirements for the safety and environmental adaptability of autonomous vehicles, the on-board environmental perception systems have become more and more complex, and the types and numbers of sensors have also increased. The risk of sensor failures is also increased, which has a serious impact on vehicle safety. Therefore, the detection of faulty sensor signals and fault-tolerant control mechanism is very important to autonomous driving safety. In our work, we established a single-track 3 DOF vehicle dynamics model and based on this model, a fault-tolerant model predictive control method was developed. Our motivation for designing this algorithm is to enable the autonomous vehicle to effectively detect and isolate the fault signal and to perform robust longitudinal path tracking motion control when the sensor failure occurs.

In order to verify the effectiveness of the method proposed in this paper, we set up a single lane changing path tracking control condition in Driving Scenario Designer. The vehicle motion state information, such as the lateral position and yaw angle, can be obtained by the GPS integrated navigation system, LIDAR perception system, and visual perception system. We assume that there is interference in the simulation environment, which causes the GPS signal to be temporarily lost. The description and analysis of simulation results show that the proposed fault-tolerant MPC algorithm can effectively detect the fault signal when the sensor failure occurs. The value of fault detection functions of fault yaw angle signal and fault lateral position signal is 10 and 100 times, respectively, more than that of normal signals. This means that we can easily detect the fault signal by setting the appropriate threshold and generate the corresponding fault detection matrix. The fault isolation is accomplished in the process of data fusion, which is one of the innovations of our work. With the fault signal isolation, the path tracking control performance of autonomous vehicle can be significantly improved, further confirming the effectiveness and robustness of the proposed algorithm in this work.

5. Conclusions

In this paper, a novel and robust fault-tolerant model predictive control algorithm was proposed, which can be used for robust vehicle lateral motion control in case of sensor failures. First, by constructing the objective function and considering dynamic constraints, the linear time-varying model predictive control algorithm for path tracking control of vehicle was designed. Second, an improved weighted data fusion algorithm was proposed for multi-sensor information fusion and fault signal isolation. Then, based on Chi-square detectors and Kalman filters, we designed a novel fault signal detection algorithm, which can produce the fault detection matrix used in data fusion algorithm for fault signal isolation. Finally, a lane changing path tracking control simulation was carried out for validation of the effectiveness and correctness of the proposed algorithm. The simulation results show that the proposed algorithm can efficiently detect the fault signal. After the fault signal isolation, the reference path can be effectively tracked by using the proposed fault-tolerant MPC algorithm. In further studies, this method can be combined with reinforcement learning to improve fault detection and isolation performance. We will explore the effectiveness of the proposed method for fault detection in electronic fuel injection system, automatic steering control system, suspension systems, etc. Meanwhile, we will extend our fault-tolerant model predictive control algorithm into the fault diagnosis fields that are out of the real driving environment, for instance, fault diagnosis in complex driving scenarios and strong environmental noise conditions. In addition, considering the longitudinal speed changes and the adaptive MPC algorithm can be used to improve the path tracking performance of motion controller. It is worth noting that we approximately use the fused multi-sensor data as the real motion states of vehicle, which means that when faults occur to most or all sensors, our proposed method is invalid. In addition, the robustness is weak by using the thresholds to detect the fault signal in our method. For solving these two problems, using the convolutional neural network to automatically extract the features of the fault detection function and detect the sensor fault could be a reliable solution.

Author Contributions: In this article, the author's contributions are shown below: Methodology, K.G.; writing—original draft preparation, Z.W.; project administration, N.A.C.; data curation, Z.W.; investigation, K.G.; resources, K.G. All authors have read and agreed to the published version of the manuscript.

Funding: The research is supported by the National Key Research and Development Program of China (Grant No. 2016YFD0700905), National Natural Science Foundation of China (Grant No. 51905095), and National Natural Science Foundation of Jiangsu Province (Grant No. BK20180401).

Conflicts of Interest: The authors declare no conflict of interest.

References

1. Xia, Y.; Fan, P.; Li, S.; Yuan, G. Lateral path tracking control of autonomous land vehicle based on ADRC and differential flatness. *IEEE Trans. Ind. Electron.* **2016**, *63*, 3091–3099. [CrossRef]
2. Alessandri, A.; Caccia, M.; Veruggio, G. Fault detection of actuator faults in unmanned underwater vehicles. *Control Eng. Pract.* **1999**, *7*, 357–368. [CrossRef]

3. Heredia, G.; Ollero, A.; Mahtani, R. Detection of Sensor Faults in Autonomous Helicopters. In Proceedings of the IEEE International Conference on Robotics & Automation, Barcelona, Spain, 18–22 April 2005.

4. Persis, C.D.; Isidori, A. A geometric approach to nonlinear fault detection and isolation. *Autom. Control. IEEE Trans. Autom. Control.* **2001**, *46*, 853–865. [CrossRef]

5. Patton, R.J.; Frank, P.M.; Clark, R.N. (Eds.) *Issues of Fault Diagnosis for Dynamic Systems*; Springer: Berlin/Heidelberg, Germany, 2010.

6. Wang, D.; Wang, W.; Shi, P. Robust fault detection for switched linear systems with state delays. *IEEE Trans. Syst. Man Cybern Part B* **2009**, *39*, 800–805. [CrossRef] [PubMed]

7. Goupil, P.; Marcos, A. The European ADDSAFE project: Industrial and academic efforts towards advanced fault diagnosis. *Control Eng. Pract.* **2014**, *31*, 109–125. [CrossRef]

8. Freeman, P.; Seiler, P.; Balas, G.J. Air data system fault modeling and detection. *Control Eng. Pract.* **2013**, *21*, 1290–1301. [CrossRef]

9. Castaldi, P.; Mimmo, N.; Simani, S. Differential geometry based active fault tolerant control for aircraft. *Control Eng. Pract.* **2014**, *32*, 227–235. [CrossRef]

10. Marzat, J.; Piet-Lahanier, H.; Damongeot, F.; Walter, E. Model-based fault diagnosis for aerospace systems: A survey. *Proc. Inst. Mech. Eng. Part G J. Aerosp. Eng.* **2012**, *226*, 1329–1360. [CrossRef]

11. Marzat, J.; Piet-Lahanier, H.; Damongeot, F. Control-based fault detection and isolation for autonomous aircraft. *Proc. Inst. Mech. Eng.* **2012**, *226*, 510–531. [CrossRef]

12. Avram, R.C.; Zhang, X.; Campbell, J. IMU Sensor Fault Diagnosis and Estimation for Quadrotor UAVs. *IFAC Pap.* **2015**, *48*, 380–385. [CrossRef]

13. Zhang, D.; Wang, Q.G.; Yu, L. Fuzzy-model-based fault detection for a class of nonlinear systems with networked measurements. *IEEE Trans. Instrum. Meas.* **2013**, *62*, 3148–3159. [CrossRef]

14. Ballesteros-Moncada, H.; Enrique, J.H.-L.; Juan, A.-M. Fuzzy model-based observers for fault detection in CSTR. *ISA Trans.* **2015**, *59*, 325–333. [CrossRef] [PubMed]

15. Chen, M.; Shi, P.; Lim, C.C. Adaptive Neural Fault-Tolerant Control of a 3-DOF Model Helicopter System. *IEEE Trans. Syst. Man Cybern. Syst.* **2017**, *46*, 260–270. [CrossRef]

16. Baghernezhad, F.; Khorasani, K. Computationally autonomous strategies for robust fault detection, isolation, and identification of mobile robots. *Neurocomputing* **2016**, *171*, 335–346. [CrossRef]

17. Yu, Z.; Qu, Y.; Zhang, Y. Fault-Tolerant Containment Control of Multiple Unmanned Aerial Vehicles Based on Distributed Sliding-Mode Observer. *J. Auton. Robot. Syst. Theory Appl.* **2019**, *93*, 163–177. [CrossRef]

18. Lu, P.; Kampen, E.J.V.; Visser, C.D. Nonlinear aircraft sensor fault reconstruction in the presence of disturbances validated by real flight data. *Control Eng. Pract.* **2016**, *49*, 112–128. [CrossRef]

19. Chamseddine, A.; Noura, H. Control and Sensor Fault Tolerance of Vehicle Active Suspension. *IEEE Trans. Control Syst. Technol.* **2008**, *16*, 416–433. [CrossRef]

20. Kim, W.; Lee, J.; Yoon, S. *Development of Mando's New Continuously Controlled Semi-Active Suspension System*; Sae World Congress & Exhibition: Detroit, MI, USA, 2005.

21. Isermann, R.M.; Schmitt, M. *A Sensor and Process Fault Detection System for Vehicle Suspension Systems*; Sae World Congress & Exhibition: Detroit, MI, USA, 2002.

22. Sébastien, V.; Morales-Menendez, R.; Lozoya-Santos, D.J. Fault Detection in Automotive Semi-Active Suspension: Experimental Results. In Proceedings of the SAE 2013 World Congress & Exhibition, Detroit, MI, USA, 16–18 April 2013.

23. Svendenius, J.; Gäfvert, M. A Semi-Empirical Tire-Model for Transient Combined-Slip Forces. *Veh. Syst. Dyn.* **2006**, *44*, 189–208. [CrossRef]

24. Kristian, H.; Florian, B.; Klaus, A. Extending the Magic Formula Tire Model for Large Inflation Pressure Changes by Using Measurement Data from a Corner Module Test Rig. *SAE Int. J. Passeng. Cars Mech. Syst.* **2018**, *11*, 103–117.

25. Wu, J.Y.; Wang, Z.P.; Zhao, Z. Influence of Tire Inflation Pressure on Vehicle Dynamics and Compensation Control on FWID Electric Vehicles. *J. Dyn. Syst. Meas. Control Trans. ASME* **2020**, *142*, 071001. [CrossRef]

26. Feng, Y.; Chao, Y.; Xing, L. Experimental Evaluation on Depth Control Using Improved Model Predictive Control for Autonomous Underwater Vehicle (AUVs). *Sensors* **2018**, *18*, 2321–2340.

27. Shi, X.; Gao, J.; Lu, Y. Simulation of Disturbance Recovery Based on MPC and Whole-Body Dynamics Control of Biped Walking. *Sensors* **2020**, *20*, 2971. [CrossRef] [PubMed]

28. Linben, L.; Zigang, L.I.; Chaoying, C. Optimal weight distribution principle used in the fusion of multi-sensor data. *J. Chin. Inert. Technol.* **2000**, *8*, 36–39.
29. Da, R. Failure detection of dynamical systems with the state chi-square test. *J. Guid. Control Dyn.* **1994**, *17*, 271–277. [CrossRef]

Review

Fault Detection, Isolation, Identification and Recovery (FDIIR) Methods for Automotive Perception Sensors Including a Detailed Literature Survey for Lidar

Thomas Goelles *, Birgit Schlager and Stefan Muckenhuber

VIRTUAL VEHICLE Research GmbH, Inffeldgasse 21a, 8010 Graz, Austria; Birgit.Schlager@v2c2.at (B.S.);
Stefan.Muckenhuber@v2c2.at (S.M.)
* Correspondence: thomas.goelles@v2c2.at

Received: 19 May 2020; Accepted: 24 June 2020 ; Published: 30 June 2020

Abstract: Perception sensors such as camera, radar, and lidar have gained considerable popularity in the automotive industry in recent years. In order to reach the next step towards automated driving it is necessary to implement fault diagnosis systems together with suitable mitigation solutions in automotive perception sensors. This is a crucial prerequisite, since the quality of an automated driving function strongly depends on the reliability of the perception data, especially under adverse conditions. This publication presents a systematic review on faults and suitable detection and recovery methods for automotive perception sensors and suggests a corresponding classification schema. A systematic literature analysis has been performed with focus on lidar in order to review the state-of-the-art and identify promising research opportunities. Faults related to adverse weather conditions have been studied the most, but often without providing suitable recovery methods. Issues related to sensor attachment and mechanical damage of the sensor cover were studied very little and provide opportunities for future research. Algorithms, which use the data stream of a single sensor, proofed to be a viable solution for both fault detection and recovery.

Keywords: automotive; perception sensor; lidar; fault detection; fault isolation; fault identification; fault recovery; fault diagnosis; fault detection and isolation (FDIR)

---

1. Introduction

Advancing the level of automation for vehicles is a major challenge in today's automotive industry. Automated vehicles are expected to provide great benefits for the driver and enable new transportation use cases and applications, e.g., [1]. Improving passenger safety is among the key arguments for the development of automated vehicles. Today, more than 1.35 million people die in road traffic crashes each year, making road traffic crashes the leading cause of death among children and young adults between 5 and 29 years of age [2]. Advanced driver assistance system (ADAS) and automated driving (AD) functions have the potential to reduce this number significantly, since most car accidents are traceable to a human error [3–5].

The Society of Automotive Engineers (SAE) classifies six levels of driving automation [6]. Currently, available vehicles provide up to SAE level 2 "partial driving automation", where an ADAS function can take over lateral and longitudinal vehicle motion control. Examples are Cadillac's Super Cruise, Mercedes-Benz's Drive Pilot, Nissan's ProPILOT Assist, Tesla's Autopilot, and Volvo's Pilot Assist. Vehicles that fulfil SAE level 3 "conditional driving automation" must provide an AD function that allows the driver to remove his attention off the road and only intervene when the system requests. In this case, the vehicle is responsible for object and event detection and proper response. Going from SAE level 2 to level 3+ implies that the responsibility for environment perception is transferred from the driver to the vehicle.

Automated vehicles work according to the SENSE-PLAN-ACT cycle [1]. SENSE represents everything related to environment perception. Within PLAN the necessary decisions are made based on the information from SENSE. ACT includes the controlling of steering, acceleration, and deceleration of the vehicle based on commands from PLAN. This chain of command illustrates how important a reliable environment perception system is for every automated vehicle. In particular, SAE level 3+ vehicles, that cannot rely on a human driver monitoring the environment constantly, have very high demands in terms of robustness and reliability. A combination of diverse and redundant sensor types is required to provide a robust environment perception during all possible environmental conditions. A sensor set including camera, radar (radio detection and ranging), and lidar (light detection and ranging) sensors is considered the best option to fulfil the SENSE demands of level 3+ vehicles [7] eventually.

Sensor faults, either caused by internal malfunctions (e.g., broken antennas) or disturbing external factors (e.g., adverse weather conditions), propose a considerable danger for each automated vehicle's anticipated functionality. This can be health and even life-threatening for passengers in the vehicle and people in the proximity of automated vehicles. A standard method to tackle sensor faults of critical systems in the aviation or military domain is sensor redundancy. However, in the automotive industry, the cost factor plays a significant role and therefore proposes a limiting factor to sensor redundancy.

To address sensor faults in automated vehicles in a cost effective manner, sensor fault detection, isolation, identification, and recovery (FDIIR) systems, e.g., [8] can be included into each individual perception sensor that is contributing to the SENSE-PLAN-ACT cycle (Figure 1). The sensor FDIIR system constantly monitors the activity of the respective perception sensor for correct operation. In case of a detected sensor fault, the system intervenes by preventing erroneous information reaching the PLAN step, and tries to recover the affected sensor (e.g., a wiper removes dirt from the optical aperture of the sensor). Additionally, FDIIR systems can provide sensor performance information to the ADAS/AD function (e.g., adverse weather conditions are currently reducing the sensors field of view (FOV) by 50 %) to increase the information content for the PLAN step.

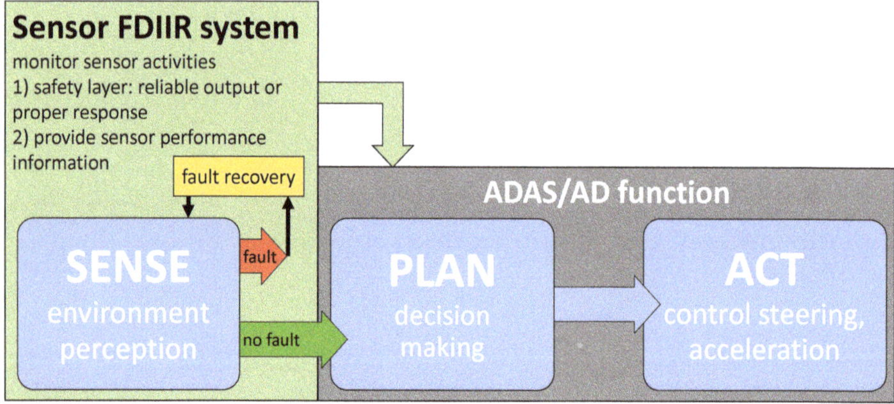

Figure 1. Schematic illustration of the sense-plan-act cycle including a sensor FDIIR (fault detection, isolation, identification, and recovery) system that monitors the sensor activity to ensure a reliable environment perception.

Two EU research projects are currently addressing faults of automotive perception sensors under unfavourable environmental conditions. The RobustSENSE project [9] aims to develop a system performance assessment module, which shall monitor the reliability of every perception sensor in the vehicle under adverse weather conditions. The DENSE project [10] approaches the problem by developing a fused sensor set that combines radar, gated short-wave infrared camera, and short-wave infrared lidar to provide a reliable perception under challenging conditions.

Sensor FDIIR systems are equally important for all types of automotive perception sensors, i.e., camera, radar, and lidar. This publication introduces a classification schema for faults, detection methods and recovery methods that applies to all perception sensor types in the automotive domain.

The literature survey and state-of-the-art analysis in this publication put the focus on FDIIR systems for lidar sensors, since lidar is a recently emerging sensor for automotive applications, and there is far less experience in the automotive industry with lidar compared to radar and camera. Today's level 2 vehicles typically rely on radar and camera sensors to perceive the environment [7]. However, lidar sensors are expected to play an essential role in level 3+ vehicles, since lidars provide a three-dimensional depth map with very high angular resolution [11,12]. High prices of mechanically spinning lidars are currently a limiting factor, but costs will decrease significantly with new technologies such as optical phased array, MEMS-based mirrors, etc. [11,13]. For example, Druml et al. [14] introduced a lidar prototype that shall eventually lead to commercially available lidar sensors with a range of more than 200 m for costs less than USD 200.

The main objectives of the presented study are the following:

- Identify all types of faults, detection methods, and recovery methods for automotive perception sensors and develop a corresponding classification schema.
- Evaluate the state of the art of FDIIR methods for automotive lidar.
- Explain, discuss, and compare the most promising existing FDIIR methods for automotive lidar.
- Identify research opportunities related to FDIIR methods for automotive lidar.

The publication is structured as follows: Section 2 introduces a new fault classification for automotive perception sensors and relates the fault classes to current and upcoming international safety standards for automated vehicles. Section 3 introduces a corresponding classification schema for fault detection methods and Section 4 introduces a corresponding classification schema for fault recovery methods. Section 5 describes the methods applied in the literature survey on lidar FDIIR methods that is presented in Section 6. Section 7 completes the paper with a discussion and conclusion and gives an outlook on future work.

A glossary of the technical terms can be found at the end of this paper.

2. Classification of Faults of Perception Sensors

Faults can occur on different levels in an automotive perception system. Ranging from the whole perception system, including the sensor fusion layer, to the individual sensor and its components. Faults on each level need different detection and recovery strategies. In this review, we focus on the individual sensor and its subcomponents, disregarding the perception algorithms that extract information based on raw data from individual sensors or perception systems consisting of several sensors. Therefore, we consider faults that occur until the raw data of the sensor is generated before any optional object detection and classification algorithm is applied (Figure 2).

Different faults occurring in perception sensors may have a similar cause. Therefore, perception sensor faults can be classified. A classification of faults is useful when it comes to the design of fault detection, fault isolation, fault identification, and fault recovery algorithms since similar faults may benefit from similar or equal algorithms. We suggest a classification into the following fault classes: defect subcomponent, mechanical damage to sensor cover, layer on sensor, mounting issue, security attack, unfavourable environmental condition, and crosstalk (see Figure 2). Table 1 lists the seven fault classes including exemplary faults and international safety standards addressing the faults.

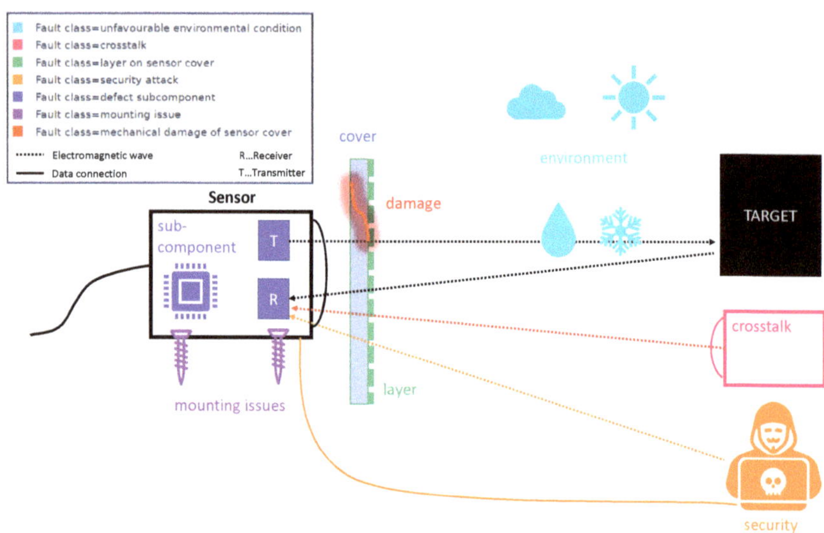

Figure 2. Schematic illustration of sensor fault classes.

Table 1. Classification of perception sensor faults including exemplary faults and related international safety standards.

Fault Class	Exemplary Fault	Safety Standard
defect subcomponent	defect transmitter defect receiver defect processing unit	ISO 26262
mechanical damage to sensor cover	deformation scratch crack hole missing cover	
layer on sensor	dirt layer water layer ice layer salt layer snow layer	
mounting issue	position change vibration	ISO PAS 21448
security attack	laser hack electronics hack	
unfavourable environmental condition	low/high temperature dust smoke rain fog hail electromagnetic interference light disturbance snow	
crosstalk	other active perception sensor	

A defect subcomponent is a faulty internal part of the sensor, e.g., a defect transmitter, a defect receiver, or a defect processing unit. A mechanical damage to the sensor cover may be scratches, cracks, or holes in the sensor cover but also a missing sensor cover or a deformation of the sensor cover count to this fault class. A layer on the sensor cover like dirt, water, ice, salt, or snow can cause that objects are not detected (false negatives), that not existing objects are detected (false positives) or that objects are misclassified. Mounting issues may be a change of the sensor's position or vibrations while driving. A further risk are security attacks as shown in, e.g., [15,16]. This may be a security attack by another active electromagnetic source, i.e., denial of service and false data injection, or an electronics hack over a wired or wireless connection to the sensor. Furthermore, a perception sensor has limitations under different unfavourable environmental conditions, which can cause, e.g., a reduction of the field of view. Other active perception sensors may cause a fault in the sensor data because of crosstalk, as described, e.g., in [17,18]. The difference to security attacks with laser hacks is that crosstalk occurs accidentally, and security attacks occur by malicious hacking.

International Safety Standards Addressing Faults of Perception Sensors

The international standard ISO 26262 [19] and the international specification ISO PAS 21448 [20] represent the state-of-the-art for safety standardization of road vehicles. FDIIR methods play an important role to fulfil the current and upcoming requirements of ISO 26262 and ISO PAS 21448 for automotive perception sensors. This section explains, how the above introduced fault classification schema (Section 2) relates to the ISO 26262 and ISO PAS 21448.

The ISO 26262 "Road vehicles—Functional safety" from 2011 includes safety measures and safety mechanisms to ensure functional safety of electrical and electronic systems of road vehicles. The standard differentiates between systematic faults and random hardware faults and provides a systematic development framework for electrical and electronic systems. The Automotive Safety Integrity Level (ASIL) is defined in this standard. This level defines which requirements and measures have to be considered and implemented to avoid an unreasonable risk [19]. Concerning FDIIR methods for automotive perception sensors, the ISO 26262 especially applies to the sensor fault class defect subcomponent listed in Table 1. This fault class affects the sensor itself and is not caused by sensor limitations due to the environment.

The ISO PAS 21448 "Road vehicles—Safety of the intended functionality" addresses limitations of systems where no fault, according to ISO 26262, occurred. The letter code PAS implies that this is a publicly available specification and not a standard at that point in time. The rework of the PAS version to an international standard has started in the beginning of 2019, and it is planned to be released as an international standard by the end of 2021. The ISO PAS 21448 especially applies to the limitations of systems that are based on sensing the environment. Such limitations can be caused by, e.g., different environmental conditions or other road users. The Annex F of the standard lists such limitations [20]. The sensor fault classes mechanical damage to sensor cover, layer on sensor cover, mounting issue, security attack, unfavourable environmental condition, and crosstalk listed in Table 1 address the issues described in the ISO PAS 21448. These sensor fault classes are caused by triggering conditions, which are combinations of sensor weaknesses in the design or specific technology or negative influencing factors of the environment.

3. Classification of FDII Methods for Perception Sensors

Fault detection, isolation, and identification (FDII) methods determine whether a fault occurred in a system, where the fault occurred, and how severe the fault is. The term "fault diagnosis" is often used as a synonym, e.g., in Chen and Patton [21], Chen et al. [22], Piltan and Kim [23].

The fault detection unit or fault detection algorithm of a perception sensor determines whether a fault occurred in the perception sensor or not. If the fault detection unit decides that a fault occurred, the fault isolation unit identifies where the fault occurred and which part of the sensor is faulty.

When the location of the fault is determined, the fault identification unit determines the severity of the fault and the level of sensor degradation.

To categorize FDII methods for perception sensors, we suggest a classification into the following classes (Table 2): comparison to sensor model, monitoring sensor output, comparison to static ground-truth, comparison to dynamic ground-truth, comparison to other sensor of same type, comparison to other sensor of different type, monitoring internal interface, and comparison of multiple interfaces.

Table 2. Classification of FDII methods for perception sensors including exemplary methods.

FDII Class	Exemplary Method
comparison to sensor model	objects detected by the sensor model compared to objects detected by the real sensor
monitoring sensor output	signal analysis and plausibility check of sensor output
comparison to static ground-truth	infrastructure detected by the sensor compared to ground-truth infrastructure in the environment
comparison to dynamic ground-truth	road users detected by another vehicle compared to road user detected by the ego-vehicle
comparison to other sensor of same type	compare objects that are detected by the sensor under observation with objects detected by another sensor of the same type (two lidar sensors)
comparison to other sensor of different type	compare objects that are detected by the sensor under observation with objects detected by another sensor of a different type (a lidar and a radar sensors)
monitoring internal interface	signal analysis and plausibility check of the output of a single sensor interface
comparison of multiple interfaces	a part of the sensor between sensor interfaces is modelled; comparison of the output of the modelled part with the output of the respective sensor interface

The class comparison to sensor model includes the comparison of real sensor measurements with the output of a sensor model (Figure 3a). In general, comparing a process with its model is a very common method in FDII literature for simple systems like bearings. Monitoring a perception sensor with this method requires very detailed knowledge about the environment and a sensor model.

The class monitoring sensor output refers to methods that use only the output data of the sensor under observation (Figure 3b). The monitoring algorithm determines whether the sensor is healthy or faulty, the fault's location, and the severity of the fault solely based on the sensor's output.

The class comparison to static ground-truth identifies faults based on comparing the sensor's output with static ground-truth data (Figure 3c). Static ground-truth may be immobile objects (houses, traffic signs, ...) stored in a map or well-defined test targets placed at known locations specifically for sensor assessment.

The class comparison to dynamic ground-truth identifies faults based on comparing the sensor's output data with dynamic ground-truth data. Dynamic ground-truth may be obtained by connected infrastructure in terms of vehicle to everything (V2X), e.g., vehicles that collect and broadcast data with their own sensors (Figure 3d). Furthermore, the position and the geometry of other vehicles may be sent from vehicle to vehicle (V2V) to check if the sensor on the ego-vehicle perceives the right position and geometry of the other vehicle.

The class comparison to other sensor of same type identifies faults based on comparing the output of two sensors of the same type, e.g., two lidar sensors, two radar sensors, or two camera sensors, with overlapping FOV (Figure 3e).

Contrary, the class comparison to other sensor of different type identifies faults based on comparing the output of two sensors of different types, e.g., lidar compared to radar, with overlapping FOV (Figure 3f). Therefore, an algorithm that transforms the data of one sensor to comparable data of the other sensor has to be applied.

The class monitoring internal interface identifies faults based on monitoring the output of a single internal sensor interface (Figure 3g). The monitoring algorithm determines whether the sensor is healthy or faulty, the location of the fault, and the severity of the fault based on the output of a single sensor interface.

The class comparison of multiple interfaces identifies faults based on comparing the output of two interfaces of one sensor (Figure 3h). The subcomponents between the two interfaces are modelled. A fault can be detected by comparing the model's output at a specific interface with the real output of the same interface.

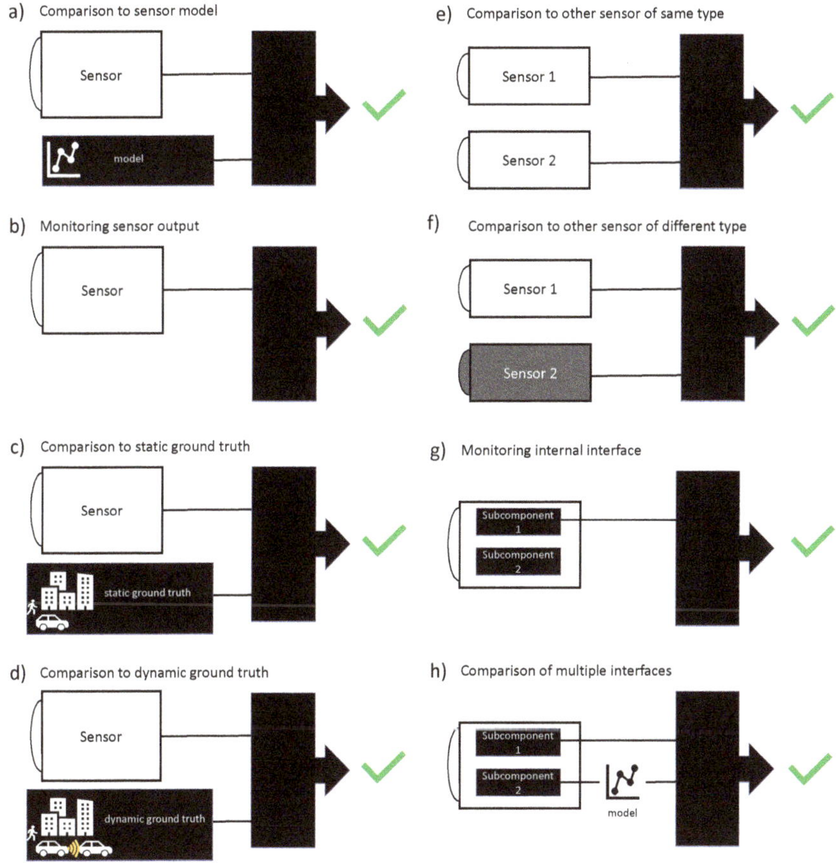

Figure 3. Schematic illustration of FDII methods.

4. Classification of Recovery Methods for Perception Sensors

If the FDII system detects a fault, a fault recovery method can be applied to eliminate or mitigate the fault and restore the functionality of the perception sensor and thereby the system that relies on the sensor.

To categorize fault recovery methods for perception sensors, we suggest a classification into the following classes: software adaption, hardware adaption, temperature regulation, and cleaning of sensor cover. Table 3 shows fault recovery classes and respective exemplary methods.

Table 3. Classification of fault recovery methods for perception sensors including exemplary methods.

Recovery Class	Exemplary Method
software adaption	filter pointcloud mitigate faults in sensor data by averaging
hardware adaption	change alignment of sensor cover adjust sensor position and rotation
temperature regulation	heating cooling
cleaning of sensor cover	air nozzle fluid nozzle wiper

Software adaption as a recovery class includes algorithms that are either applied on the sensor output data to mitigate faults directly or that are applied in the processing chain that uses the sensor data, for example, filtering of the point cloud for unwanted signals from precipitation.

Similarly to software adaption, it is also possible to apply hardware adaption, if required. Hardware adaption can be adjusting the sensor's alignment after a change of the sensor's position, or aligning the sensor cover if only a part of the sensor cover is faulty.

A further recovery class is temperature regulation. This can be heating if the sensor is covered by an ice layer, for example, or cooling to prevent overheating of the sensor.

The last recovery class, cleaning of sensor cover, is required in case of a disturbing layer on the sensor cover. For example, wipers may be used to remove dirt or water from the sensor cover.

5. Literature Survey Methodology

This section explains the methods used for the literature survey, presented in the following Section 6. Relevant publications addressing faults, fault detection methods, and recovery methods for automotive lidars were identified, analysed and then classified in a quantitative approach based on the PRISMA scheme, e.g., [24,25]. The presented literature survey methodology can also be applied for literature studies on other perception sensors.

Figure 4 shows the literature search procedure adopted for identifying relevant studies. We selected the interdisciplinary databases ISI Web of Science (also known as Web of Knowledge), the engineering database of IEEE, and the patent search engines of Google and depatisnet. Further, we also included a general Google search to cover grey literature and websites in addition to the research articles, conferences, and patents. The search was conducted between August and December of 2019; therefore we included sources published between 1900 and 31 December 2019.

After the initial search, we removed duplicates and missing full texts. In the next stage, we assessed the documents for eligibility. Here, we excluded records that were outside the focus of this study, i.e., records which did not mention any of the following: a fault, a detection method, or a recovery method. In terms of recovery methods, the focus was put on publications, where the sensor itself can perform the recovery method. This means that the following recovery solutions were excluded: recovery in the form of functional redundancy, an adaption of sensor fusion, direct or hardware redundancy, or fall back to a safe state like "limp home" and shut down.

In the next stage the literature was categorized according to the classification schema introduced in Tables 1–3 and the results were stored in a spreadsheet. Followed by filtering oft records which did not explicitly deal with lidar, leaving 95 records which were further analysed.

The records were stored in a BiBteX file and were analysed together with the spreadsheet file with Python. The bibliography processor Pybtex (https://pybtex.org) was used to parse the BibTeX files. The BibTeX key was used to link the BiBteX file with the spreadsheet. For further details, please refer to the supplements, which include the BibTeX file, the spreadsheet, and the saved Web of Science search queries.

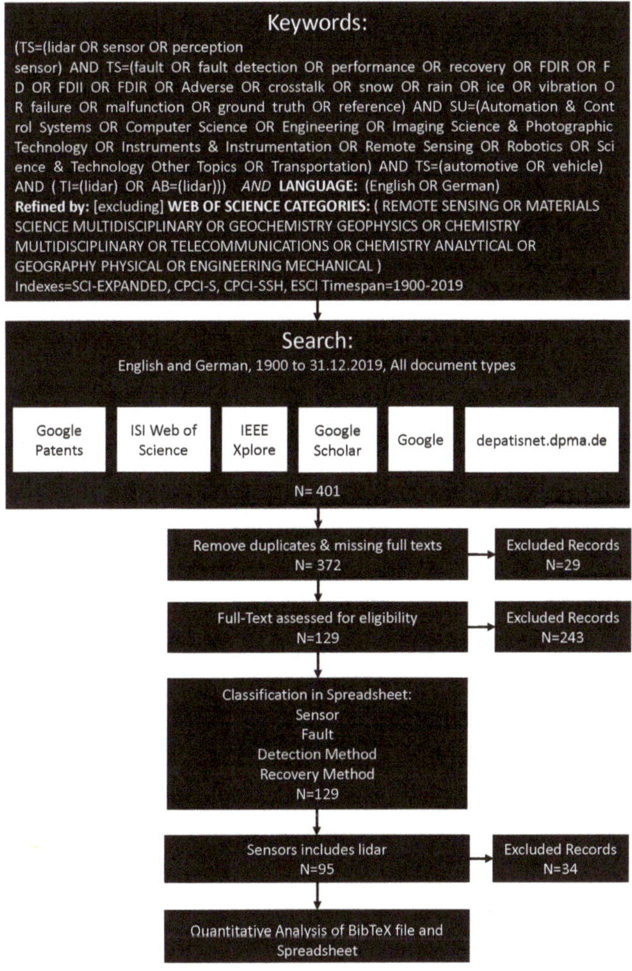

Figure 4. Review process for FDIIR methods of automotive lidar sensors.

6. Literature Survey on FDIIR Methods for Automotive Lidar

This section provides an overview of the state-of-the-art of fault detection, isolation, identification, and recovery methods for automotive lidar and their faults. We identified and collected 95 highly relevant publications. Out of those publications the majority were conference papers, followed by patents and journal articles, as Figure 5a shows. Of those publications, the majority were published in the last five years, as can be seen in panel Figure 5b. Since 2014, there has been a steady increase in publications until 2019. This increase is most likely due to the recent focus of the automotive industry on lidar technology.

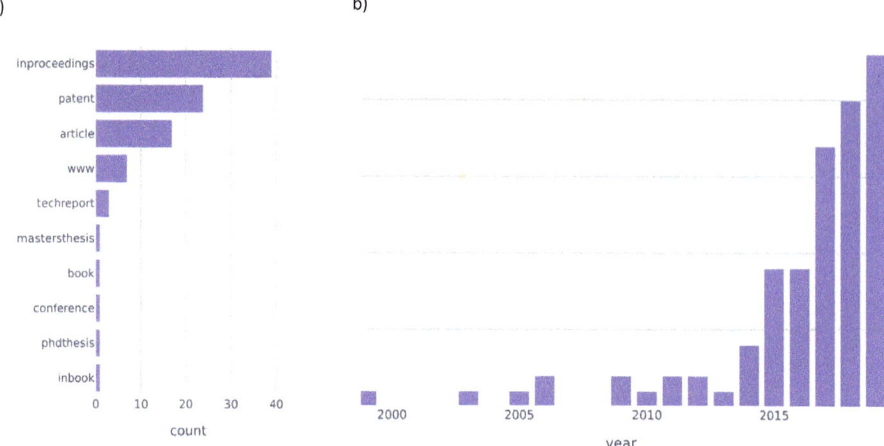

Figure 5. Overview of BibTeX publications which were analysed (N = 95). (**a**) shows the counts of BibTeX classes. (**b**) shows the number of publications per year.

6.1. Fault Classes of Automotive Lidar

First, we classified the literature according to the fault classes. Table 4 and Figure 6 give an overview of the mentioned faults and fault classes in the literature. Faults belonging to the class unfavourable environmental condition are mentioned in 46 documents, making it the class with the most entries. Followed by layer on sensor cover and security attack. Mounting issues are covered in two documents and mechanical damage to sensor cover only in one. The total number of records is larger than the number of publications because one publication can present several fault classes.

Table 4. Literature review of fault classes, FDII and recovery methods for automotive lidar.

Fault Class	Literature
defect subcomponent	[26–28]
mechanical damage to sensor cover	[29]
layer on sensor cover	[29–38]
mounting issue	[39–41]
security attack	[15,16,42–47]
unfavourable environmental condition	[16,28,30,41,48–88]
crosstalk	[17,18,63,89–94]
FDII Class	
comparison to sensor model	[43,95,96]
monitoring sensor output	[18,26,33,38,41,43,46,47,53,60,84,90,97–103]
comparison to static ground-truth	[104,105]
comparison to dynamic ground-truth	[104–107]
comparison to other sensor of same type	[38,108–110]
comparison to other sensor of different type	[27,38,43,46,52,53,62,108–112]
monitoring internal interface	-
comparison of multiple interfaces	[113]
Recovery Class	
software adaption	[18,44,47,70,73–75,90,99,103,112,114–116]
hardware adaption	[39]
temperature adaption	[30,36,38,67,86]
cleaning of sensor cover	[30,32,34,35,37,38]

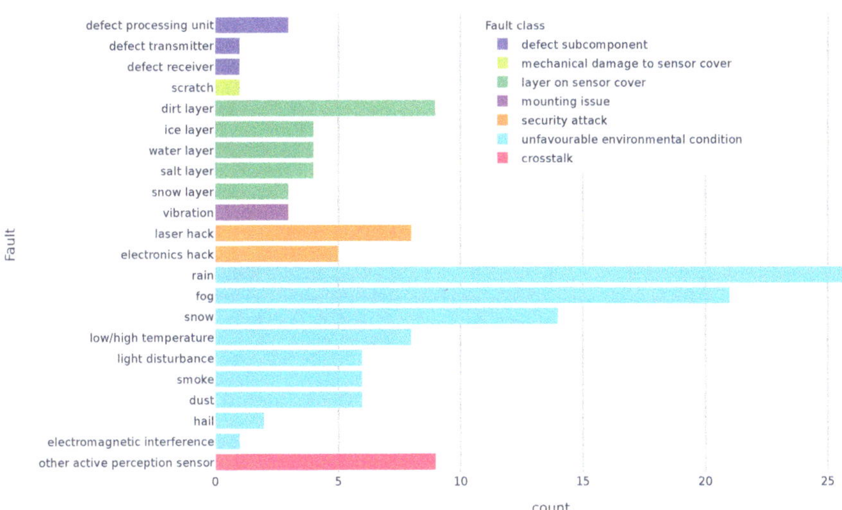

Figure 6. Counts of mentioned faults in the literature.

6.2. FDII Classes and Realizations for Automotive Lidar

Figure 7 shows the number of mentions per FDII class according to the definitions in Section 3. The most common method of FDII is monitoring the output of a single sensor. Detecting faults by comparing different sensors has been described 12 times, which makes it the second most common method. FDII by monitoring internal interface was never mentioned.

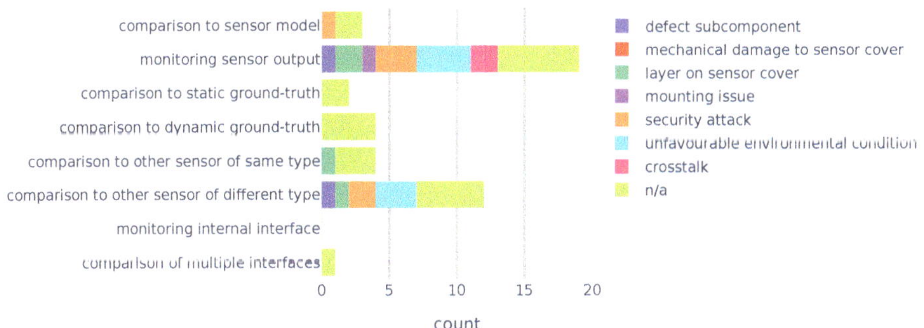

Figure 7. FDII classes count with colour coding according to the mentioned fault class in the publication. Some papers present fault detection methods without mentioning a specific fault, as indicated by the "n/a" category.

In the comparison to sensor models method, residuals are calculated between the output of the sensor and a sensor model. Faults are indicated when the residual is above a threshold value. For example, a general observer scheme with extended Kalman filters is described by Huang and Su [95]. Several publications describe sensor fault detection based on monitoring sensor output with different realizations. For example, Tascione and Bode [101] detects lidar faults by checking if spin rate, laser power, laser frequency, and photodetector alignment are in its tolerance ranges. Jokela et al. [84]

evaluate the standard deviation of range measurements under different environmental conditions. Some publications like Leslar et al. [99] and Segata et al. [26] describe outlier detection. Filters or machine learning approaches may be applied to sensor data to suppress outliers. For example, a supervised learning method is mentioned by Breed [98], which used pattern recognition algorithms and neural networks. Furthermore, James et al. [33] used deep neural networks to classify sensor contaminations. An alternative method used for lidar fault detection is the comparison to static ground-truth which is described in Zhu and Ferguson [105] and Bruns and Venator [104]. As described in these publications, static ground-truth may be landmarks, road markings, or terrain maps. Contrary to using static ground-truth as reference data, some publications describe how to implement the comparison to dynamic ground-truth. Dynamic ground-truth is typically received from other vehicles in the vicinity over V2V or by servers over V2X as described by Mehrdad [106] and Ghimire and Prokhorov [107]. The three patents described in Zhu et al. [108], Zhu et al. [110] and Kim et al. [109] utilize the comparison to sensor of same type. A distance function is used to decide whether an error occurred or not. For example, in Kim et al. [109] a fault is detected if the Mahalanobis distance exceeds a specified threshold. Kim et al. [109] detects the fault first and then recalibrates the sensor whereas Zhu et al. [108] and Zhu et al. [110] describe no further reactions after fault detection. Another method is the comparison to a different type of sensor, which is typically more complex, since the data format has to be transformed before comparison. Dannheim et al. [52] use a camera and a lidar to detect and classify unfavourable environmental conditions like snow, rain, and fog. Daniel et al. [62], as another example, uses a radar, a lidar, and a stereo camera under foggy conditions. Other publications like Choi et al. [46] and Guo et al. [43] describe how to suppress security attacks by comparing between sensors of different types.

6.3. Recovery Methods for Automotive Lidar

In terms of recovery methods for automotive lidar, software adaption was used the most and for a variety of faults (see Figure 8). Algorithms were used to mitigate security hacks, crosstalk, and filter out effects of unfavourable environmental conditions. In addition, software adaption methods were presented without going into detail about the fault, indicated as the "n/a" category in Figure 8. Hardware adaption was only used once to stabilize a lidar on a tractor [39]. Temperature adaption was used to recover from faults caused by a layer on the sensor cover and to keep the temperature within specifications. Naturally, cleaning of sensor cover recovery methods were applied solely when the fault was caused by an unwanted layer on the sensor cover.

The recovery methods used for each individual exemplary fault shown in Table 1, are shown in Figure 9. The majority of publications, which discussed one or more faults, did not cover a suitable recovery method. Especially, faults of the class unfavourable environmental condition and defect subcomponent lacked suitable recovery methods.

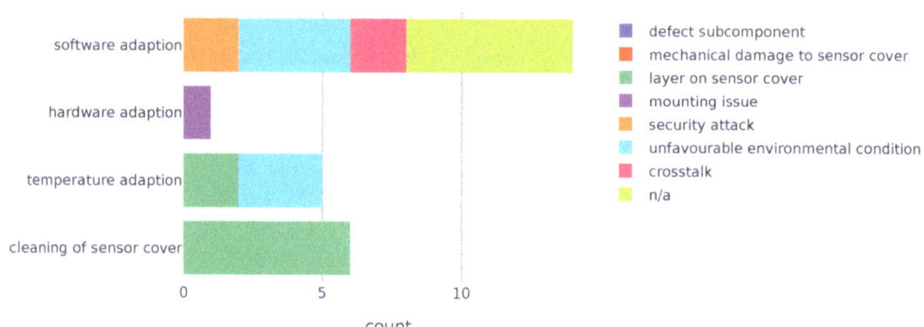

Figure 8. Recovery class count colour coded by fault classes.

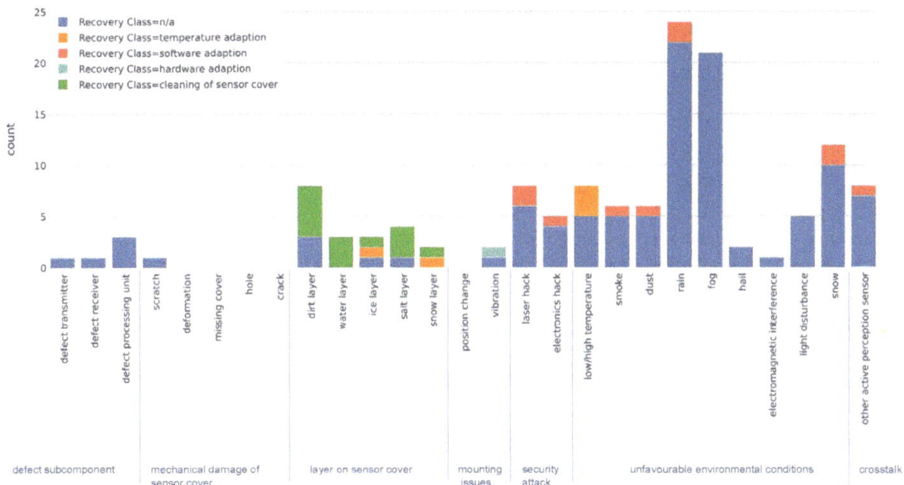

Figure 9. Exemplary faults (Table 1) and their associated recovery class count.

Defect subcomponents are faults that are directly addressed by the manufacturer, and information is not readily available. Only some documentations of data protocols reveal hints about the internals. For example, in Ibeo [28], where error and warnings indicate internal fault detection.

Mechanical damage to the sensor cover by scratches was described in Rivero et al. [29] without a recovery method. However, cover related issues mostly dealt with a layer on the sensor cover.

Disruptive layers on the cover were mostly removed mechanically by air nozzles, spraying fluid, and wiping. Other publications approached the problem by shaking the unit [34] or by heating of the cover [36].

Only vibration was discussed as a mounting issue in the literature in Periu et al. [39]. They introduced an active stabilization mechanism to mitigate vibration effects on an agricultural vehicle. Changes of the sensor's position were not covered in the literature.

Security attacks were addressed via prevention or software adaption. Dutta [47] developed an algorithm based on the least square estimator to mitigate attacks such as a denial of service and false data injection. Petit et al. [44] mitigated attacks with, besides redundancy and probing multiple times, algorithms based on increasing the difficulty for the hacker to hit the right time-window for attacks.

From the fault class unfavourable environmental condition rain and fog were mentioned the most (Figure 8). Nevertheless, only two publications [74,75] applied an algorithm to mitigate its negative effects. Both of them used a filter to minimize reflections from the particles in the air. Peynot et al. [75] use a Bayes filter and Zhang et al. [74] use a histogram and particle filter. Besides algorithms, temperature adaption methods are used to recover from environmental hazards. These are used to keep the sensor temperature within the specified range. Sick [86] used heating, while Wilson and Sinha [67] and Tegler [30] used cooling to keep the temperature within specifications.

Crosstalk by another automotive lidar was investigated by Kim et al. [17] and Zhang et al. [18]. They presented an adaptive algorithm to distinguish between target echo and crosstalk. There are also several patents, e.g., [89,91–94,117], which address methods for preventing crosstalk.

7. Discussion and Conclusions

7.1. Limitations of the Literature Study

Literature that might have used a different terminology could have been overlooked during the search, as well as methods buried deep in the text of patents. Furthermore, the interpretation of the classification of the FDII class is bound to some subjective judging. Nevertheless, the remaining 95 records have been read and interpreted carefully and analysed quantitatively.

7.2. Faults and Fault Classes

The presented lists of fault classes and, in particular, the exemplary faults are certainly not complete, but they are meant to cover all major issues which are of importance to reach the next step in terms of reliable environment perception. Out of our defined fault classes, unfavourable environmental conditions have been studied by far the most for lidar sensors, likely due to its frequent occurrence and its strong impact on the lidar perception principle.

In an alternative classification approach, faults could be classified according to the underlying physics. This would potentially allow common fault identification, isolation, and recovery methods per class. For example, the faults snow, dust, rain, smoke, and fog all can be described by scattering and multiple scattering. In more detail, the relevant description depends on the relationship between the particle size and the laser wavelength. Depending on the laser source, the wavelength is between 0.8 and 1.5 micrometers. Mie scattering can be used if the particle is of similar size compared to the wavelength, e.g., [118–120]. If the particle is smaller than the laser wavelength, Rayleigh scattering can be used. Otherwise, if the particle is much larger than the wavelength like it is the case for hail, non-selective scattering and absorption formulation is suitable. Scattering and absorption are not only relevant for precipitation, but also for issues related to the sensor cover.

7.3. Fault Detection, Isolation, and Identification

Fault detection, isolation, and identification methods based on monitoring the data of a single sensor are attractive since they are independent of other input sources. They can be directly implemented on the sensor firmware and are capable of detecting faults throughout all fault classes.

Third-party development of fault detection, isolation, and identification methods based on internal interfaces are hindered by the manufactures which are very protective of their intellectual property. This is especially the case for automotive lidar as the market is rather new, and competition is high.

7.4. Recovery Methods

Software adaption is the most commonly described recovery class and often implemented in the firmware of the sensor by the manufacturers. Algorithms for recovery can be applied directly to the raw data, before object detection.

7.5. Research Opportunities

Based on the presented survey, the following research opportunities can be identified:

- Mounting issues and mechanical damage to the sensor cover have been studied very little.
- Comparison to static and dynamic ground truth are very promising FDII methods. Still, these methods require more research to evaluate which and how ground truth information can be accessed by the FDIIR system.
- Although unfavourable environmental conditions have been discussed in many publications, possible recovery methods for this fault class were typically absent.
- Literature on defective subcomponents is sparse and no suitable recovery methods were described. This is most likely because commercial lidar units are provided as "black box", because manufacturers are very protective of their intellectual properties. This calls for an open standard in order to access interfaces between subcomponents.
- Due to similar reasons, fault detection, isolation, and identification methods that access internal interfaces are mentioned only once in the literature.
- Detailed investigations of first principles of root causes for faults could lead to fault detection and recovery methods that apply for a wide range of faults.

- Hardware adaption has been studies very little and might be well suited for other faults related to mounting issues.

7.6. Closing Remarks

The introduced classification schema for faults, detection methods, and recovery methods, as well as the literature survey methodology and many of the presented fault detection and recovery methods, are not restricted to the automotive domain, but may also be applied to other domains such as aviation and maritime. The presented FDIIR classification and methods can be relevant for any automated system that requires a very reliable environment perception system (i.e., automated drones, automated trains, automated ships, automated airplanes, …).

In the last five years, an increasing interest in automotive perception sensors, especially lidar, and associated faults and recovery methods has been observed. However, there is still much to be done to offer reliable perception for ADAS/AD functions. Fault detection, isolation, identification, and recovery methods for automotive lidar are still at their infancy.

Author Contributions: Conceptualization, T.G., B.S., S.M.; methodology, T.G., B.S., S.M.; software, T.G.; formal analysis, T.G., B.S.; investigation, T.G., B.S.; data curation, T.G., B.S.; writing–original draft preparation, T.G., B.S., S.M.; writing–review and editing, T.G., B.S., S.M.; visualization, T.G., S.M.; supervision, T.G., S.M.; project administration, T.G., S.M.; funding acquisition, S.M.; All authors have read and agreed to the published version of the manuscript.

Funding: This research received financial support within the COMET K2 Competence Centers for Excellent Technologies from the Austrian Federal Ministry for Climate Action (BMK), the Austrian Federal Ministry for Digital and Economic Affairs (BMDW), the Province of Styria (Dept. 12) and the Styrian Business Promotion Agency (SFG).

Acknowledgments: The publication was written at Virtual Vehicle Research GmbH in Graz, Austria. The authors would like to acknowledge the financial support within the COMET K2 Competence Centers for Excellent Technologies from the Austrian Federal Ministry for Climate Action (BMK), the Austrian Federal Ministry for Digital and Economic Affairs (BMDW), the Province of Styria (Dept. 12) and the Styrian Business Promotion Agency (SFG). The Austrian Research Promotion Agency (FFG) has been authorised for the programme management. The authors would furthermore like to express their thanks to Helmut Martin for a helpful conversation about ISO standards. The security icon by Vitaly Gorbachev is available at flaticon.com.

Conflicts of Interest: The authors declare no conflict of interest. The funders had no role in the design of the study; in the collection, analyses, or interpretation of data; in the writing of the manuscript, or in the decision to publish the results.

Abbreviations

AD	Automated Driving
ADAS	Advanced Driver-Assistance Systems
ASIL	Automotive Safety Integrity Level
FDIIR	Fault Detection Isolation Identification Recovery
FOV	Field Of View
MEMS	MicroElectroMechanical Systems
SAE	Society of Automotive Engineers
V2V	Vehicle-to-vehicle communication
V2X	Vehicle-to-everything communication

References

1. Watzenig, D.; Horn, M. (Eds.) *Automated Driving—Safer and More Efficient Future Driving*; Springer International Publishing: Cham, Switzerland, 2017. [CrossRef]
2. World Health Organisation. *Global Status Report on Road Safety 2018*; Technical Report; World Health Organisation: Geneva, Switzerland, 2018.
3. Anderson, J.; Kalra, N.; Stanley, K.; Sorensen, P.; Samaras, C.; Oluwatola, T. *Autonomous Vehicle Technology: A Guide for Policymakers*; RAND Corporation: Santa Monica, CA, USA, 2014.

4. Fagnant, D.J.; Kockelman, K. Preparing a nation for autonomous vehicles: Opportunities, barriers and policy recommendations. *Transp. Res. Part A Policy Pract.* **2015**, *77*, 167–181. [CrossRef]

5. Thomas, P.; Morris, A.; Talbot, R.; Fagerlind, H. Identifying the causes of road crashes in Europe. *Ann. Adv. Automot. Medicine. Assoc. Adv. Automot. Med. Annu. Sci. Conf.* **2013**, *57*, 13–22.

6. SAE International. *Taxonomy and Definitions for Terms Related to Driving Automation Systems for on-Road Motor Vehicles*; SAE International: Warrendale, PA, USA, 2018. [CrossRef]

7. Marti, E.; de Miguel, M.A.; Garcia, F.; Perez, J. A Review of Sensor Technologies for Perception in Automated Driving. *IEEE Intell. Transp. Syst. Mag.* **2019**, *11*, 94–108. [CrossRef]

8. Isermann, R. *Fault-Diagnosis Systems: An Introduction from Fault Detection to Fault Tolerance*; Springer: Berlin, Germany, 2006.

9. RobustSENSE. Available online: http://www.robustsense.eu/ (accessed on 12 January 2020).

10. Dense. Available online: https://dense247.eu (accessed on 17 October 2019).

11. Hecht, J. Lidar for Self-Driving Cars. *Opt. Photonics News* **2018**, *29*, 26–33. [CrossRef]

12. Thakur, R. Scanning LIDAR in Advanced Driver Assistance Systems and Beyond: Building a road map for next-generation LIDAR technology. *IEEE Consum. Electron. Mag.* **2016**, *5*, 48–54. [CrossRef]

13. Warren, M.E. Automotive LIDAR Technology. In Proceedings of the 2019 Symposium on VLSI Circuits, Kyoto, Japan, 9–14 June 2019; pp. C254–C255. [CrossRef]

14. Druml, N.; Maksymova, I.; Thurner, T.; Lierop, D.; Hennecke, M.; Foroutan, A. 1D MEMS Micro-Scanning LiDAR. In Proceedings of the International Conference on Sensor Device Technologies and Applications (SENSORDEVICES), Venice, Italy, 16–20 September 2018.

15. Petit, J.; Shladover, S.E. Potential cyberattacks on automated vehicles. *IEEE Trans. Intell. Transp. Syst.* **2014**, *16*, 546–556. [CrossRef]

16. Shin, H.; Kim, D.; Kwon, Y.; Kim, Y. Illusion and Dazzle: Adversarial Optical Channel Exploits against Lidars for Automotive Applications. In *Lecture Notes in Computer Science*; Springer International Publishing: Cham, Switzerland, 2018; pp. 445–467. [CrossRef]

17. Kim, G.; Eom, J.; Park, Y. An Experiment of Mutual Interference between Automotive LIDAR Scanners. In Proceedings of the 2015 12th International Conference on Information Technology—New Generations, Las Vegas, NV, USA, 13–15 April 2015; pp. 680–685. [CrossRef]

18. Zhang, F.; Liu, Q.; Gong, M.; Fu, X. Anti-dynamic-crosstalk method for single photon LIDAR detection. In Proceedings of the SPIE 10605, LIDAR Imaging Detection and Target Recognition 2017, Bellingham, WA, USA, 23–25 July 2017; Lv, Y., Su, J., Gong, W., Yang, J., Bao, W., Chen, W., Shi, Z., Fei, J., Han, S., Jin, W., Eds.; International Society for Optics and Photonics: Bellingham, WA, USA, 2017; Volume 10605, p. 1060503. [CrossRef]

19. ISO26262. *26262: Road Vehicles-Functional Safety*; International Standard; International Organization for Standardization: Geneva, Switzerland, 2011.

20. ISO/PAS 21448:2019. *Road Vehicles—Safety of the Intended Functionality*; Technical Report; International Organization for Standardization: Geneva, Switzerland, 2019.

21. Chen, J.; Patton, R. *Robust Model-based Fault Diagnosis for Dynamic Systems*, 1st ed.; Springer: New York, NY, USA, 1999; Volume 3. [CrossRef]

22. Chen, X.; Wang, Z.; Zhang, Z.; Jia, L.; Qin, Y. A Semi-Supervised Approach to Bearing Fault Diagnosis under Variable Conditions towards Imbalanced Unlabeled Data. *Sensors* **2018**, *18*, 2097. [CrossRef]

23. Piltan, F.; Kim, J.M. Bearing Fault Diagnosis by a Robust Higher-Order Super-Twisting Sliding Mode Observer. *Sensors* **2018**, *18*, 1128. [CrossRef] [PubMed]

24. Liberati, A.; Altman, D.G.; Tetzlaff, J.; Mulrow, C.; Gøtzsche, P.C.; Ioannidis, J.P.A.; Clarke, M.; Devereaux, P.J.; Kleijnen, J.; Moher, D. The PRISMA Statement for Reporting Systematic Reviews and Meta-Analyses of Studies That Evaluate Health Care Interventions: Explanation and Elaboration. *PLoS Med.* **2009**, *6*, 1–28. [CrossRef] [PubMed]

25. Yunusa-Kaltungo, A.; Labib, A. A hybrid of industrial maintenance decision making grids. *Prod. Plan. Control* **2020**, 1–18. [CrossRef]

26. Segata, M.; Cigno, R.L.; Bhadani, R.K.; Bunting, M.; Sprinkle, J. A LiDAR Error Model for Cooperative Driving Simulations. In Proceedings of the 2018 IEEE Vehicular Networking Conference (VNC), Taipei, Taiwan, 5–7 December 2018; pp. 1–8. [CrossRef]

27. Sun, X.H. Method and Apparatus for Detection and Ranging Fault Detection and Recovery. U.S. Patent 010203408B, 12 February 2019.

28. Ibeo. *Ethernet Data Protocol Ibeo LUX and Ibeo LUX Systems*; Technical Report; Ibeo: Hamburg, Germany, 2009.

29. Rivero, J.R.V.; Tahiraj, I.; Schubert, O.; Glassl, C.; Buschardt, B.; Berk, M.; Chen, J. Characterization and simulation of the effect of road dirt on the performance of a laser scanner. In Proceedings of the 2017 IEEE 20th International Conference on Intelligent Transportation Systems (ITSC), Yokohama, Japan, 16–19 October 2017; pp. 1–6. [CrossRef]

30. Tegler, E. Available online: https://www.popularmechanics.com/cars/car-technology/a29309664/bugs-self-driving-cars/ (accessed on 17 September 2019).

31. Trierweiler, M.; Caldelas, P.; Gröninger, G.; Peterseim, T.; Neumann, C. Influence of sensor blockage on automotive LiDAR systems. In Proceedings of the 2019 IEEE SENSORS, Montreal, QC, Canada, 27–30 October 2019; pp. 1–4. [CrossRef]

32. Jalopnik. Available online: https://jalopnik.com/google-has-a-wiper-for-its-lidar-to-handle-bird-shit-an-1795346982 (accessed on 5 October 2019).

33. James, J.K.; Puhlfürst, G.; Golyanik, V.; Stricker, D. Classification of LIDAR Sensor Contaminations with Deep Neural Networks. In Proceedings of the Computer Science in Cars Symposium (CSCS) 2018, Munich, Germany, 13–14 September 2018; p. 8. [CrossRef]

34. Green, A.; Bidner, D.K. Lidar Windscreen Vibration Control. U.S. Patent 20190359179A1, 28 November 2019.

35. Zhao, C.; Gopalan, S. Automotive Image Sensor Surface Washing and Drying System. U.S. Patent 20190202410A, 4 July 2019.

36. Wideye. Available online: https://www.wideye.vision/use-cases-value/lidar-covers/ (accessed on 17 September 2019).

37. Doorley, G.; Karplus, P.T.H.; Avram, P. Control for Passive Wiper System. U.S. Patent 2017/0151933 A1, 1 June 2017.

38. McMichael, R.; Kentley-Klay, T.D.; Schabb, D.E.; Thakur, A.; Torrey, J. Sensor Obstruction Detection and Mitigation Using Vibration and/ot Heat. U.S. Patent WO 2019246029A1, 19 December 2019.

39. Periu, C.; Mohsenimanesh, A.; Laguë, C.; McLaughlin, N. Isolation of Vibrations Transmitted to a LIDAR Sensor Mounted on an Agricultural Vehicle to Improve Obstacle Detection. *Can. Biosyst. Eng.* **2013**, *55*, 233–242. [CrossRef]

40. Hama, S.; Toda, H. Basic Experiment of LIDAR Sensor Measurement Directional Instability for Moving and Vibrating Object. In Proceedings of the 6th Annual International Conference on Material Science and Environmental Engineering, Chongqing, China, 23–25 November 2018; IOP Conference Series-Materials Science and Engineering; Wang, K., Ed.; IOP Publishing Ltd.: Bristol, UK, 2019; Volume 472. [CrossRef]

41. McManamon, P. *Field Guide to Lidar*; SPIE Field Guides; SPIE: Bellingham, WA, USA, 2015; Volume FG36.

42. Stottelaar, B.G. Practical Cyber-Attacks on Autonomous Vehicles. Master's Thesis, University of Twente, Enschede, The Netherlands, 2015.

43. Guo, P.; Kim, H.; Virani, N.; Xu, J.; Zhu, M.; Liu, P. RoboADS: Anomaly Detection Against Sensor and Actuator Misbehaviors in Mobile Robots. In Proceedings of the 2018 48th Annual IEEE/IFIP International Conference on Dependable Systems and Networks (DSN), Luxembourg, 25–28 June 2018; pp. 574–585. [CrossRef]

44. Petit, J.; Stottelaar, B.; Feiri, M.; Kargl, F. Remote attacks on automated vehicles sensors: Experiments on camera and lidar. *Black Hat Eur.* **2015**, *11*, 2015.

45. Thing, V.L.L.; Wu, J. Autonomous Vehicle Security: A Taxonomy of Attacks and Defences. In Proceedings of the 2016 IEEE International Conference on Internet of Things (iThings) and IEEE Green Computing and Communications (GreenCom) and IEEE Cyber, Physical and Social Computing (CPSCom) and IEEE Smart Data (SmartData), Chengdu, China, 15–18 December 2016; pp. 164–170. [CrossRef]

46. Choi, H.; Lee, W.C.; Aafer, Y.; Fei, F.; Tu, Z.; Zhang, X.; Xu, D.; Deng, X. Detecting Attacks Against Robotic Vehicles: A Control Invariant Approach. In Proceedings of the 2018 ACM SIGSAC Conference on Computer and Communications Security, CCS '18, Toronto, ON, Canada, 15–19 October 2018; ACM: New York, NY, USA, 2018; pp. 801–816. [CrossRef]

47. Dutta, R.G. Security of Autonomous Systems Under Physical Attacks: With Application to Self-Driving Cars. Ph.D. Thesis, University of Central Florida, Orlando, FL, USA, 2018.

48. Zang, S.; Ding, M.; Smith, D.; Tyler, P.; Rakotoarivelo, T.; Kaafar, M.A. The Impact of Adverse Weather Conditions on Autonomous Vehicles: How Rain, Snow, Fog, and Hail Affect the Performance of a Self-Driving Car. *IEEE Veh. Technol. Mag.* **2019**, *14*, 103–111. [CrossRef]

49. Filgueira, A.; González-Jorge, H.; Lagüela, S.; Díaz-Vilariño, L.; Arias, P. Quantifying the influence of rain in LiDAR performance. *Measurement* **2017**, *95*, 143–148. [CrossRef]

50. Hasirlioglu, S.; Doric, I.; Kamann, A.; Riener, A. Reproducible Fog Simulation for Testing Automotive Surround Sensors. In Proceedings of the 2017 IEEE 85th Vehicular Technology Conference (VTC Spring), Sydney, NSW, Australia, 4–7 June 2017; pp. 1–7. [CrossRef]

51. Kim, B.K.; Sumi, Y. Performance evaluation of safety sensors in the indoor fog chamber. In Proceedings of the 2017 IEEE Underwater Technology (UT), Busan, Korea, 21–24 February 2017; pp. 1–3. [CrossRef]

52. Dannheim, C.; Icking, C.; Mader, M.; Sallis, P. Weather Detection in Vehicles by Means of Camera and LIDAR Systems. In Proceedings of the 2014 Sixth International Conference on Computational Intelligence, Communication Systems and Networks, Tetova, Macedonia, 27–29 May 2014; pp. 186–191. [CrossRef]

53. RobustSENSE. *Deliverable D5.1 System Performance Assessment Specification*; Technical Report; RobustSENSE: Ulm, Germany, 2016.

54. Hasirlioglu, S.; Riener, A. Introduction to Rain and Fog Attenuation on Automotive Surround Sensors. In Proceedings of the 2017 IEEE 20th International Conference on Intelligent Transportation Systems (ITSC), Yokohama, Japan, 16–19 October 2017. [CrossRef]

55. Mäyrä, A.; Hietala, E.; Kutila, M.; Pyykönen, P. Spectral attenuation in low visibility artificial fog: Experimental study and comparison to literature models. In Proceedings of the 2017 13th IEEE International Conference on Intelligent Computer Communication and Processing (ICCP), Cluj-Napoca, Romania, 7–9 September 2017; pp. 303–308. [CrossRef]

56. Mäyrä, A.; Hietala, E.; Kutila, M.; Pyykönen, P.; Tiihonen, M.; Jokela, T. Experimental study on spectral absorbance in fog as a function of temperature, liquid water content, and particle size. In *Optics in Atmospheric Propagation and Adaptive Systems*; International Society for Optics and Photonics: Bellingham, WA, USA, 2017; Volume 10425, p. 104250G. [CrossRef]

57. Heinzler, R.; Schindler, P.; Seekircher, J.; Ritter, W.; Stork, W. Weather Influence and Classification with Automotive Lidar Sensors. In Proceedings of the 2019 IEEE Intelligent Vehicles Symposium (IV), Paris, France, 9–12 June 2019; IEEE: Piscataway, NJ, USA, 2019; pp. 1527–1534. [CrossRef]

58. Aho, A.T.; Viheriälä, J.; Mäkelä, J.; Virtanen, H.; Ranta, S.; Dumitrescu, M.; Guina, M. High-power 1550 nm tapered DBR laser diodes for LIDAR applications. In Proceedings of the 2017 Conference on Lasers and Electro-Optics Europe European Quantum Electronics Conference (CLEO/Europe-EQEC), Munich, Germany, 25–29 June 2017. [CrossRef]

59. Goodin, C.; Carruth, D.; Doude, M.; Hudson, C. Predicting the Influence of Rain on LIDAR in ADAS. *Electronics* **2019**, *8*, 89. [CrossRef]

60. Ruiz-Llata, M.; Rodriguez-Cortina, M.; Martin-Mateos, P.; Bonilla-Manrique, O.E.; Ramon Lopez-Fernandez, J. LiDAR design for Road Condition Measurement ahead of a moving vehicle. In Proceedings of the 2017 16th IEEE Sensors, IEEE Sensors, Glasgow, UK, 29 October–1 November 2017; pp. 1062–1064. [CrossRef]

61. Rosique, F.; Navarro, P.J.; Fernández, C.; Padilla, A. A Systematic Review of Perception System and Simulators for Autonomous Vehicles Research. *Sensors* **2019**, *19*, 648. [CrossRef]

62. Daniel, L.; Phippen, D.; Hoare, E.; Stove, A.; Cherniakov, M.; Gashinova, M. Low-THz radar, lidar and optical imaging through artificially generated fog. In Proceedings of the International Conference on Radar Systems 2017, Belfast, UK, 23–26 October 2017; pp. 1–4. [CrossRef]

63. Rasshofer, R.H.; Spies, M.; Spies, H. Influences of weather phenomena on automotive laser radar systems. *Adv. Radio Sci.* **2011**, *9*, 49–60. [CrossRef]

64. Xique, I.J.; Buller, W.; Fard, Z.B.; Dennis, E.; Hart, B. Evaluating Complementary Strengths and Weaknesses of ADAS Sensors. In Proceedings of the 2018 IEEE 88th Vehicular Technology Conference (VTC-Fall), Chicago, IL, USA, 27–30 August 2018; pp. 1–5.

65. Kutila, M.; Pyykonen, P.; Ritter, W.; Sawade, O.; Schaufele, B. Automotive LIDAR sensor development scenarios for harsh weather conditions. In Proceedings of the 2016 IEEE 19th International Conference on Intelligent Transportation Systems (ITSC), Rio de Janeiro, Brazil, 1–4 November 2016; pp. 265–270. [CrossRef]

66. Phillips, T.G.; Guenther, N.; McAree, P.R. When the Dust Settles: The Four Behaviors of LiDAR in the Presence of Fine Airborne Particulates. *J. Field Robot.* **2017**, *34*, 985–1009. [CrossRef]

67. Wilson, J.; Sinha, P. Thermal Considerations For Designing Reliable and Safe Autonomous Vehicle Sensors. Available online: https://www.fierceelectronics.com/components/thermal-considerations-for-designing-reliable-and-safe-autonomous-vehicle-sensors (accessed on 25 September 2019).

68. Hasirlioglu, S.; Riener, A.; Huber, W.; Wintersberger, P. Effects of exhaust gases on laser scanner data quality at low ambient temperatures. In Proceedings of the 2017 IEEE Intelligent Vehicles Symposium (IV), Los Angeles, CA, USA, 11–14 June 2017; pp. 1708–1713. [CrossRef]

69. Hasirlioglu, S.; Kamann, A.; Doric, I.; Brandmeier, T. Test methodology for rain influence on automotive surround sensors. In Proceedings of the 2016 IEEE 19th International Conference on Intelligent Transportation Systems (ITSC), Rio de Janeiro, Brazil, 1–4 November 2016; pp. 2242–2247. [CrossRef]

70. Moorehead, S.; Simmons, R.; Apostolopoulos, D.D.; Whittaker, W.R.L. Autonomous Navigation Field Results of a Planetary Analog Robot in Antarctica. In Proceedings of the International Symposium on Artificial Intelligence, Robotics and Automation in Space, Noordwijk, The Netherlands, 1–3 June 1999.

71. Pfeuffer, A.; Dietmayer, K. Optimal Sensor Data Fusion Architecture for Object Detection in Adverse Weather Conditions. *arXiv* **2018**, arXiv:1807.02323.

72. Sick. Available online: https://lidarnews.com/articles/top-5-considerations-for-choosing-lidar-for-outdoor-robots/ (accessed on 25 September 2019).

73. Waterloo Autonomous Vehicles Lab. Available online: http://wavelab.uwaterloo.ca/?weblizar_portfolio=real-time-filtering-of-snow-from-lidar-point-clouds (accessed on 25 September 2019).

74. Zhang, C.; Ang, M.H.; Rus, D. Robust LIDAR Localization for Autonomous Driving in Rain. In Proceedings of the 2018 IEEE/RSJ International Conference on Intelligent Robots and Systems (IROS), Madrid, Spain, 1–5 October 2018; pp. 3409–3415. [CrossRef]

75. Peynot, T.; Underwood, J.; Scheding, S. Towards reliable perception for Unmanned Ground Vehicles in challenging conditions. In Proceedings of the 2009 IEEE/RSJ International Conference on Intelligent Robots and Systems, St. Louis, MO, USA, 10–15 October 2009; pp. 1170–1176. [CrossRef]

76. Bijelic, M.; Gruber, T.; Ritter, W. A Benchmark for Lidar Sensors in Fog: Is Detection Breaking Down? In Proceedings of the 2018 IEEE Intelligent Vehicles Symposium (IV), Changshu, China, 26–30 June 2018. [CrossRef]

77. Hasirlioglu, S.; Riener, A. A Model-based Approach to Simulate Rain Effects on Automotive Surround Sensor Data. In Proceedings of the 2018 21st International Conference on Intelligent Transportation Systems (ITSC), Maui, HI, USA, 4–7 November 2018; pp. 2609–2615. [CrossRef]

78. Ijaz, M.; Ghassemlooy, Z.; Minh, H.L.; Rajbhandari, S.; Perez, J. Analysis of Fog and Smoke Attenuation in a Free Space Optical Communication Link. In Proceedings of the 2012 International Workshop on Optical Wireless Communications (IWOW), Pisa, Italy, 22 October 2012; pp. 1–3.

79. Fersch, T.; Buhmann, A.; Koelpin, A.; Weigel, R. The Influence and of Rain on Small Aperture and LiDAR Sensors. In Proceedings of the German Microwave Conference (GeMiC) 2016, Bochum, Germany, 14–16 March 2016.

80. Hadj-Bachir, M.; Souza, P. *LIDAR Sensor Simulation in Adverse Weather Condition for Driving Assistance Development*; hal-01998668; Hyper Articles en Ligne (HAL): Lyon, France, 2019.

81. Kutila, M.; Pyykönen, P.; Holzhüter, H.; Colomb, M.; Duthon, P. Automotive LiDAR performance verification in fog and rain. In Proceedings of the 2018 21st International Conference on Intelligent Transportation Systems (ITSC), Maui, HI, USA, 4–7 November 2018; pp. 1695–1701. [CrossRef]

82. Hasirlioglu, S.; Doric, I.; Lauerer, C.; Brandmeier, T. Modeling and simulation of rain for the test of automotive sensor systems. In Proceedings of the 2016 IEEE Intelligent Vehicles Symposium (IV), Gotenburg, Sweden, 19–22 June 2016; pp. 286–291. [CrossRef]

83. Schönhuber, M. Available online: https://www.distrometer.at/2dvd/ (accessed on 13 September 2019).

84. Jokela, M.; Kutila, M.; Pyykönen, P. Testing and Validation of Automotive Point-Cloud Sensors in Adverse Weather Conditions. *Appl. Sci.* **2019**, *9*, 2341. [CrossRef]

85. Michaud, S.; Lalonde, J.F.; Giguere, P. Towards Characterizing the Behavior of LiDARs in Snowy Conditions. In Proceedings of the 7th Workshop on Planning, Perception and Navigation for Intelligent Vehicles, Hamburg, Germany, 28 September 2015.

86. Sick. Available online: https://www.sick.com/ag/en/detection-and-ranging-solutions/2d-lidar-sensors/lms1xx/c/g91901 (accessed on 18 October 2019).

87. O'Brien, M.E.; Fouche, D.G. Simulation of 3D Laser Radar Systems. *Linc. Lab. J.* **2005**, *15*, 37–60.

88. McKnight, D.; Miles, R. Impact of reduced visibility conditions on laser based DP sensors. In Proceedings of the Dynamic Positioning Conference, Houston, TX, USA, 14–15 October 2014.

89. Lingg, A.J.; Beck, S.W.; Stepanian, J.G.; Clifford, D.H. Method and Apparatus Crosstalk and Multipath Noise Reduction in A Lidar System. U.S. Patent 2019/0339393A1, 7 November 2019.

90. Diehm, A.L.; Hammer, M.; Hebel, M.; Arens, M. Mitigation of crosstalk effects in multi-LiDAR configurations. In Proceedings of the Electro-Optical Remote Sensing XII, Berlin, Germany, 12–13 Septenber 2018; Kamerman, G., Steinvall, O., Eds.; SPIE: Washington, DC, USA, 2018; Volume 10796. [CrossRef]

91. Retterath, J.E.; Laumeyer, R.A. Methods and Apparatus for Array Based LiDAR Systems With Reduced Interference. U.S. Patent 10,203,399, 12 February 2019.

92. Denham, M.; Gilliland, P.B.; Goldstein, B.M.; Musa, O. Crosstalk Mitigation Circuit for Lidar Pixel Receivers. U.S. Patent 2019/0317196A1, 17 October 2019.

93. Hall, D.S. High Definition LiDAR System. U.S. Patent 7969558B2, 28 June 2011.

94. Eichenholz, J.M.; LaChapelle, J.G. Detection of Crosstalk and Jamming Pulses With Lidar System. U.S. Patent 16/178,049, 2 May 2019.

95. Huang, W.; Su, X. Design of a Fault Detection and Isolation System for Intelligent Vehicle Navigation System. *Int. J. Navig. Obs.* **2015**, *2015*, 1–19. [CrossRef]

96. Qin, S.J.; Guiver, J.P. Sensor Validation Apparatus and Method. U.S. Patent 6594620B1, 15 July 2003.

97. Song, J.; Fry, G.; Wu, C.; Parmer, G. CAML: Machine Learning-based Predictable, System-Level Anomaly Detection. In Proceedings of the 1st Workshop on Security and Dependability of Critical Embedded Real-Time Systems, in Conjunction with IEEE Real-Time Systems Symposium, Porto, Portugal, 29 November 2016.

98. Breed, D.S. System and Method for Vehicle Diagnostics. U.S. Patent 7103460B1, 5 September 2006.

99. Leslar, M.; Wang, J.; Hu, B. A Comparison of Two New Methods of Outlier Detection for Mobile Terrestrial Lidar Data. *Proc. Int. Arch. Photogramm. Remote Sens. Spatial Inf. Sci.* **2010**, *38*, 78–84.

100. Khalastchi, E.; Kaminka, G.; Lin, R.; Kalech, M. Anomaly Detection Methods, Devices and Systems. U.S. Patent 2014/0149806A1, 29 May 2014.

101. Tascione, D.; Bode, M. Autonomous Vehicle Diagnostic System. U.S. Patent 2018/0050704A1, 22 February 2018.

102. Abt, T.L.; Hirsenkorn, K.; Isert, C.; Parolini, L.; Radler, S.; Rauch, S. Method, System, and Computer Program Product for Determining a Blockage of a Sensor of a Plurality of Sensors of an Ego Vehicle. EP 3511740A1, 17 July 2019.

103. Mori, D.; Sugiura, H.; Hattori, Y. Adaptive Sensor Fault Detection and Isolation using Unscented Kalman Filter for Vehicle Positioning. In Proceedings of the 2019 IEEE Intelligent Transportation Systems Conference (ITSC), Auckland, New Zealand, 27–30 October 2019; pp. 1298–1304. [CrossRef]

104. Bruns, E.; Venator, M. Verfahren und Steuervorrichtung zum Erkennen einer Fehlfunktion zumindest eines Umfeldsensors eines Kraftfahrzeugs. DE 102018205322A1, 10 April 2019.

105. Zhu, J.; Ferguson, D.I. System to Optimize Sensor Parameters in an Autonomous Vehicle. U.S. Patent 2018/0329423A1, 15 November 2018.

106. Mehrdad, S.K. Verfahren Zum Verarbeiten Von Sensordaten, Anordnung Zur Verarbeitung Von Sensordaten, Anordnung Zum Verarbeiten Einer Anfrage Eines Ego-Fahrzeugs Für Ein Weiteres Fahrzeug, Rechner Zum Verarbeiten Von Sensordaten Sowie Rechner Zum Verarbeiten Einer Anfrage Eines Ego-Fahrzeugs Für Ein Weiteres Fahrzeug. DE 102018207658A1, 21 November 2019.

107. Ghimire, S.; Prokhorov, D. Collaborative Multi-Agent Vehicle Fault Diagnostic System & Associated Methodology. WO 2012/148514 A1, 1 November 2012.

108. Zhu, J.; Dolgov, D.A.; Urmson, C.P. Cross-Validating Sensors of an Autonomous Vehicle. U.S. Patent 9555740B1, 31 January 2017.

109. Kim, H.S.; Seo, M.W.; Yang, J.W.; Kim, S.Y.; Yang, D.H.; Ko, B.C.; Seo, G.W. Apparatus and Method for Failure Diagnosis and Calibration of Sensors for advanced Driver Assistance Systems. U.S. Patent 10026239B2, 17 July 2018.

110. Zhu, J.; Dolgov, D.; Urmson, C.P. Cross-Validating Sensors of an Autonomous Vehicle. U.S. Patent 9221396B1, 29 December 2015.
111. Hartung, J.J.; Brink, P.; Lamb, J.; Miller, D.P. Autonomous Vehicle Platform and Safety Architecture. U.S. Patent 2017139411A1, 18 May 2017.
112. Gao, B.; Coifman, B. Vehicle identification and GPS error detection from a LIDAR equipped probe vehicle. In Proceedings of the 2006 IEEE Intelligent Transportation Systems Conference, Toronto, ON, Canada, 17–20 September 2006; pp. 1537–1542. [CrossRef]
113. Egnor, D.T.; Zbrozek, A.; Schultz, A. Methods and Systems for Compensating for Common Failures. U.S. Patent 9,195.232 B1, 24 November 2015.
114. Ning, X.; Li, F.; Tian, G.; Wang, Y. An efficient outlier removal method for scattered point cloud data. *PLoS ONE* **2018**, *13*, e0201280. [CrossRef]
115. Shin, B. *Fault Tolerant Control and Localization for Autonomous Driving: Systems and Architecture*; Technical Report UCB/EECS-2016-83; University of California at Berkeley: Berkeley, CA, USA, 2016.
116. Sazara, C.; Nezafat, R.V.; Cetin, M. Offline reconstruction of missing vehicle trajectory data from 3D LIDAR. In Proceedings of the 2017 IEEE Intelligent Vehicles Symposium (IV), Los Angeles, CA, USA, 11–14 June 2017; pp. 792–797. [CrossRef]
117. Avlas, L.N.; Berseth, E.N. Systems and methods for mitigating optical crosstalk in a light ranging and detection system. U.S. Patent 2020/0064452A1, 27 February 2020.
118. El-Nayal, M.K.; Aly, M.M.; Fayed, H.A.; AbdelRassoul, R.A. Adaptive free space optic system based on visibility detector to overcome atmospheric attenuation. *Results Phys.* **2019**, *14*, 102392. [CrossRef]
119. Colomb, M.; Duthon, P.; Laukkanen, S. *Characteristics of Adverse Weather Conditions*; Technical Report, DENSE247.eu; DENSE Consortium: Ulm, Germany, 2017.
120. Hespel, L.; Riviere, N.; Huet, T.; Tanguy, B.; Ceolato, R. Performance evaluation of laser scanners through the atmosphere with adverse condition. In *Electro-Optical Remote Sensing, Photonic Technologies, and Applications V*; Kamerman, G.W., Steinvall, O., Bishop, G.J., Gonglewski, J.D., Lewis, K.L., Hollins, R.C., Merlet, T.J., Eds.; International Society for Optics and Photonics: Bellingham, WA, USA, 2011; Volume 8186, p. 818606. [CrossRef]

Article

A Novel Intelligent Fault Diagnosis Method for Rolling Bearing Based on Integrated Weight Strategy Features Learning

Jun He [1], Ming Ouyang [1,*], Chen Yong [1], Danfeng Chen [1], Jing Guo [1] and Yan Zhou [2]

[1] College of Automation, Foshan University, Foshan City 528000, Guangdong Province, China;
    hejun_723@fosu.edu.cn (J.H.); cheny@fosu.edu.cn (C.Y.); cdf2017@fosu.edu.cn (D.C.);
    guojing@fosu.edu.cn (J.G.)
[2] College of Computer Science, Foshan University, Foshan City 528000, Guangdong Province, China;
    zhouyan791266@163.com
*   Correspondence: ouymouym@outlook.com; Tel.: +86-134-3311-0891

Received: 2 February 2020; Accepted: 19 March 2020; Published: 23 March 2020

Abstract: Intelligent methods have long been researched in fault diagnosis. Traditionally, feature extraction and fault classification are separated, and this process is not completely intelligent. In addition, most traditional intelligent methods use an individual model, which cannot extract the discriminate features when the machines work in a complex condition. To overcome the shortcomings of traditional intelligent fault diagnosis methods, in this paper, an intelligent bearing fault diagnosis method based on ensemble sparse auto-encoders was proposed. Three different sparse auto-encoders were used as the main architecture. To improve the robustness and stability, a novel weight strategy based on distance metric and standard deviation metric was employed to assign the weights of three sparse auto-encodes. Softmax classifier is used to classify the fault types of integrated features. The effectiveness of the proposed method is validated with extensive experiments, and comparisons with the related methods and researches on the widely-used motor bearing dataset verify the superiority of the proposed method. The results show that the testing accuracy and the standard deviation are 99.71% and 0.05%.

Keywords: stacked auto-encoder; weighting strategy; deep learning; bearing fault diagnosis

---

1. Introduction

With the upgrading of industrial capacity, the connection between machine equipment is increasingly inseparable. Once unexpected faults happen in a machine, it may indirectly effect the reliability of other connected machineries [1]. These failures will cause heavy economic loss, and even more seriously, they could be life-threatening [2]. Therefore, the automatic, accurate, and timely recognition of the health conditions of machine equipment is highly necessary [3,4].

In the past few years, intelligent fault diagnosis methods have attracted great attentions and widely adopted in the condition monitoring systems [5,6]. Generally, intelligent fault diagnosis methods can be divided three main steps: (1) signals acquisition; (2) feature extraction and selection; (3) fault classification [7,8]. After a literature review, it can be found that a tremendous amount of researches have focused on how to extract discriminative features from collected signals based on abundant signal processing technologies [9,10], such as time-domain [11,12], frequency-domain [13], time-frequency-domain statistics analytical methods [14], or other waveform transform methods [15,16]. To classify the extracted features, a few artificial intelligence methods (ANN, SVM, etc.) are applied. For instance, Fu et al. [17] proposed a novel hybrid approach coupling variational mode decomposition and SVM to identify rolling bearing fault types. Ali et al. [15] used empirical mode decomposition to extract 10 time-domain statistical features and an artificial neural network is used to identify the health conditions of rolling bearing. He et al. [18] proposed an ensemble error minimized learning machine method to recognize rolling bearing faults, empirical mode decomposition technology is adopted to extract the ensemble time-domain features. However, although these traditional intelligent methods did work and achieved an accurate diagnosis result, they still have two deficiencies: (1) the features are usually manually extracted depending on prior knowledge and diagnostic expertise, which accorded to a specific fault type and probably unsuitable for other faults [19,20]; (2) In real industries, the collected signals are usually exposed to environmental noises, which cause the signals to be complex and non-stationary, and signal processing technologies need to be employed to filter the collected signals to obtain the effective features [3,21]. Consequently, there is an urgent need to develop new intelligent fault diagnosis methods to accomplish fault diagnosis tasks automatically.

As an emerging research field, deep learning has a powerful ability to extract the representative features from the collected signals, which makes it has the potential to overcome the shortcomings of the traditional intelligent diagnosis methods [22,23]. The advantage of deep learning is that can automatically learn discriminative features and classified faults, which removes the requirements of manual feature extraction and prior knowledge from the diagnosis model. After more than ten years of development, deep learning has been gradually applied to the field of fault diagnosis. For example, Liu et al. [24] presented a fault diagnosis method for rolling bearings based on convolution neural network (CNN) in which the step k is used to discretize the vibration signal and the discrete sequence as the input data of CNN. Jia et al. [25] used the normalized sparse AEs to constitute local connection network, and the model can learn to avoid similar, repeated features and overcome the problem of feature change. Shao et al. [26] proposed an improved convolution deep belief network method based on compressed sensing technology, this method used compressed data as the input of the model and obtained less time consumption of the fault diagnosis. A novel cross-domain fault diagnosis method was proposed by Li et al. [27] whereby multiple deep generative neural networks were employed to generate corresponding-domain fake samples, and faults in different domains could be discriminated well. Long et al. [28] used a competitive swarm optimizer and a local search algorithm to optimize the weights of echo state networks for decreasing the affect caused by random selection of input weights and reservoir weights. Although the above researches are successfully applied in fault diagnosis, there still exist shortcomings in that these intelligent diagnosis methods based on deep learning mainly focus on the research of the individual learning model. Due to complexity of the collected vibration data and even there are exiting the imbalance between different data [29], the generalization can seldom perform well consistently when used individual deep learning model. This problem derives from the limitation of individual deep learning models for the fault diagnosis of complicate mechanical equipment [30]. Ensemble learning is another method of machine learning that can effectively deal with this problem, ensemble learning uses several models and an integration strategy to maximize the strengths of individual models and achieve better results than an individual model [31,32]. Among them, the integration strategy plays an important role in the ensemble learning, and directly affect the accuracy of the diagnosis results. Therefore, it is meaningful to study ensemble learning models.

In this paper, a novel ensemble learning method based on multiple stacks sparse AEs is proposed for bearing intelligent fault diagnosis. The proposed method is mainly included three steps: Firstly, three stack sparse AEs with different weights are used to extract the representative features from the raw vibration signals. Secondly, a feature integrated strategy based on distance and standard deviation metrics is designed to fine tune the extracted features, which improves the robustness and stability of the diagnosis result. Finally, the softmax classifier is used to classify the fault types based on the integrated features. Experimental results show that the proposed method can get rid of the dependence of manual design algorithm to extract features, and overcome the limitations of an individual deep learning model, which is superior compared with other similar intelligent diagnosis methods. In brief, the contributions of this paper are summarized as follows:

(1) A novel ensemble deep learning method-based multiple stacks sparse AE is proposed for bearing intelligent fault diagnosis. This method is a segmented adaptive feature extraction procedure and can automatically classify the health status of the rolling machinery. Since the proposed method can process three segments of signals at the same time, it is more suitable for processing massive data in the fields of condition monitoring and fault diagnosis.

(2) A feature integrated strategy is designed to assign the weight of each feature. The strategy is composed of distance weight and variance weight, which can decrease the distance of intra-class and increase the distance of inter-class, improving the robustness and stability of fault diagnosis.

(3) A common motor bearing dataset is used to verify the proposed method. In the course of research, the selection of several key parameters and effects of segments and training samples on the diagnosis performance were studied. In addition, this method is compared with different methods and relative similar studies, the results show the superiority of the proposed method.

The remainder parts are organized as follows. In Section 2, the theory of the stack sparse AEs and softmax classifier are briefly introduced. In Section 3, the proposed method is described in detail. Section 4 demonstrates the experiment results on a popular rolling bearing. Conclusions are given in Section 5.

2. Stack Sparse Auto-Encoders and Softmax Classifier

2.1. Stack Sparse Auto-Encoders

In this section, we will briefly introduce the standard stack auto-encoder (SAE). As an unsupervised learning model, SAE has wide application in pattern recognition fields [33]. It consists of several auto-encoders, each of which is a symmetrical three-layer neural network, including encoder network and decoder. The network parameters can be initialized by minimizing the reconstruction error between the input data and the output data [34,35]. Further, the expected SAE can be obtained through layer by layer training, the structure of auto-encoder (AE), and the training process of SAE, as shown in Figure 1.

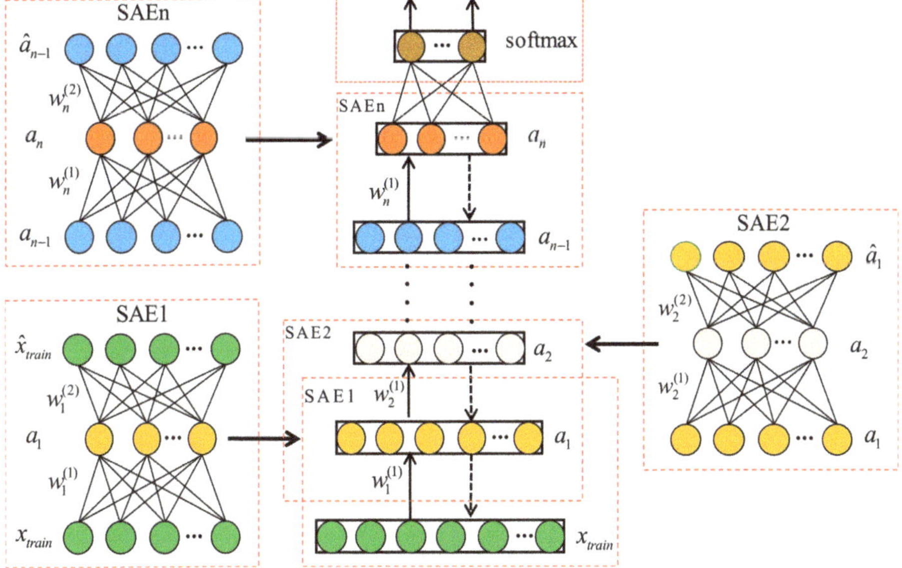

Figure 1. AE structure and the training process of SAE.

Suppose a n-dimensional unlabeled training sample is $x = \{x_1, x_2, \ldots, x_n\} \in \mathcal{R}^{1 \times n}$, the training process of AE is a representation that transform the input sample x into a hidden layer vector a_1, the vector a_1 can be denoted as $a_1 = \{a^1, a^2, \ldots, a^s\} \in \mathcal{R}^{1 \times s}$, the calculation procedures is as follows:

$$a_1 = f\left(w_1^{(1)} x + b_1^{(1)}\right) \tag{1}$$

where $w_1^{(1)}$ is the weight matrix, $b_1^{(1)}$ and $f(\cdot)$ are the offset vector and the activation function, respectively. Sigmoid [36,37] as the activation function used to train AE given as follows:

$$f(z) = 1/(1 + e^{-z}) \tag{2}$$

Then, the hidden vector a_1 will be decoded and reconstructed as the vector $\hat{x}$ by the Equation (3), the vector $\hat{x}$ can be denoted as $\hat{x} = \{\hat{x}_1, \hat{x}_2, \ldots, \hat{x}_n\} \in \mathcal{R}^{1 \times n}$. Equation (3) gives as follows:

$$\hat{x} = f(w_1^{(2)} a_1 + b_1^{(2)}) \tag{3}$$

where the $w_1^{(2)}$ and $b_1^{(2)}$ are the parameters of hidden layer to output layer. This works as the Equation (1).

The aims of training process is to obtain the approximation optimal value of parameter w and b through minimized the reconstruction errors.

For a sample set $\{x^m\}_{m=1}^M$ with M samples, its reconstruction cost function can be expressed as follow:

$$J_1(w, b) = \frac{1}{M} \sum_{m=1}^{m} L(x^m, \hat{x}^m) \tag{4}$$

where $L(x^m, \hat{x}^m)$ is the reconstruction error square, which is given as Equation (5)

$$L(x^m, \hat{x}^m) = \left\| x^m - \hat{x}^m \right\|^2 \tag{5}$$

2.2. Sparse Auto-Encoder

In the training process of AE, training samples usually contain a lot of redundant information, which means that the training samples only contain a small amount of useful information, and the hidden neurons are not all activated to represent the information of input data, especially when the dimension of input data is less than the number of hidden neurons. Therefore, for each AE, a sparse constraint is adopted to limit the number of activated neurons in the hidden layer [37,38]. Kullback–Leibler (KL) divergence, as a constraint condition usually used in AE training, can be expressed as follows:

$$KL(\rho \| \hat{\rho}_j) = \rho \log \frac{\rho}{\hat{\rho}_j} + (1 - \rho) \log \frac{1 - \rho}{1 - \hat{\rho}_j} \tag{6}$$

where ρ and $\hat{\rho}_j$ are the sparse factor and average activated number of jth hidden neurons, respectively, and $KL(\rho \| \hat{\rho}_j)$ denotes the discrepancy ρ and $\hat{\rho}_j$.

To sum up the above conclusions, it can minimize the cost function of each AE to get optimal pre-training parameters w and b. So, the cost function can be rewritten as follows:

$$J(W, b) = \frac{1}{M} \sum_{m=1}^{M} L(x^m, \hat{x}^m) + \beta \cdot \sum_{j=1}^{s} KL(\rho \| \hat{\rho}_j) \tag{7}$$

where β is the dilution penalty factor.

2.3. Softmax Classifier

Softmax classifier is a linear classifier that commonly used in multi-classification tasks, whose output is the probability value of each class [39]. Given a training sample set $\{x^m\}_{m=1}^M$ and $x^m \in \Re^{1 \times n}$, its corresponding sample label set is $\{y^m\}_{m=1}^M$ with $y^m \in \{1, 2, \ldots, K\}$. For each given sample x^m, softmax classifier will compute the values $p(y^m = k | x^m)$, which is the probability of each class. Therefore, for each different input sample, the output is always a K dimension vector of probability, and the position of the maximum probability determines the class of the sample, which can be expressed by the following hypothetical functions

$$
h_\theta(x^m) = \begin{bmatrix} p(y^m = 1 | x^m; \theta) \\ p(y^m = 2 | x^m; \theta) \\ \vdots \\ p(y^m = K | x^m; \theta) \end{bmatrix} = \frac{1}{\sum\limits_{k=1}^{K} e^{\theta_k^T x^m}} \begin{bmatrix} e^{\theta_1^T x^m} \\ e^{\theta_2^T x^m} \\ \vdots \\ e^{\theta_K^T x^m} \end{bmatrix}
\tag{8}
$$

where $\theta = [\theta_1, \theta_2, \ldots, \theta_K]^T$ is the parameter of Softmax classifier, h_θ is the normalized probability. The optimization of model parameters can be achieved by minimizing the cost function $J(\theta)$.

$$
J(\theta) = -\frac{1}{M} \left[\sum_{m=1}^{M} \sum_{k=1}^{K} I\{y^m = k\} \log \frac{e^{\theta_k^T x^m}}{\sum\limits_{k=1}^{K} e^{\theta_k^T x^m}} \right]
\tag{9}
$$

where $I\{\cdot\}$ is an indicator function, when the condition is true, the function return 1 otherwise return 0.

3. Proposed Fault Diagnosis Method

In this section, the proposed bearing fault diagnosis method is presented. First, three different sparse auto-encoders are constructed and used to extract the features from the raw vibration signal in Section 3.1. The weight strategy is described in Section 3.2. In Section 3.3, the feature integration is introduced. Softmax classifier is used to classify the health condition of the integrated features, and the detailed process is shown in Figure 2.

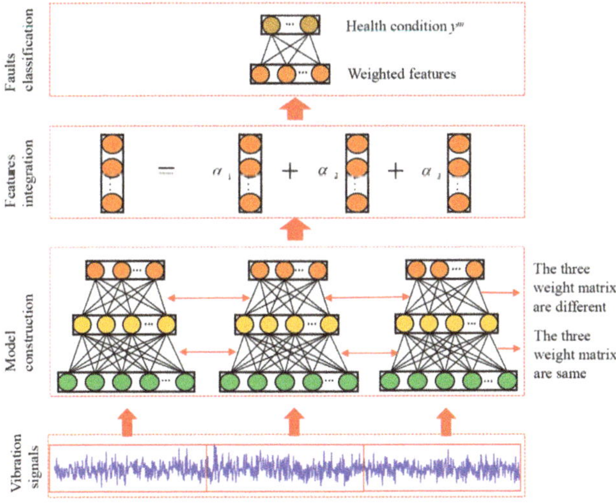

Figure 2. Illustration of the proposed.

3.1. Ensemble Auto-Encoders Construction

In order to construct three different sparse auto-encoder models, we divide the original vibration signal into three segments, and each segment uses a SAE to extract features. Assuming the input dimension of SAE is N_{in}, when training individual SAE, we randomly select N_t training samples from the data set, which are obtained by overlapping sampling method. Each training sample consists of three N_{in} segments, which means that there are three segment samples that can be used to training in each N_t. The details of training process are shown in the Figure 3.

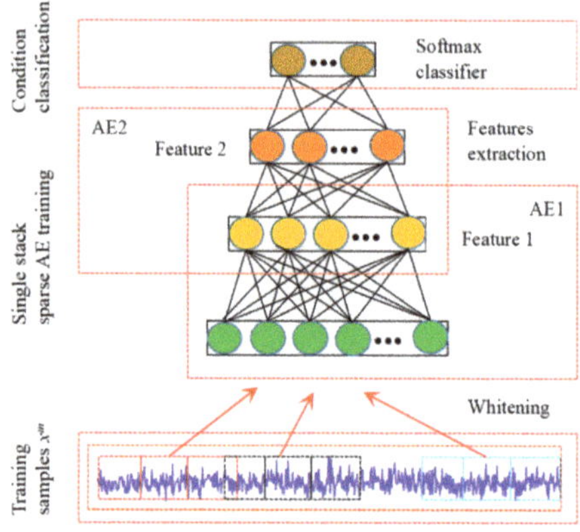

Figure 3. Illustration of the SAE training process.

When the training of individual SAE is completed, removing softmax classifier, and reproduce the parameters of the feature extraction part to other two SAEs. Then, keeping the weight $w_1^{(1)}$ unchanged, and add a small variety to the weight $w_2^{(1)}$ of other two SAEs. This not only can extract representative features from raw data and increase the diversity between features, but it is also beneficial when the input samples are similar. Therefore, the proposed model is very concise, greatly reducing training time and increasing the practicality. In addition, in order to improve the robustness of the model, Gaussian white noise is added to the training samples.

3.2. Weighting Strategy

A common ensemble strategy is voting method, which has been wildly applied in different ensemble learning models [40,41]. The voting method includes majority voting and weighting voting [31,32]. The majority voting is used to directly calculate the average value of features, and the advantages of this method are convenience and intelligibility. When machines work in a stable environment or without noise interference, majority voting can get good results for the mechanical fault diagnosis. Different with the majority voting, weighting voting assigns different weight for each feature. Obviously, the majority voting is a special case of the weighting voting. When the working environment of the machine changes or the signal contains a lot of noise, the weighting voting has better performance than the majority voting. There are other ensemble methods for integration features, such as the learning method which outputs features to form a new data set, and learning with a new model [42,43].

In this paper, in order to improve the robustness and stability of the proposed method, we select the weight voting method to design an integration strategy. As show in Figure 4, assuming that the output features of the three SAEs are $\{a_n^1, a_n^2, a_n^3\}$, their distances to the expectation $\bar{a}_n$ are d_1, d_2, d_3, respectively. When the distance is larger, it means that the feature deviates from the category, and the lower accuracy will be obtained for fault diagnosis. Therefore, we select the distance metric of the three features to the expectation to measure the weight, such as, the larger the distance, the smaller the weight, and vice versa. Suppose the weights of the three SAEs based on distance metric are $\alpha' = [\alpha'_1, \alpha'_2, \alpha'_3]^T$, mathematically, it can be written as:

$$\alpha' = \frac{1}{\sum\limits_{i=1}^{N_s} \left\| a_n^i - \bar{a}_n \right\|_2} \begin{bmatrix} \sum\limits_{i=2}^{N_s} \left\| a_n^i - \bar{a}_n \right\|_2 \\ \sum\limits_{i=1, i \neq 2}^{N_s} \left\| a_n^i - \bar{a}_n \right\|_2 \\ \sum\limits_{i=1, i \neq 3}^{N_s} \left\| a_n^i - \bar{a}_n \right\|_2 \end{bmatrix} \tag{10}$$

where N_s denotes the number of input segments; n is the nth hidden layer; $\|\cdot\|_2$ is the Euclidean distance; denominator $\sum\limits_{i=1}^{N_s} \left\| a_n^i - \bar{a}_n \right\|_2$ is to normalize the weight distribution.

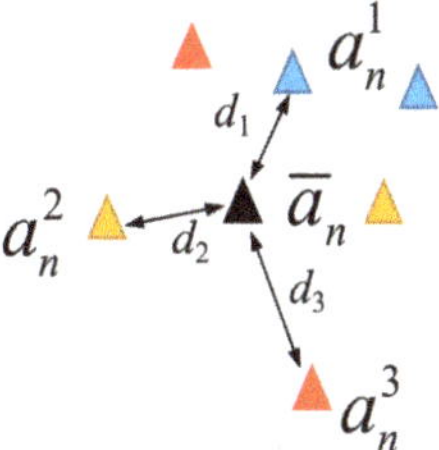

▲ ▲ ▲   Represent the three features respectively

▲   Represent the average feature

Figure 4. Illustration of the weight selection.

It should be noticed that, although the distance metric can constrain the feature deviation from the average feature on the same faulty category, when the feature itself has a large deviation, the distance metric may not have a good function. Based on this, we introduce the second weight measure condition: standard deviation metric. Standard deviation can reflect the degree of data fluctuation, the larger the standard deviation gets, the greater the data fluctuation is. So, the standard deviation can reflect the stability of features. Suppose that the weights of the three SAEs based on standard deviation metric are $\alpha'' = [\alpha''_1, \alpha''_2, \alpha''_3]^T$, they are defined by:

$$\alpha'' = \frac{1}{\sum\limits_{i=1}^{N_s} \rho_i} \begin{bmatrix} \sum\limits_{i=2}^{N_s} \rho_i \\ \sum\limits_{i=1, i \neq 2}^{N_s} \rho_i \\ \sum\limits_{i=1, i \neq 3}^{N_s} \rho_i \end{bmatrix} \tag{11}$$

where ρ denotes the standard deviation of each feature; Denominator $\sum\limits_{i=1}^{N_s} \rho_i$ is to normalize the weight distribution.

Now we have two feature-related weight vectors, distance metric weight and standard deviation metric weight. To implement an excellent integration strategy, we assume the target weight $\alpha = [\alpha_1, \alpha_2, \alpha_3]^T$, it is defined as follow:

$$\alpha = \lambda \alpha' + \gamma \alpha'' \tag{12}$$

where λ and γ are two hyper-parameters by user-specifying, which the limits of the values are between 0 and 1 and their sum is 1. In the proposed method, the two hyper-parameters will be studied in detail for the effect of diagnosis performance.

3.3. Feature Integration

After the above analysis, the weight vector $\alpha = [\alpha_1, \alpha_2, \alpha_3]^T$ can be determined for each sample x^m. Meanwhile, three feature vectors $[f_1, f_2, f_3]$ are extracted from the input sample x^m by the three SAEs. The final object features f^m are aggregated using the weight strategy, which is written as follows:

$$f^m = \alpha_1 f_1 + \alpha_2 f_2 + \alpha_3 f_3 \tag{13}$$

This weighted strategy is beneficial that can decrease the influence of the random features caused by ambient noise and interference. Also, the weighted way enhances the discriminative features that these features are complementary and improves the stability due to having the weight constraint term. The detailed process of the proposed method given as Figure 5.

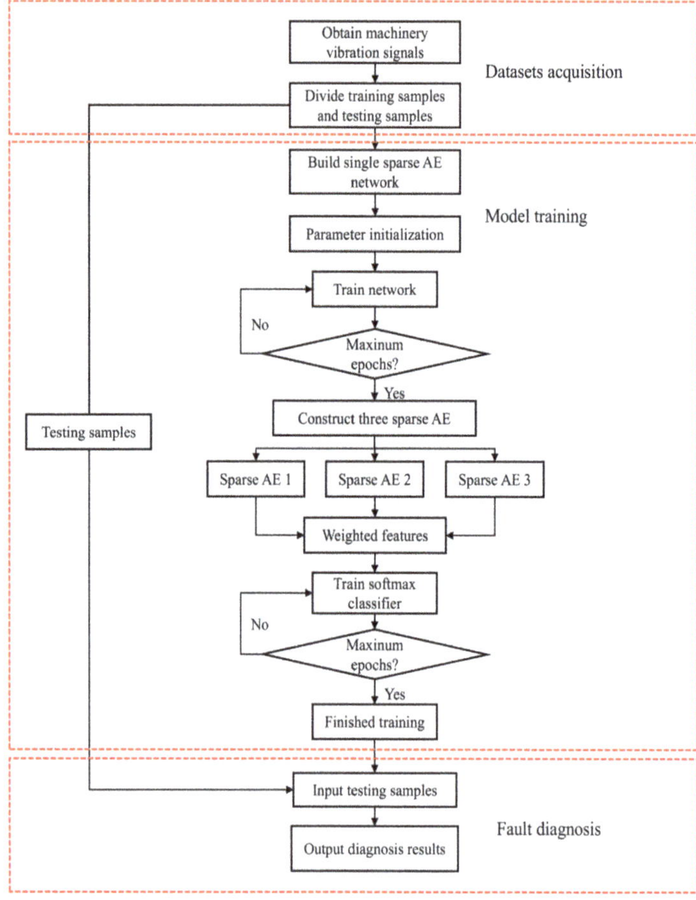

Figure 5. Flow chat of the proposed bearing fault diagnosis method.

4. Experiment and Analysis

4.1. Dataset Description

The bearing dataset provided by Case Western Reserve University [44] is analyzed in this section. As show in Figure 6, the test rig main consists of a 2-horsepower (hp) motor, a torque converter/encoder, a dynamometer and a control circuit. The vibration data were collected from the drive end of a motor under four different conditions: normal condition, inner race fault (IF), roller fault (RF), and outer race fault (OF). Single point faults were introduced of the motor with fault diameters of 0.18 mm, 0.36 mm, and 0.54 mm, respectively. The bearing data were all collected under four load conditions (0, 1, 2, and 3 hp) with the sampling frequency of 12 kHz.

Figure 6. Bearing platform used for experiment.

These vibration data compose the motor bearing dataset, which is used to verify the effectiveness of the proposed method. These data contain ten bearing health conditions under four loads, where the same health condition under different loads is defined as one class. The details of the experimental condition are summarized in Table 1. In this experiment, the first 120,000 points of the vibration data are selected as the preprocessed data under each condition. These preprocessed data are divided into training set and test set.

Table 1. Bearing data information used to experiment in this proposed.

Fault Type	Fault Size (mm)	Load(hp)	Label
Normal	0.0	0,1,2,3	1
IF	0.18	0,1,2,3	2
IF	0.36	0,1,2,3	3
IF	0.53	0,1,2,3	4
RF	0.18	0,1,2,3	5
RF	0.36	0,1,2,3	6
RF	0.53	0,1,2,3	7
OF	0.18	0,1,2,3	8
OF	0.36	0,1,2,3	9
OF	0.53	0,1,2,3	10

4.2. Compare Studies

In order to verify the superiority of the proposed method, three methods were selected to compare with the proposed method, namely, Support Vector Machine (SVM), Back-Propagation Neural Network (BPNN) with two hidden layers, and the individual stack sparse AE with two hidden layers. They are widely used in fault diagnosis of rotating machinery. The input data is raw vibration data, and the comparison of diagnosis performance of the three methods under 20 experiments is shown in Figure 7.

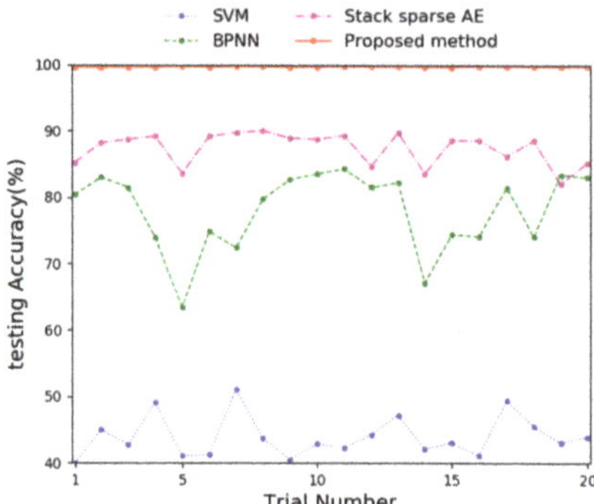

Figure 7. Comparison of different method with 20 experiments.

In the Figure 7, the results show that the proposed method has the highest diagnostic accuracy and the smallest fluctuations. Compared with the proposed method, the individual stack sparse AE has smaller diagnosis accuracy and greater diagnosis fluctuation, which indicates that the proposed ensemble method has better diagnosis performance than an individual stack sparse AE. Of course, the individual stack sparse AE is better than the other two diagnosis methods. Since the BPNN is not pre-trained like the individual stack sparse AE, it is under fitting, this proves that AE can reduce the number of training samples. SVM has minimal diagnosis accuracy, because SVM is not suitable for processing high-dimensional data, usually, it needs to preprocess the original vibration data and transform them into statistical features. The specific diagnosis results are summarized in Table 2.

Table 2. Experimental results of average accuracy and standard deviation of various methods.

Method	Average Accuracy	Standard Deviation
SVM	43.99%	3.09%
BPNN	78.07%	5.91%
SAE	87.40%	2.44%
Our method	99.71%	0.05%

From the perspective of the average accuracy in Table 2, the proposed method shows the best marks with the highest average accuracy (99.71%), while the SVM has the worst diagnosis performance (43.99%), and the individual stack sparse AE only gets intermediate testing accuracy (87.40%). In addition, from the perspective of stability, the proposed method has the optimum performance, with the smallest standard deviation (0.05%), and the BPNN with two hidden layer has the worst performance, with the largest standard deviation (5.91%).

For further proving the superiority of the proposed method, we also compared with other similar studies used the same dataset. In [45], a method adopting 15 stack sparse AEs to extract bearing features was proposed. The 15 stack sparse AEs use different activation function, and the extracted features are integrated with an accuracy threshold. This proposed method classifies the health conditions of 12 motor bearings at 0 hp, and finally obtained an average test accuracy of 97.18% and a standard deviation of 0.11%. Sun et al. [46] proposed a method based on compressed sensing theory. Their method combined with stack sparse AEs to extract features from the compressed data which were used to represent seven bearing health conditions under the load 2. The fault recognition rate of this method is 97.47% and the standard deviation is 0.43% in the bearing database. Lei et al. [47] proposed a bearing diagnosis method to integrate 12 sparse filter networks. The method used a simple average weighted combination strategy to process 12 local features that extracted from raw vibration data and white Gaussian noise is added during training. The method achieved 99.66% diagnosis accuracy and 0.19% standard deviation under 10 fault types and 4 different loads. In method [48], a dynamic weighted average method is designed to aggregate these learned features. This method used three different deep auto-encoder to extract the features, and the accuracy of k-fold cross-validation is used as a metric to assign the weights of the three deep auto-encoders. They obtained the accuracy of 99.69% and standard deviation of 0.24%. Comparing with the above methods, the proposed method in this paper achieved the highest accuracy of fault identification and the smallest standard deviation. The results of the above comparison are displayed in Table 3.

Table 3. Performance comparison with various studies.

Method	Load(hp)	No. of Health Condition	Testing Accuracy	Standard Deviation
[45]	0	12	97.18%	0.11%
[46]	2	7	97.41%	0.43%
[47]	0,1,2,3	10	99.66%	0.19%
[48]	0,1,2,3	10	99.69%	0.24%
Proposed	0,1,2,3	10	99.71%	0.05%

4.3. Visualization of Learned Representation

In this section, to qualitatively illustrate the effectiveness of the proposed fault diagnosis method, we visualize the features using four methods. The other three methods are sparse AEs with two hidden layers, the proposed model without whitening method, and the weight average method, respectively. The visual features are extracted from testing sample by the four methods, and the experiment conducted under the condition of noise for a better visual comparison of the result.

A technique called 't-SNE' is used to map the extracted features into a two-dimensional space to achieve visualization of high-dimensional data [49]. This technique has two processes, firstly, the principal component analysis (PCA) is used to reduce the dimension of the features to 50. Then, a technology called 't-SNE' is used to represent the 50-dimensional data as two-dimensional planar data.

Figure 8 is the feature visualization of individual sparse AEs with two hidden layers. It can be seen from the figure that the individual sparse AE method performs aggregation poorly on different fault types. this method cannot correctly diagnose the bearing fault, and only 82% of the test accuracy is obtained.

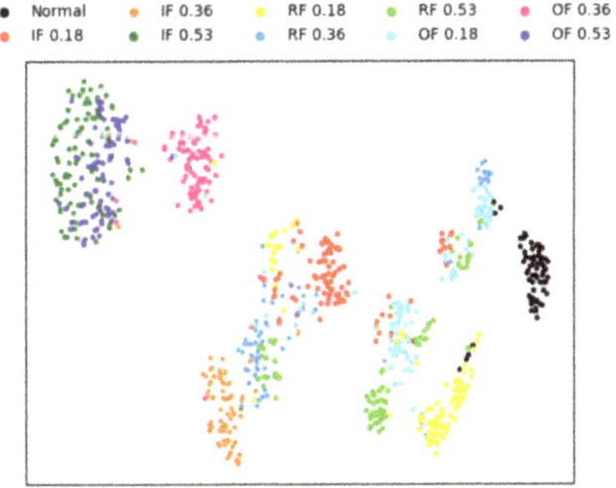

Figure 8. Visualization of sparse AE for the learned features.

Figure 9 is the feature visualization of the proposed method without whitening. Comparing with Figure 8, it can be noticed that most of the testing data are clustered in their own category and different types of faults are scattered in different regions. In Figure 9, there are only less intersection between the different fault classes, and the mainly error of fault diagnosis is concentrated in IF 0.18, that is mean that the proposed method cannot completely classify IF 0.18. The distances between different classes are far away, which also shows that the proposed method is robust. The final test accuracy of this method is 96.93%.

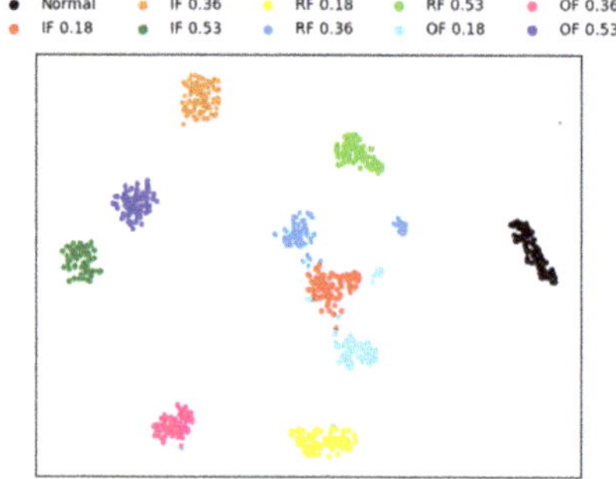

Figure 9. Visualization of without whitening for the learned features.

Figures 10 and 11 shows the feature visualization of the weight average method and our method, respectively. Figures 10 and 11 are very similar, which verifies that the features extracted by the three SAEs are similar, weighting strategy only a fine-tuning operation . Although the difference is not great, the proposed method has better performance than the weight average method, with the fault identification accuracy of 97.78% and 98.23% respectively obtained.

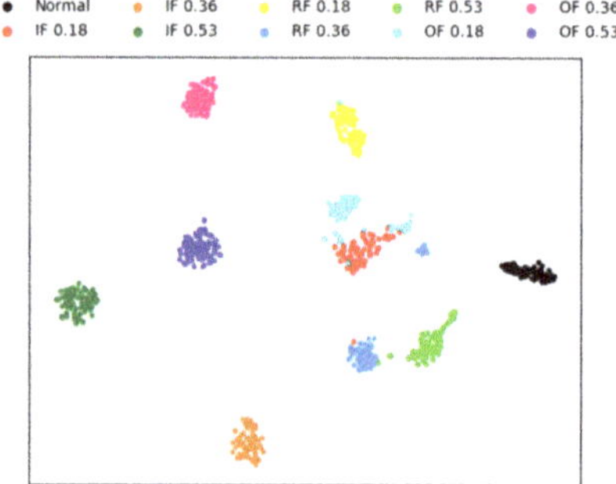

Figure 10. Visualization of weight average for the learned features.

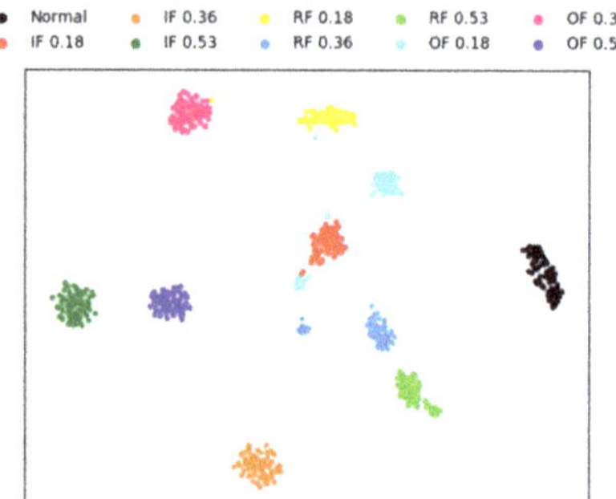

Figure 11. Visualization of proposed method for the learned features.

4.4. Parameters Selection of the Proposed Method

There are several key parameters need to determine in the proposed method, such as: the input dimension of SAE, the number of hidden layer neurons and sparse parameters ρ, etc. Next, we will respectively investigate the selection of these parameters. In addition, in order to reduce the influence of the randomness, 20 trials are repeated for each experiment. The environment of all experiments are 4G RAM and python 3.6.

First, we investigate the selection of the input dimension. We select a certain number of samples to train the proposed method, where 40,000 samples are sampled from the bearing dataset, and the rest samples are used for testing. For each trial of different input dimension, we always keep other parameters unchanged. The diagnosis results are displayed in Figure 12, wherein the positive error bars show the standard deviations and the point of time are the average time. It can be seen that when the input dimensions are increasing from 100 to 300, the accuracies are going higher, and when the input dimension is 300, the standard deviation is the smallest. When the input dimension is greater than 300, the average test accuracy only decreases slightly, but the time consumption is growing linearly. Therefore, considering the results from the experiment, we choose 300 as the input dimension.

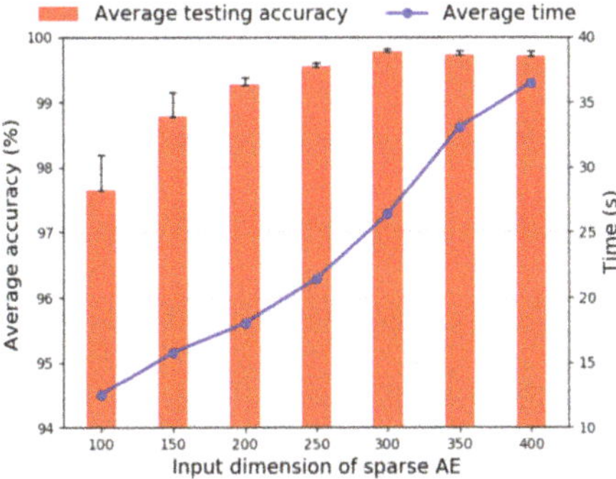

Figure 12. Diagnosis accuracy of various input dimension.

Next we investigate the number of the first hidden layer neurons. As shown Figure 13, the fault recognition accuracy increases gradually, and standard deviation is also reduced as the number of neurons in the first hidden layer increases from 50 to 200. When the number of neurons is greater than 200, the accuracy is stable and corresponding standard deviations are higher. The average time is also increasing. So, we choose 200 as the number of the first hidden layer neurons.

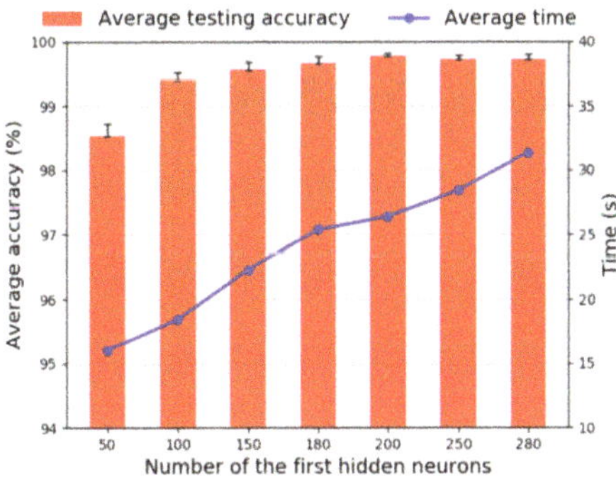

Figure 13. Diagnosis accuracy of neurons in the first layer.

Then, we investigate the number of the second hidden layer neurons. Generally, the number of neurons in the second hidden layer is less than the first hidden layer. Therefore, the number of neurons we studied was between 40 and 200, and the diagnosis results are shown in Figure 14. It can be seen that the accuracy varies only slightly in the whole neural unit interval. When the number of neurons is 100, the average testing accuracy is highest and standard deviation is smallest. Although time consumption has increased, the increasing values are far from acceptable. Therefore, we choose 100 as the number of the second hidden layer neurons.

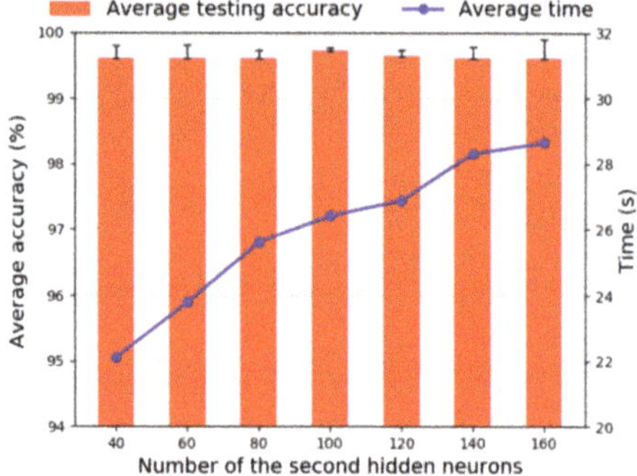

Figure 14. Diagnosis accuracy of neurons number in the second layer.

Afterward we investigate the selection of sparse parameter ρ. The sparse parameter plays an important role in the process of achieving high accuracy. In general, it is a small value close to zero. According to the general experiment results, the selection of sparse parameter varying from 0.05 to 0.5 is studied. Figure 15 shows the average diagnosis accuracy whit different sparse parameters. It can be seen from the figure that, as the value of sparse factor is 0.15, the highest average test accuracy and smallest standard deviation are obtained. Therefore, 0.15 is chosen as the value of the sparse parameter.

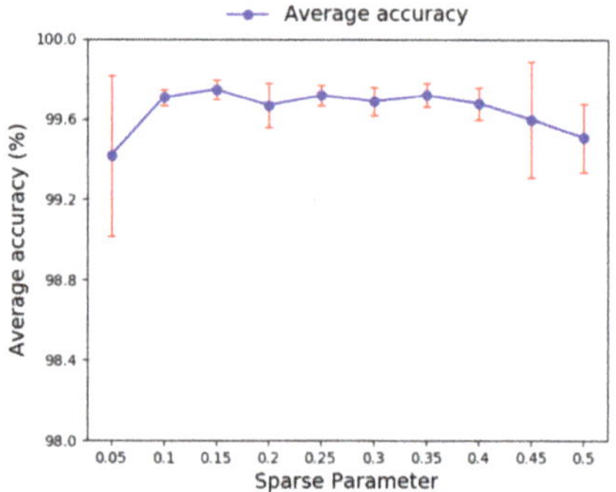

Figure 15. Diagnosis accuracy of different sparse parameters.

Finally, we investigate the selection of parameters λ and γ. These two parameters are the hyper-parameters of distance metric and standard metric. The diagnosis results are shown in Figure 16, where the two hyper-parameters λ and γ are selected as [0,0.1,0.2,0.3,0.4,0.5,0.6,0.7,0.8,0.9,1] and [1,0.9,0.8,0.7,0.6,0.5,0.4,0.3,0.2,0.1,0], respectively. It can be known that when the model is determined, the average diagnosis accuracy slightly changes as the two parameters correspond to different values, which means that the two hyper-parameters only have the function of fine-tuning. The highest accuracy can be obtained when the parameter λ is between 0.4 and 0.5. Therefore, it is reasonable that λ chooses 0.4 or 0.5. Furthermore, when the parameter γ is greater than λ, the average accuracy is higher, which indicates that the proposed method prefers to choose the standard deviation metric and the distance metric has a smaller impact.

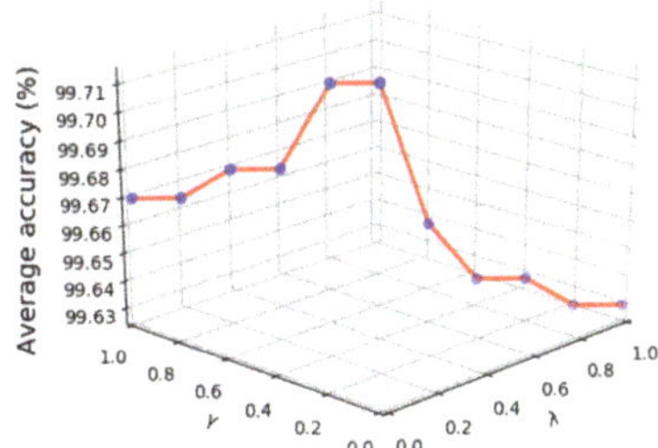

Figure 16. The relationship between average accuracy and parameters λ, γ.

In summary, the detailed parameters of the proposed method are as follows, the input dimension of sparse AE is 300, the number of two hidden layer neurons is 200, 100, respectively, and the value of sparsely parameter ρ is 0.15, the two hyper-parameters λ γ are 0.5. A wider selection of these parameters in the proposed method are listed in Table 4.

Table 4. Key parameters of the proposed method.

Parameters Description	Value
The dimension of Sparse AE	300
The number of the hidden layers	2
The number of the first hidden neurons	200
The number of the second hidden neurons	100
Learning rate	0.007
Sparse parameter	0.15
Sparse penalty factor	2
Batch size	100
Hyper-parameters (λ, γ)	0.5

4.5. Effect of Segments and Training Samples

The proposed model involves different number of segments and training samples, i.e., a different number of segments for training input and the percentage of training samples for training the proposed model will both significantly impact the diagnosis accuracy and time consumption of the proposed method. Therefore, we study the effect of different number of segments and training samples.

(1) Effect of segments: Different segments determine the structure and diagnosis performance of the model. In this study, signals with different segments will be used as the input for the proposed framework. In order to quantitatively evaluate the effect of the input segments on the classification performance, different segments ranging from one to four are studies. Figure 17 shows the diagnosis accuracy and training time choosing various segments. It is easily observed that when the segment number goes from one to four, the superior diagnosis performance is obtained. The results indicate that the more segments are used, the proposed model can achieve better and more stable performance, because these extracted features from different segments are rich and complementary, it is helpful for classification. Furthermore, a significant accuracy increase and standard deviation decrease from one segment to two segments can be noticed. More segments can achieve better diagnosis performance, however, in reality, it does not mean that more segments are always beneficial, from the comprehensive consideration of the model complexity and computational cost, choosing three segments are reasonable. Table 5 lists the diagnosis performance of different segments corresponding to Figure 17. This result validates that the proposed method can extract more discriminative and stable features from raw vibration signals.

(2) Effect of training samples: In general, as more samples are used to train the model, the higher accuracy can be achieved. The diagnosis results using different percentage of training samples are shown in Figure 18. It can be seen that when the training samples goes larger, the average test accuracy

is higher, and the standard deviation is smaller. However, the time consumption is increasing linearly. It means that the selection of training samples is a trade-off between the diagnosis accuracy and the time consumption. The same is true for Figures 12–14 and Figure 17. In Figure 18, when the proportion of training samples is 40%, the average test accuracy is 99.71% and the standard deviation is only 0.05%, which means that our proposed method achieves very high diagnosis accuracy and has good stability.

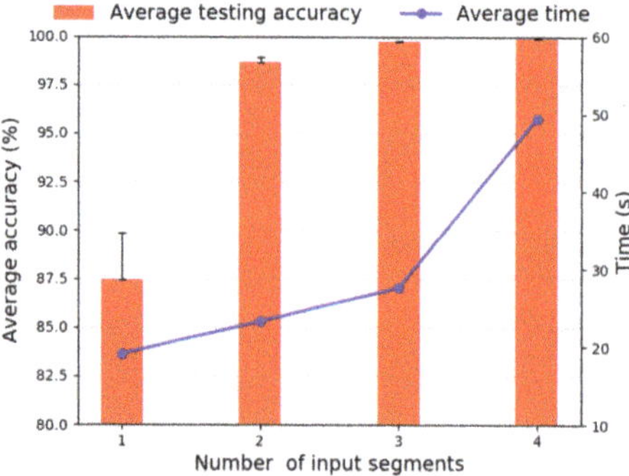

Figure 17. Diagnosis performance with different segments from 1 to 4.

Table 5. Diagnosis accuracy and time consuming for different segments.

Segments	Average Accuracy	Standard Deviation	Training Time (s)	Testing Time (s)
1	87.40%	2.44%	19.11	0.28
2	98.62%	0.23%	23.27	0.36
3	99.71%	0.05%	27.62	0.44
4	99.88%	0.04%	49.44	0.50

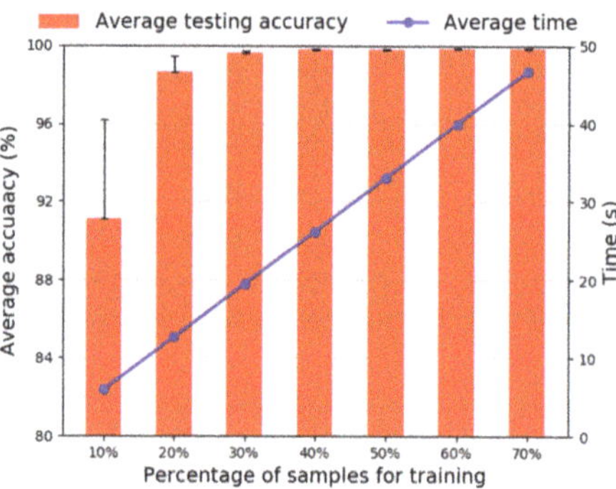

Figure 18. Diagnosis accuracy of different percentage of training samples.

4.6. Robustness Against Environmental Noises

In the actual industrial production process, noise is everywhere. The raw vibration signals are collected often contain a lot of noise, which has complex variability. For all possible noise, we can't get all the label samples corresponding to noises. So, in this section, we will study the effect of noise on diagnosis performance by adding Gaussian white noise. The robustness of the proposed method against environmental noise is verified by adding noise to the test data based on the original experiments. Specifically, the noise data is generated by adding Gaussian white noise with different signal-to-noise ratio (SNR) to the test data. The signal-to-noise ratio is defined as

$$SNR = 10 \log_{10}\left(\frac{P_{signal}}{P_{noise}}\right) \tag{14}$$

where P_{signal} and P_{nosie} represent the power of the original signal and added noise, respectively, the unit of SNR is dB. In this study, we evaluate the proposed method adding noisy signals with different SNR ranging from 0 dB to 8 dB. The results are shown in Figure 19.

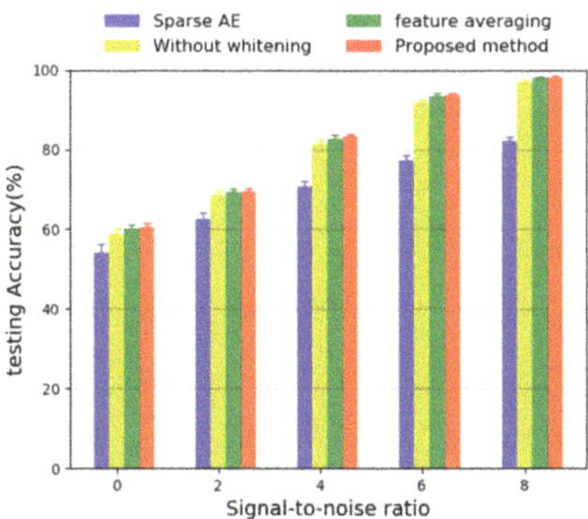

Figure 19. Diagnosis results with environmental noises by different methods.

It can be seen from the figure that when the SNR increases from 0 dB to 8 dB, the test accuracy of the four methods is increasing. Among them, at each SNR, the proposed method has the highest accuracy and the smallest deviation, the second is the method that using feature averaging method and next is the method without adding Gaussian white noise to the training data, the individual sparse AE with two hidden layer gets worst performance. Compared with the individual sparse AE, it is obvious that the proposed method has better anti-noise performance. In addition, it can be noticed that the proposed method is only slightly better (it is about 0.5%) than the feature average method; it infers that the input data has the same distribution, also, the feature average method can be considered as a sample mean filter, thus random noises will be filtered to some extent. Maybe, when the vibration signals are collected by multiple sensors or the distribution of input data is different, the proposed method may achieve better performance.

5. Concluding Remarks

In this paper, a novel bearing fault diagnosis method based on ensemble stack sparse auto-encoder was proposed. A common bearing data set is used, and a large number of experiments are carried out to verify the effectiveness of the proposed method. This paper studies the selection of several key parameters and the influence of segments and training samples on the diagnosis performance. By a comparison with other methods and related studies using the same data set, the superiority of the proposed method is proved. Additionally, the robustness of the proposed method against environmental noises is demonstrated under different levels of noise.

Future research will be extended to other complex models and other fault diagnosis problems such as using a CNN model and remaining useful-life prediction for rolling bearings. In addition, although the proposed method in this paper obtained a high accuracy of fault recognition, it did not achieve satisfactory results in a noisy environment, and there is still much room for improvement. This is one of the future research directions on how to improve the anti-noise performance of the model.

Author Contributions: Writing—original draft, J.H.; data curation, M.O.; supervision, C.Y.; funding acquisition, D.C., J.G. and Y.Z. All authors have read and agreed to the published version of the manuscript.

Funding: The research is supported by National Natural Science Foundation of China (Grant No.61703104, 61803087, 61972091), by Guangdong Natural Science Fund (Grant No. 2017A030310580, 2017A030313388), by Key Project of University of Guangdong Province (Grant No. 2019KZDXM007).

Conflicts of Interest: The authors declare no conflict of interest.

References

1. Lei, Y.; Li, N.; Guo, L.; Li, N.; Yan, T.; Lin, J. Machinery health prognostics: A systematic review from data acquisition to RUL prediction. *Mech. Syst. Signal Process.* **2018**, *104*, 799–834. [CrossRef]
2. Lu, C.; Wang, Z.-Y.; Qin, W.-L.; Ma, J. Fault diagnosis of rotary machinery components using a stacked denoising autoencoder-based health state identification. *Signal Process.* **2017**, *130*, 377–388. [CrossRef]
3. Li, X.; Zhang, W.; Ding, Q. A robust intelligent fault diagnosis method for rolling element bearings based on deep distance metric learning. *Neurocomputing* **2018**, *310*, 77–95. [CrossRef]
4. Xu, Y.; Sun, Y.; Wan, J.; Liu, X.; Song, Z. Industrial big data for fault diagnosis: Taxonomy, review, and applications. *IEEE Access* **2017**, *5*, 17368–17380. [CrossRef]
5. Gao, Z.; Cecati, C.; Ding, S.X. A survey of fault diagnosis and fault-tolerant techniques—Part I: Fault diagnosis with model-based and signal-based approaches. *IEEE Trans. Ind. Electron.* **2015**, *62*, 3757–3767. [CrossRef]
6. Gao, Z.; Ding, S.X.; Cecati, C. Real-time fault diagnosis and fault-tolerant control. *IEEE Trans. Ind. Electron.* **2015**, *62*, 3752–3756. [CrossRef]
7. Jia, F.; Lei, Y.; Lin, J.; Zhou, X.; Lu, N. Deep neural networks: A promising tool for fault characteristic mining and intelligent diagnosis of rotating machinery with massive data. *Mech. Syst. Signal Process.* **2016**, *72*, 303–315. [CrossRef]
8. Liu, R.; Yang, B.; Zio, E.; Chen, X. Artificial intelligence for fault diagnosis of rotating machinery: A review. *Mech. Syst. Signal Process.* **2018**, *108*, 33–47. [CrossRef]
9. Feng, Z.; Ma, H.; Zuo, M.J. Vibration signal models for fault diagnosis of planet bearings. *J. Sound Vib.* **2016**, *370*, 372–393. [CrossRef]
10. Rai, A.; Upadhyay, S. A review on signal processing techniques utilized in the fault diagnosis of rolling element bearings. *Tribol. Int.* **2016**, *96*, 289–306. [CrossRef]
11. Bustos, A.; Rubio, H.; Castejón, C.; García-Prada, J.C.J. EMD-based methodology for the identification of a high-speed train running in a gear operating state. *Sensors* **2018**, *18*, 793. [CrossRef] [PubMed]
12. Yan, R.; Gao, R.X.; Chen, X. Wavelets for fault diagnosis of rotary machines: A review with applications. *Signal Process.* **2014**, *96*, 1–15. [CrossRef]
13. Bustos, A.; Rubio, H.; Castejon, C.; Garcia-Prada, J.C. Condition monitoring of critical mechanical elements through Graphical Representation of State Configurations and Chromogram of Bands of Frequency. *Measurement* **2019**, *135*, 71–82. [CrossRef]
14. Li, X.; Yang, Y.; Pan, H.; Cheng, J.; Cheng, J.J.M.; Theory, M. Non-parallel least squares support matrix machine for rolling bearing fault diagnosis. *Mech. Mach. Theory* **2020**, *145*, 103676. [CrossRef]
15. Ali, J.B.; Fnaiech, N.; Saidi, L.; Chebel-Morello, B.; Fnaiech, F. Application of empirical mode decomposition and artificial neural network for automatic bearing fault diagnosis based on vibration signals. *Appl. Acoust.* **2015**, *89*, 16–27.
16. Lee, J.; Wu, F.; Zhao, W.; Ghaffari, M.; Liao, L.; Siegel, D. Prognostics and health management design for rotary machinery systems—Reviews, methodology and applications. *Mech. Syst. Signal Process.* **2014**, *42*, 314–334. [CrossRef]
17. Fu, W.; Shao, K.; Tan, J.; Wang, K.J. Fault diagnosis for rolling bearings based on composite multiscale fine-sorted dispersion entropy and SVM with hybrid mutation SCA-HHO algorithm optimization. *IEEE Access* **2020**, *8*, 13086–13104. [CrossRef]
18. He, C.; Liu, C.; Wu, T.; Xu, Y.; Wu, Y.; Chen, T. Medical rolling bearing fault prognostics based on improved extreme learning machine. *J. Comb. Optim.* **2019**, 1–22. [CrossRef]

19. Kang, M.; Kim, J.; Kim, J.-M.; Tan, A.C.; Kim, E.Y.; Choi, B.-K. Reliable fault diagnosis for low-speed bearings using individually trained support vector machines with kernel discriminative feature analysis. *IEEE Trans. Power Electron.* **2014**, *30*, 2786–2797. [CrossRef]

20. Cao, H.; Fan, F.; Zhou, K.; He, Z. Wheel-bearing fault diagnosis of trains using empirical wavelet transform. *Measurement* **2016**, *82*, 439–449. [CrossRef]

21. Huang, S.; Tan, K.K.; Lee, T.H. Fault diagnosis and fault-tolerant control in linear drives using the Kalman filter. *IEEE Trans. Ind. Electron.* **2012**, *59*, 4285–4292. [CrossRef]

22. Wen, L.; Li, X.; Gao, L.; Zhang, Y. A new convolutional neural network-based data-driven fault diagnosis method. *IEEE Trans. Ind. Electron.* **2017**, *65*, 5990–5998. [CrossRef]

23. Khan, S.; Yairi, T. A review on the application of deep learning in system health management. *Mech. Syst. Signal Process.* **2018**, *107*, 241–265. [CrossRef]

24. Liu, R.; Meng, G.; Yang, B.; Sun, C.; Chen, X. Dislocated time series convolutional neural architecture: An intelligent fault diagnosis approach for electric machine. *IEEE Trans. Ind. Inform.* **2016**, *13*, 1310–1320. [CrossRef]

25. Jia, F.; Lei, Y.; Guo, L.; Lin, J.; Xing, S. A neural network constructed by deep learning technique and its application to intelligent fault diagnosis of machines. *Neurocomputing* **2018**, *272*, 619–628. [CrossRef]

26. Shao, H.; Jiang, H.; Zhang, H.; Duan, W.; Liang, T.; Wu, S. Rolling bearing fault feature learning using improved convolutional deep belief network with compressed sensing. *Mech. Syst. Signal Process.* **2018**, *100*, 743–765. [CrossRef]

27. Li, X.; Zhang, W.; Ding, Q. Cross-domain fault diagnosis of rolling element bearings using deep generative neural networks. *IEEE Trans. Ind. Electron.* **2018**, *66*, 5525–5534. [CrossRef]

28. Long, J.; Zhang, S.; Li, C. Evolving Deep Echo State Networks for Intelligent Fault Diagnosis. *IEEE Trans. Ind. Inform.* **2019**. [CrossRef]

29. Mao, W.; He, L.; Yan, Y.; Wang, J. Online sequential prediction of bearings imbalanced fault diagnosis by extreme learning machine. *Mech. Syst. Signal Process.* **2017**, *83*, 450–473. [CrossRef]

30. Ma, S.; Chu, F. Ensemble deep learning-based fault diagnosis of rotor bearing systems. *Comput. Ind.* **2019**, *105*, 143–152. [CrossRef]

31. Rokach, L. Ensemble-based classifiers. *Artif. Intell. Rev.* **2010**, *33*, 1–39. [CrossRef]

32. Polikar, R. Ensemble based systems in decision making. *IEEE Circuits Syst. Mag.* **2006**, *6*, 21–45. [CrossRef]

33. Qi, Y.; Shen, C.; Wang, D.; Shi, J.; Jiang, X.; Zhu, Z. Stacked sparse autoencoder-based deep network for fault diagnosis of rotating machinery. *IEEE Access* **2017**, *5*, 15066–15079. [CrossRef]

34. Bengio, Y.; Courville, A.; Vincent, P. Representation learning: A review and new perspectives. *IEEE Trans. Pattern Anal. Mach. Intell.* **2013**, *35*, 1798–1828. [CrossRef] [PubMed]

35. Hinton, G.E.; Osindero, S.; Teh, Y.-W. A fast learning algorithm for deep belief nets. *Neural Comput.* **2006**, *18*, 1527–1554. [CrossRef]

36. Makhzani, A.; Frey, B. K-sparse autoencoders. *arXiv* **2013**, arXiv:1312.5663.

37. Ng, A. Sparse autoencoder. *CS294A Lect. Notes* **2011**, *72*, 1–19.

38. Erhan, D.; Bengio, Y.; Courville, A.; Manzagol, P.-A.; Vincent, P.; Bengio, S. Why does unsupervised pre-training help deep learning? *J. Mach. Learn. Res.* **2010**, *11*, 625–660.

39. Tao, S.; Zhang, T.; Yang, J.; Wang, X.; Lu, W. Bearing fault diagnosis method based on stacked autoencoder and softmax regression. In Proceedings of the 2015 34th Chinese Control Conference (CCC), Hangzhou, China, 28–30 July 2015; pp. 6331–6335.

40. Zheng, J.; Pan, H.; Cheng, J. Rolling bearing fault detection and diagnosis based on composite multiscale fuzzy entropy and ensemble support vector machines. *Mech. Syst. Signal Process.* **2017**, *85*, 746–759. [CrossRef]

41. Li, S.; Liu, G.; Tang, X.; Lu, J.; Hu, J. An ensemble deep convolutional neural network model with improved DS evidence fusion for bearing fault diagnosis. *Sensors* **2017**, *17*, 1729. [CrossRef]

42. Zhou, F.; Hu, P.; Yang, S.; Wen, C. A Multimodal Feature Fusion-Based Deep Learning Method for Online Fault Diagnosis of Rotating Machinery. *Sensors* **2018**, *18*, 3521. [CrossRef]

43. Jiang, G.; He, H.; Yan, J.; Xie, P. Multiscale convolutional neural networks for fault diagnosis of wind turbine gearbox. *IEEE Trans. Ind. Electron.* **2018**, *66*, 3196–3207. [CrossRef]

44. Smith, W.A.; Randall, R.B. Rolling element bearing diagnostics using the Case Western Reserve University data: A benchmark study. *Mech. Syst. Signal Process.* **2015**, *64*, 100–131. [CrossRef]

45. Shao, H.; Jiang, H.; Lin, Y.; Li, X. A novel method for intelligent fault diagnosis of rolling bearings using ensemble deep auto-encoders. *Mech. Syst. Signal Process.* **2018**, *102*, 278–297. [CrossRef]
46. Sun, J.; Yan, C.; Wen, J. Intelligent bearing fault diagnosis method combining compressed data acquisition and deep learning. *IEEE Trans. Instrum. Meas.* **2017**, *67*, 185–195. [CrossRef]
47. Lei, Y.; Jia, F.; Lin, J.; Xing, S.; Ding, S.X. An intelligent fault diagnosis method using unsupervised feature learning towards mechanical big data. *IEEE Trans. Ind. Electron.* **2016**, *63*, 3137–3147. [CrossRef]
48. Zhang, Y.; Li, X.; Gao, L.; Chen, W.; Li, P. Intelligent fault diagnosis of rotating machinery using a new ensemble deep auto-encoder method. *Measurement* **2020**, *151*, 107232. [CrossRef]
49. Maaten, L.V.d.; Hinton, G. Visualizing data using t-SNE. *J. Mach. Learn. Res.* **2008**, *9*, 2579–2605.

Article

Scan-Chain-Fault Diagnosis Using Regressions in Cryptographic Chips for Wireless Sensor Networks

Hyunyul Lim, Minho Cheong and Sungho Kang *

Electrical and Electronic Engineering Department, Yonsei University, Seoul 03722, Korea;
lim8801@soc.yonsei.ac.kr (H.L.); cmh9292@soc.yonsei.ac.kr (M.C.)
* Correspondence: shkang@yonsei.ac.kr; Tel.: +82-21-232-775

Received: 15 July 2020; Accepted: 20 August 2020; Published: 24 August 2020

Abstract: Scan structures, which are widely used in cryptographic circuits for wireless sensor networks applications, are essential for testing very-large-scale integration (VLSI) circuits. Faults in cryptographic circuits can be effectively screened out by improving testability and test coverage using a scan structure. Additionally, scan testing contributes to yield improvement by identifying fault locations. However, faults in circuits cannot be tested when a fault occurs in the scan structure. Moreover, various defects occurring early in the manufacturing process are expressed as faults of scan chains. Therefore, scan-chain diagnosis is crucial. However, it is difficult to obtain a sufficiently high diagnosis resolution and accuracy through the conventional scan-chain diagnosis. Therefore, this article proposes a novel scan-chain diagnosis method using regression and fan-in and fan-out filters that require shorter training and diagnosis times than existing scan-chain diagnoses do. The fan-in and fan-out filters, generated using a circuit logic structure, can highlight important features and remove unnecessary features from raw failure vectors, thereby converting the raw failure vectors to fan-in and fan-out vectors without compromising the diagnosis accuracy. Experimental results confirm that the proposed scan-chain-diagnosis method can efficiently provide higher resolutions and accuracies with shorter training and diagnosis times.

Keywords: cryptography; wireless sensor networks; machine learning; scan-chain diagnosis

---

1. Introduction

Wireless sensor networks (WSNs) are composed of several sensor nodes that are deployed in the area to be monitored, through certain topologies and for certain purposes. Through some suitable methods and their respective information exchanges, WSNs collaboratively perceive physical world information and collect and collate the information of perceived objects within the network coverage area [1,2]. Because of these characteristics, WSNs have been widely used for various environmental, health, military, and commercial applications, such as intelligent transportation, smart homes, industrial monitoring, logistics, and healthcare systems [3,4]. However, the rapid deployment of WSNs has created critical problems in privacy and security [5–7]. Consequently, cryptography is generally implemented to ensure the security and integrity of information and data in WSNs. Cryptography systems can be divided into symmetric-key and asymmetric-key algorithms. Symmetric-key algorithms utilize only one secret key to cipher and decipher information, whereas asymmetric-key algorithms utilize public and secret keys to encrypt and decrypt information. The public key is freely available to anyone, whereas the private key is maintained secure and is only known to its owner. These cryptography algorithms require numerous modular multiplication and exponentiation operations, which are computationally expensive. Therefore, specific circuits for cryptographic algorithms that require high computational power have been designed [8].

Cryptographic circuits must be rigorously tested to guarantee the accuracy of cryptographic algorithms. Therefore, a scan structure is generally inserted to test cryptographic circuits. Scan structures are widely used in very-large-scale integration (VLSI) circuits as a design-for-test, as they increase the fault coverage and diagnosability by enhancing the controllability and observability of the digital circuit logic. The normal operation of scan chains in the circuits is guaranteed for accurate scan testing. However, if a fault occurs in scan chains, the test cannot be performed, and the yield decreases. In particular, several defects occurring in the initial manufacturing process are expressed as scan-chain faults. Therefore, the fault phenomenon propagates, widely reducing the accuracy of fault diagnosis. Hence, 10–30% of yield loss is caused by faults in the scan chain [9,10].

In addition, the cryptographic chips are generally protected by the secure scan-chain architecture to defense the scan channel attack. The secure technique generally uses the secret-key management policy to encrypt the scan-chain content during testing. It is required that inserting the encryption hardware composed of the internal registers in the scan-chain architectures. The secure scan architecture of the cryptographic circuit reduces controllability and observability [11,12], reducing the accuracy of scan-chain diagnosis. Consequently, scan-chain diagnosis has become a crucial issue in semiconductor manufacturing for cryptographic circuits.

Therefore, several scan-chain-diagnosis methods have been studied. Special-tester-based diagnostic methods [13–16] use a tester to control the scan operation and physical failure analysis equipment to identify the defective scan cells and locations of defective dies. These methods provide high resolution and accuracy. However, these special testers are expensive, and their operations are time consuming. Hardware-based diagnostic methods [17–20] modify a scan-chain structure or scan-cell design to enhance the diagnosability. They can easily identify a defective scan cell from a scan chain. However, these methods are not typically used due to additional hardware overheads and diagnostic complexity when a fault occurs in the additional control logic.

Finally, software-based diagnostic methods [21–25] apply algorithmic diagnosis by analyzing the observed responses of the patterns, which comprise shift-in, capture, and shift-out operations, from all scan chains. Such software-based scan-chain diagnoses are typically used because they can increase the yield of manufacturing without requiring additional hardware overheads and costs for special testers. However, these methods do not provide satisfactory results for scan-chain diagnosis. They require a significant amount of fault simulations to search for candidate scan cells in a failure case. Therefore, high computation power is required for executing several fault simulations, which becomes time-consuming for a scan chain with numerous cells. Furthermore, these methods cannot achieve sufficient resolution and accuracy.

Machine learning has been widely used in the field of fault diagnosis [26–30] as well as in scan-chain diagnosis to achieve sufficient resolution and accuracy [31–33]. For example, in [33], multistage artificial neural networks (ANNs) were used to diagnose scan-chain faults. A coarse global neural network (CGNN) was used to select several suspected scan cells (affine group) from among all scan-chain cells, and a refined local neural network (RLNN) was used to identify the final suspected scan cell in the affine group. The CGNN used a vector called an integer failure vector (IFV), a bitwise summation of all the binary failure vectors, as training and inference. In addition, the RLNN was trained using a cascaded failure vector, which was created by concatenating all the binary failure vectors. The use of such ANNs for scan-chain diagnosis was effective in improving the resolution and accuracy. However, multistage ANNs have some shortcomings due to their characteristics. First, several ANNs are required for the inference: one CGNN for each scan chain and several RLNNs for each affine group on only one scan chain. Moreover, the vector used for the RLNN is extremely long; hence, the network size is increased. Therefore, multistage ANNs require significant amounts of training time.

Therefore, this article proposes a novel scan-chain diagnosis method involving regressions and various filters. The controllability and observability from the logic circuit structure are applied to the fan-in and fan-out filters; thus, the fault-affected and fault-affecting cell information from the capture sequence are collected by the fan-in and fan-out filters, respectively. Therefore, raw failure vectors are compressed by applying filters, and the values with fault effects remain only at the new vectors. This can reduce the number of models for a chain and the number of dimensions of the models without losing the resolution and accuracy of scan-chain diagnosis. Therefore, the proposed regression reduces the training and diagnosis times as well as improves the resolution and accuracy of scan-chain diagnosis.

The remainder of this article is organized as follows. The motivations for introducing the fan-in and fan-out filters are presented in Section 2. The regressions and filters are described in Section 3. The simulation results are presented in Section 4. Finally, the conclusions are provided in Section 5.

2. Motivations

In this section, the motivations for introducing the fan-in and fan-out filters are presented. First, a software-based scan-chain diagnosis is described in Section 2.1. Second, the motivation for using machine-learning algorithms in scan-chain diagnosis is described in Section 2.2. Subsequently, the concept of sensitive scan cells is presented in Section 2.3.

2.1. Software-Based Scan-Chain Diagnosis

Software-based diagnostic methods [21–25] apply software-based algorithms to identify faulty cells. Such diagnosis uses test patterns to diagnose scan-chain faults. Generally, these test patterns can be categorized into the following three categories:

1. A chain pattern: It comprises only shift-in and shift-out, without a single capture process. This pattern aims at verifying whether the chain fails, and it can categorize various scan-chain defects into three common types—the stuck-at (SA) fault (stuck-at 1, stuck-at 0), slow (slow-to-rise, slow-to-fall, slow), and fast (fast-to-rise, fast-to-fall, fast) types.
2. A scan automatic test pattern generation (ATPG) pattern: It comprises shift-in, shift-out, and multiple-capture processes. The purpose of this pattern is to test the circuit logic.
3. A special chain diagnostic pattern: It is generated solely for scan-chain diagnosis.

The software-based diagnostic methods use three test patterns and failure logs to diagnose scan-chain faults. First, a flush test is conducted to find the defective scan chain. Next, a fault is injected in each scan cell, and a fault simulation is executed for each cell to collect its simulated fault response in a software-based simulation environment. Subsequently, the capture process of the loaded values (test stimulus) is simulated, and the simulated captured values are compared with the observed values (test response). After executing all the simulations, the cell with the best match is selected as the suspected scan cell. If it cannot be identified in the defective chain, special chain diagnostic patterns are used to obtain a higher resolution and accuracy.

Figure 1 illustrates an example to explain the software-based scan-chain diagnosis. To simplify the explanation, each scan cell is indexed from the scan-out to the scan-in cells. The scan cells between the scan-in and scan cells are called the upstream scan cells, and those between the scan-out and scan cells are called the downstream scan cells. Assume that a stuck-at-1 (SA1) fault exists in Cell 2 in the defective chain.

First, all the simulated loaded values of the defective chain are 0s. Next, the simulated loaded values are changed to (0 0 1 1 1) because the test stimulus values of the upstream scan cells of the defective cell, Cell 3, are polluted by the defective cell during the shift sequence. After the fault simulation, the simulated captured values of the defective chain are (0 1 1 0 0). Subsequently, suppose the observed values on the automatic test equipment (ATE) are (1 1 1 0 0). The observed values of the downstream scan cells of the defective cells are polluted by the defective cell during the shifting sequence. The observed value at Cell 1 should be polluted by the SA1 fault. Thus, the captured value on the ATE can be (0 1 1 0 0). Therefore, the SA1 fault may be in Cell 3. Furthermore, the lower bound of the defective chain is at Cell 3 because a 0 value cannot be observed at the lower bound of the SA1 fault cell.

The software-based algorithms run the fault simulation as described above for all test patterns. The candidate cells are determined from the lower and upper bounds of the defective chain. Then, the scores are calculated by comparing the observed and simulated observed values of each candidate cell, thereby confirming the location of the defective cell. However, these methods do not provide satisfactory results for intermittent faults. The intermittent fault diagnosis accuracy of the simulation-based scan-chain diagnosis using manufacturing test patterns is 52% on the target circuit [34]. In addition, the accuracy of the simulation-based scan-chain diagnosis using signal profiling is 57% [35]. Furthermore, these methods cannot achieve sufficient accuracy.

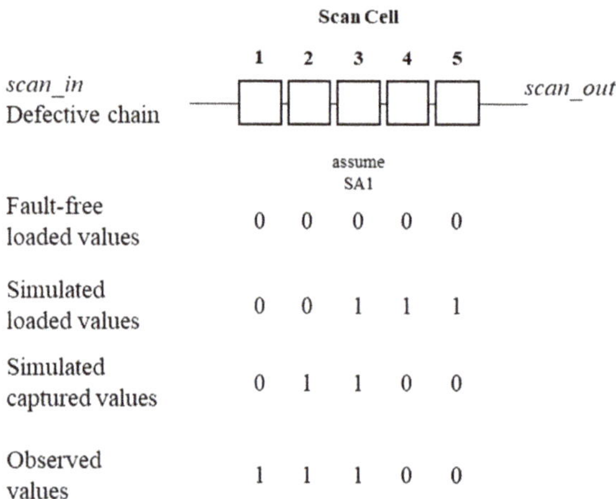

Figure 1. Example of software-based scan-chain diagnosis.

2.2. Machine Learning for Fault Diagnosis

Machine learning is an effective technique that enables systems to learn automatically from experience without being explicitly programmed. It has been applied in various fields, such as speech recognition, robotics, medical diagnosis, and computer vision to solve numerous problems [28]. The main idea of machine learning is to train a machine through example data to solve a cognitive task.

Machine learning is expressed by a function $f : x \rightarrow y$ that has an unknown intricate closed-form mathematical expression, as shown in Figure 2. The problems encountered when using machine learning are expressed as a function f using N datasets (x_i, y_i), $i = 1, 2, 3, \cdots, N$. In machine learning, x_i is called a feature, and y_i is called a target. By learning these datasets, machine learning can be used to create a function f that can infer a correct target even when an unseen feature is generated. Because of the characteristics of training and inference, machine learning can be applied in several fields where a large volume of data is available to train the models.

Scan-chain diagnosis is also suitable in terms of machine learning because a considerable number of labeled failure log datasets are generated during manufacturing. However, machine learning might not operate as intended if training is performed with only raw data. Although machine learning appears to be an all-round contributor in big-data processing, the intended results can only be achieved by understanding the problem accurately. For a specified target, some domain knowledge and decision-making processes are required to select a few effective features from among several possible features.

Owing to the recent proliferation of machine learning, various machine-learning techniques have been used for scan-chain diagnosis. However, previous scan-chain diagnosis methods using

ANNs and unsupervised machine learning [31–33] merely train simple ANNs based on raw failure log datasets. Therefore, the input vector is long, and several ANNs must be used in only one scan chain; this results in long training diagnosis times. However, if the effects of the fault are compressed by specific filters generated based on the characteristics of the circuit structure, the fault can be diagnosed using a short vector.

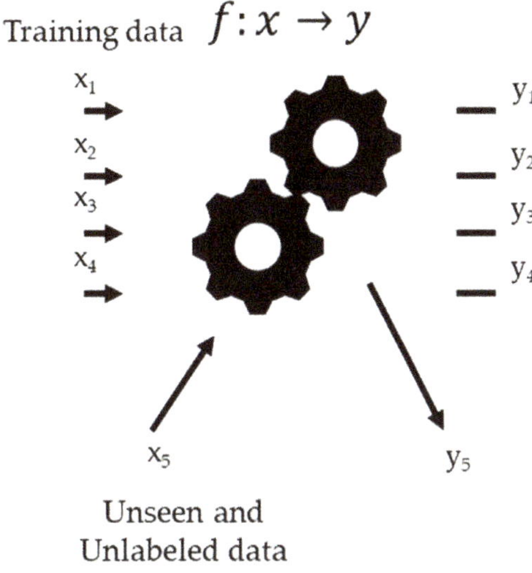

Figure 2. Scope of machine learning.

2.3. Sensitive Cells in Scan-Chain Diagnosis

In the general logic scan test process, test patterns are shifted into the circuit logic and shifted out to the scan-chain cells as a response in the capture process. Therefore, if an error is injected in a scan cell in the scan shift process, it can be spread to other cells through the circuit logic. Hence, errors will occur at the connected cells through the circuit logic. These connected cells of the scan cell are called sensitive cells and are classified into fan-in and fan-out cells. The fan-in cells affect the target cell, whereas the fan-out cells are affected by the target cells. For example, as shown in Figure 3, the circuit contains two scan chains, i.e., 1 and 2, with five scan cells from Cell 1 to Cell 5. Assume that an SA1 fault occurs in Cell 2. Cell 3 of the defective chain is connected to Cell 3 of the fault-free chain and Cell 2 of the defective chain through a NOR gate. Therefore, if an error is injected into Cell 3 of the defective chain from the shift process, it is highly likely that an error occurs in Cell 3 of the fault-free chain and Cell 2 of the defective chain after the capture process. Therefore, Cell 3 of the fault-free chain and Cell 2 of the defective chain are the fan-out cells of Cell 3 of the defective chain, thereby making Cell 3 of the defective chain the fan-in cell of Cell 3 of the fault-free chain and Cell 2 of the defective chain.

The concept of a sensitive cell has been studied previously. Scan-chain reordering [36,37] and stitching [38] have been proposed to consider logic dependency and controllability between scan cells, where scan-chain diagnosis was improved using a circuit structure. In particular, using the fan-in and fan-out dependencies between scan cells, these methods distributed sensitive cells to other chains, thereby obtaining more clues from failure logs. Therefore, software-based scan-chain diagnosis, such as full-masking scan-chain diagnosis, achieved a higher accuracy on reordered scan chains based on sensitive cells than the conventional scan-chain diagnosis.

This concept of sensitive cells can be used for scan-chain diagnosis using ANNs as well. As the impact of a fault in each cell spreads to the sensitive cells of each cell, filters that can contain this impact from the sensitive cells can be used to generate reduced input vectors without compromising accuracy.

Therefore, if a raw failure vector is edited using these filters, a shorter input vector can be constructed for the ANN to diagnose a fault with sufficient accuracy.

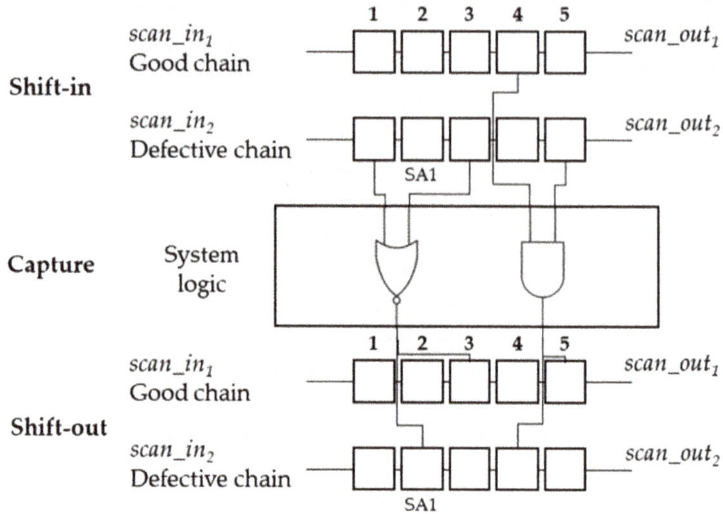

Figure 3. Sensitive cells in scan testing.

3. Proposed Methodology

In this section, the key idea of the study is presented. First, the overall flow of the proposed method and the dataset preparation are described in Section 3.1. Section 3.2 describes the fan-in and fan-out filters that can compress the raw failure log to shorter vectors by highlighting the tendency of the failure. Finally, the selection of training and loss functions is described in Section 3.2.3.

3.1. Overall Flow and Dataset Preparation

Figure 4 shows the overall flowchart of the proposed methodology. The left-hand side of the flowchart shows the test pattern generation process in which failure log datasets are generated on the generated test patterns. A target circuit is synthesized, and scan chains are inserted into the target circuit using a Synopsys design compiler. Next, an ATPG is executed on the scan-chain insertion circuit under test. Subsequently, the test patterns for the target circuit are produced, and failure log datasets are generated to obtain the training and test datasets for the target circuit. Moreover, a fault model is defined similar to the SA fault, slow, and fast models. Furthermore, an intermittent fault simulation is executed to produce the failure log datasets of the defined fault model.

The right-hand side of the flowchart shows the failure feature vector generation process in which the ANN of each scan chain is constructed with the failure feature vectors. First, during test pattern generation, a hierarchical cell report and scan-chain report are generated, and thereby, fan-in and fan-out filters are generated. Next, failure features are extracted using the failure log datasets, fan-in filters, and fan-out filters, to generate failure feature vectors. Subsequently, ANNs are trained for each chain using the failure feature vectors. Finally, the scan-chain diagnosis using the ANNs infer the defective scan cells in each scan chain.

The failure log datasets—comprising failure logs and output labels (the position of the defective cell)—are generated using the following three parameters: (1) fault model, (2) fault location, and (3) fault probability.

1. Fault model: A modeled fault that is one of the following fault types: SA0, SA1, slow-to-rise, slow-to-fall, fast-to-rise, fast-to-fall, fast, or slow.
2. Fault location: A modeled fault is generally assumed to occur in the input wire or output wire of the scan cells or scan cell logic. Therefore, the output of the scan cell can be affected by the modeled faults. The fault location is the location of the scan cell, and it is labeled by the chain and cell numbers. For example, if the third cell in the second chain has a fault, the output label becomes (2, 3).
3. Fault probability: A modeled fault may occur during the processing time. We determined the probability of the fault occurring during the generation of the failure log dataset. For example, assume that a scan chain has seven cells. If the probability of an SA0 fault occurring in Cell 3 is 20%, then each test stimulus at Cells 4–7 may have a 20% probability of failure. By contrast, the test response of Cells 1, 2, and 3 may have a 20% probability of failure.

The fault probability is determined as 10%, 20%, 30%, 40%, 50%, 60%, 70%, 80%, 90%, and 100%. Depending on the probability, the failures are injected in the test stimulus and the test response, as shown below.

1. Perform ATPG and obtain the standard test interface language (STIL) file that contains the test patterns of the target circuit.
2. Inject errors to the test stimulus of the target scan chain with the determined fault probability.
3. Perform fault simulation with the failure-injected STIL file and obtain the failure log datasets through the errors in the test stimulus.
4. Inject errors in the test response of the target chain with the determined fault probability.

In the failure log datasets, various failure cases can be generated even for the same probability. Hence, various vectors can be obtained in the same group of probabilities through 10 iterations.

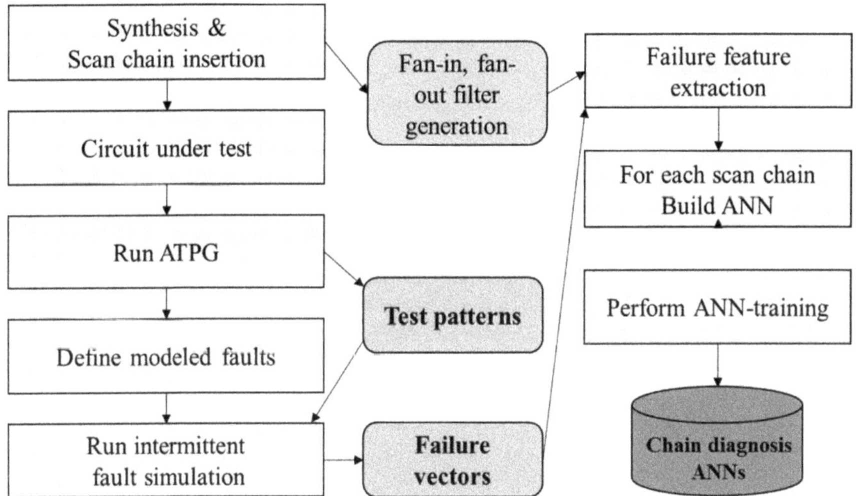

Figure 4. Flowchart of the proposed methodology. ATPG: automatic test pattern generation; ANN: artificial neural networks.

3.2. Failure Feature Extraction

There are two kinds of failures due to the scan-chain faults: test stimulus failure and test response failure. The test stimulus failures are the polluted values of the test stimulus during the shift sequence. Because the value of the test stimulus is loaded serially into the scan chain through a scan-in port-by-shift sequence, the test stimulus values of upstream scan cells of the defective cell are polluted by the defective cell during the shift sequence. This test stimulus failure can cause multiple failures in the fault-free chain through the connected combinational circuits during the capture sequence. Meanwhile, the test response failures are the failures in the test response of the defective chain during shift sequence. In the shift sequence, the state of the scan chain by the capture sequence (test response) is shifted out through scan-out port. Therefore, the test response values of the downstream cells of the defective cell are polluted by the defective cell during the shift sequence. This test response failure can cause only one failure. Therefore, test stimulus failure can cause more failures than test response failure.

Therefore, a fault that occurs near the scan-in cell can cause more test stimulus failures than the test response failures, increasing the number of failures in the failure log. On the other hand, a fault that occurs near the scan-out cell can cause fewer errors in the test stimulus and increase the number of errors in the test response of the defective scan chain, decreasing the number of failures in the failure log.

Figure 5 shows some relevant examples. To simplify the explanation, defective cells are marked by gray boxes; the cells in which errors occur at the test stimulus are marked by the dot patterned boxes; the cells in which failure is observed at the test response are marked by diagonally lined boxes. As shown in the upper part of Figure 5a, assume that an SA fault exists in Cell 10 in the defective chain. This SA fault will affect the test stimulus of Cells 11 and 12 in the defective chain. Therefore, errors will only spread to sensitive Cells 10, 11, and 12 in the fault-free chains. In contrast, assume that an SA fault exists in Cell 2 in the defective chain, as illustrated at the bottom of Figure 5a. This SA fault affects the test stimulus in Cells 2–12 in the defective chain. Hence, several more errors appear in the fault-free chains at the scan output compared with the case in Cell 10. However, the number of failures in the defective chain decreases as the defective scan cell approaches the scan-in cell and increases as the defective scan cell approaches the scan-out cell, as shown in Figure 5b.

These effects are more prominent in the sensitive cells of each cell. Therefore, a shorter vector that contains all the failure information would be generated by vectorizing the tendencies of the failures based on an investigation of all the sensitive cells in each chain. Consequently, filters are necessary to generate a vector that can train these tendencies from the raw failure log datasets.

The sensitive cells are categorized into two types. The first type includes the fan-out cells, which are the affected cells in all the chains that are connected to the cell of the defective chain. Therefore, if an error occurs at a cell, the observed responses of one or more of the fan-out cells will contain errors. For example, as shown in Figure 3, if an error occurs in the test stimulus at Cell 0 in Chain 1, the test response of Cell 0 in Chain 1 and that of Cell 1 in Chain 2 may lead to failures in these cells. Therefore, Cell 0 and Cell 1 in Chain 2 are the fan-out cells of Cell 0 in Chain 1, in contrast to the fan-in cells. Similar to this example, the fan-out filters identify the fan-out cells in the defective chain.

The second type is the fan-in cells, which are the affected scan cells at the defective chain that are connected to the cell of the fault-free chain. If there is an error in the response of a cell in a chain other than the detective chain, this error must have originated from the error stimulus of a cell in the defective chain. The fan-in cells are the cells observed by backtracking the cells that are connected. Therefore, an error occurring at a cell results in errors in the test stimulus of one or more of the fan-in cells of the cell. For example, Cell 1 in Chain 1 and Cell 0 in Chain 2 are connected to an AND gate cell, which is connected to Cell 0 in Chain 1 and Cell 1 in Chain 2. Consequently, Cell 1 in Chain 1 and Cell 0 in Chain 2 are the fan-in cells of Cell 0 in Chain 1 and Cell 1 in Chain 2.

The fan-in and fan-out filters are generated by analyzing the circuit logic structure to extract the failure feature from the failure logs. First, the hierarchical cell and scan-chain reports are analyzed. The hierarchical cell report details the connection of each cell, such as the input pins, output pins, net driver pins, and net load pins, whereas the scan-chain report shows the name of the scan cell belonging to each scan chain. Accordingly, the fan-in and fan-out filters are generated by searching all the fan-in and fan-out cells of each cell.

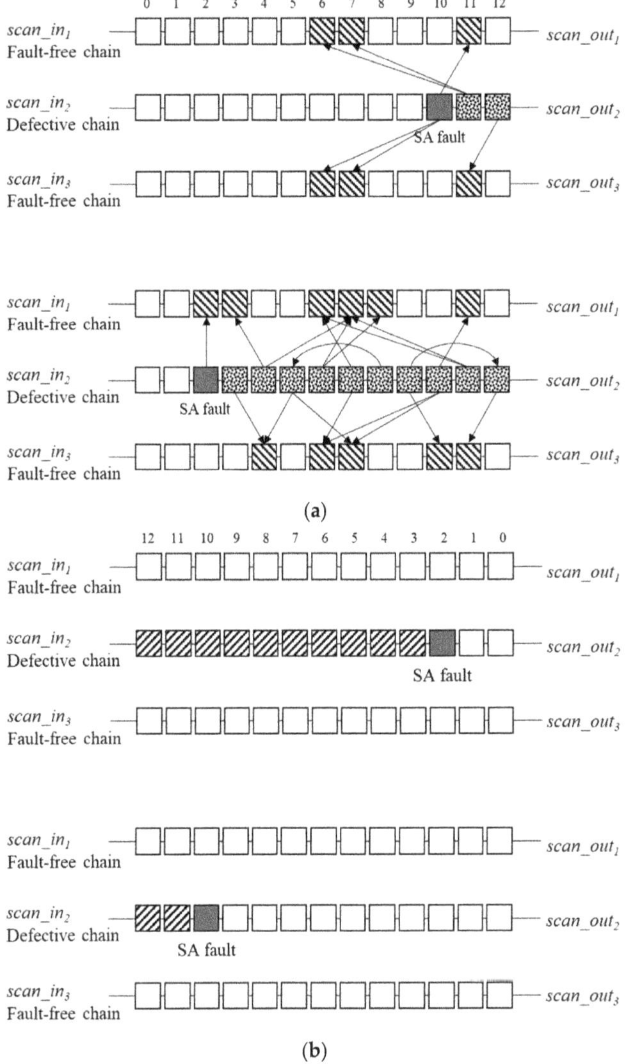

Figure 5. Failure tendencies: (**a**) failure tendency caused by shift-in; (**b**) failure tendency caused by shift-out.

3.2.1. Fan-In Filters

The fan-in filter accumulates the number of errors affected by the fan-in cells in the defective chain. Hence, each element of a vector reflects the number of errors affected by each cell in the defective chain. Therefore, the length of the fan-in vector is the same as that of the defective chain. First, the fan-in filters detect the fan-in cells of all the scan cells in the defective chain. Subsequently, the number of errors in the fan-in cells of each cell is applied to the element reflecting the cell of the fan-in vector.

To apply the number of errors, failure log datasets are analyzed. If an error exists at the location of a cell in the failure log dataset, this information is applied in the elements of the fan-in cells in the fan-in vectors, if no error occurs in the test response of the fan-in cells. Because, if a failure occurs on the test stimulus of the defective chain, failure does not occur again on the test response of the chain except the defective cell. Therefore, each element of fan-in vectors reflects the number of errors in the fault-free chain caused by each cell in the defective chain. This operation is performed for all failure log datasets. By adding all vectors investigated in this manner, a fan-in vector is created.

An example of the fan-in filters is demonstrated in Figure 6 and Table 1. The target circuit comprises three scan chains, each containing seven cells. For ease of explanation, Cell B in Chain A is referred to as $S_A C_B$. If an SA0 fault occurs at Cell 5 in scan chain 2 ($S_2 C_5$), then errors occur at Cells 1, 2, and 6 in scan chain 1 ($S_1 C_1$, $S_1 C_2$, $S_1 C_6$), Cells 1–5 in scan chain 2 ($S_2 C_1$, $S_2 C_2$, $S_2 C_3$, $S_2 C_4$, $S_2 C_5$), and Cells 1, 2, and 6 in scan chain 3 ($S_3 C_1$, $S_3 C_2$, $S_3 C_6$) in the observed responses of test pattern 0.

An analysis of the circuit structure in Table 1 shows that $S_1 C_1$ is affected by $S_2 C_1$, $S_2 C_3$, and $S_2 C_6$; $S_1 C_2$ is affected by $S_2 C_5$; $S_1 C_6$ is affected by $S_2 C_4$, $S_2 C_5$, and $S_2 C_7$; $S_3 C_1$ is affected by $S_2 C_2$ and $S_2 C_5$; $S_3 C_1$ is affected by $S_2 C_5$ and $S_2 C_6$; $S_3 C_2$ is affected by $S_2 C_7$. In pattern 0, the failure at the $S_1 C_1$ is applied to the fan-in vector of $S_2 C_1$, $S_2 C_3$, and $S_2 C_6$. There are failures on the test responses of $S_2 C_1$ and $S_2 C_3$. Therefore, only the element of $S_2 C_6$ counts, <0 0 0 0 0 1 0>. The failure of $S_1 C_2$ is applied to $S_2 C_6$, <0 0 0 0 0 2 0>. The failure of $S_1 C_6$ is applied to $S_2 C_7$, <0 0 0 0 0 2 1>. The failure of $S_3 C_1$ is applied to $S_2 C_2$ and $S_2 C_5$, but there are failures on the test response of $S_2 C_2$ and $S_2 C_5$, <0 0 0 0 0 2 1>. The failure of $S_3 C_2$ is applied to $S_2 C_5$ and $S_2 C_6$, but there is failure on the test response of $S_2 C_5$, <0 0 0 0 0 3 1>. The failure of $S_3 C_6$ is applied to $S_2 C_7$, <0 0 0 0 0 3 2>. Therefore, the fan-in vector of scan chain 2 becomes <0 0 0 0 0 3 2>. This fan-in vector is generated through all the failure log datasets. Subsequently, test patterns 0, 1, 2, and 3 acquire vectors <0 0 0 0 0 3 2>, <0 0 1 1 1 3 3>, <0 1 0 2 0 3 4>, and <0 1 0 0 1 3 2>, respectively. By adding all the vectors, the fan-in vector becomes <0 2 1 3 2 12 11>.

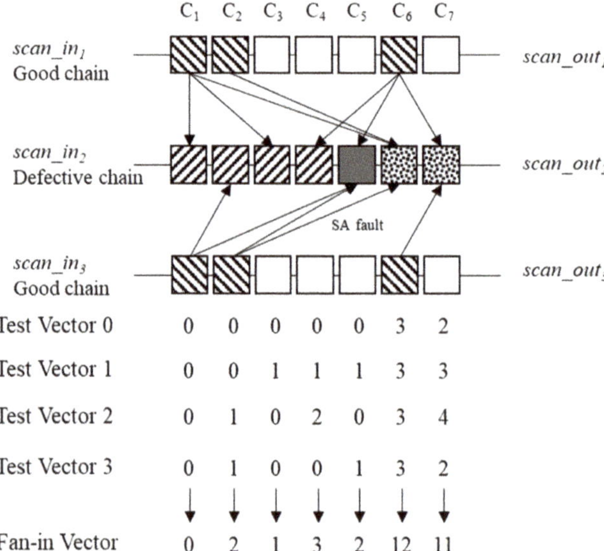

Figure 6. Example of fan-in filter.

Table 1. Fan-in connected cells of example in Figure 6.

Cell	Fan-In Connected Cells
$S_1 C_1$	$S_2 C_1$, $S_2 C_3$, and $S_2 C_6$
$S_1 C_2$	$S_2 C_5$
$S_1 C_6$	$S_2 C_4$, S2C_5, and $S_2 C_7$
$S_3 C_1$	$S_2 C_2$ and $S_2 C_5$
$S_3 C_2$	$S_2 C_5$ and $S_2 C_6$
$S_3 C_6$	$S_2 C_7$

3.2.2. Fan-Out Filters

If a fault occurs near the scan-in cell, the test stimulus will contain more errors than when a fault occurs near the scan-out cell. Therefore, the failure in the test response would have spread more than that in the case where it is close to the scan-out cell.

Hence, the fan-out filter accumulates the number of errors at the fan-out cells in all the chains. Therefore, each element of a vector reflects the number of errors in the fan-out cells. Accordingly, the length of the fan-out vector is the same as the number of cells. First, the fan-out filters detect the fan-out cells of all the scan cells. Subsequently, the number of errors in the fan-out cells of each cell is applied to the element reflecting the cell of the fan-out vector.

To obtain the features from the failure log dataset, the fan-out filters track the spreading cells of the shift-out failures from all chains using a circuit structure similar to that of the fan-in filter. Resembling the fan-in vector, each element of a vector reflects the number of errors in the fan-out cells in the defective chain. The failure log datasets are analyzed to calculate the number of errors in the fan-out cells. The fan-out cells of each cell are searched, and the total number of errors that exist at all fan-out cells is accumulated in the elements of the cell in the fan-out vectors. Therefore, the element of the fan-out vectors reflecting a cell becomes the sum of the number of errors at all the fan-out cells of the cell. Moreover, all the failure log datasets are summed. This operation is performed for all the scan chains. Subsequently, a fan-out vector is created by adding all the vectors investigated thus.

Figure 7 and Table 2 demonstrate the fan-out filters, where the target circuit comprises three scan chains, each containing seven cells. If an SA fault occurs at Cell 3 in scan chain 2 (S_2C_3), then test pattern 0 acquires the failure at Cells 3, 4, and 7 in scan chain 1 (S_1C_3, S_1C_4, S_1C_7); Cells 3, 4, 5, and 7 in scan chain 2 (S_2C_3, S_2C_4, S_2C_5, S_2C_6, S_2C_7); and Cells 5 and 7 in scan chain 3 (S_3C_5, S_3C_7).

In this circuit structure as shown in Table 2, S_2C_2 is affected by S_2C_3; S_2C_3 is affected by S_1C_2 and S_2C_3; S_2C_4 is affected by S_3C_5; S_2C_5 is affected by S_2C_3, S_1C_4, S_1C_7, and S_3C_7; S_2C_6 is affected by S_3C_2, S_3C_4, and S_3C_5; and S_2C_7 is affected by S_1C_7, S_2C_7, and S_3C_7. In this case, in pattern 0, the fan-out vector of S_2C_1 is 0 due to the absence of a fan-out cell in Cell 0; that of S_2C_2 and S_2C_3 is 1 due to failure at the fan-out cell S_2C_3; that of S_2C_4 is 1 due to failure at the fan-out cell S_3C_5. Finally, the fan-out vector of scan chain 2 becomes 0 1 1 1 3 1 3. These fan-out vectors are generated using all the failure log datasets. Subsequently, test patterns 0, 1, 2, and 3 acquire vectors <0 1 1 1 3 1 3>, <0 0 2 1 2 3 3>, <0 1 2 2 4 3 4>, and <0 1 3 0 5 3 2>, respectively. By adding all the vectors, the fan-out vector of scan chain 2 becomes <0 3 7 4 13 10 12>.

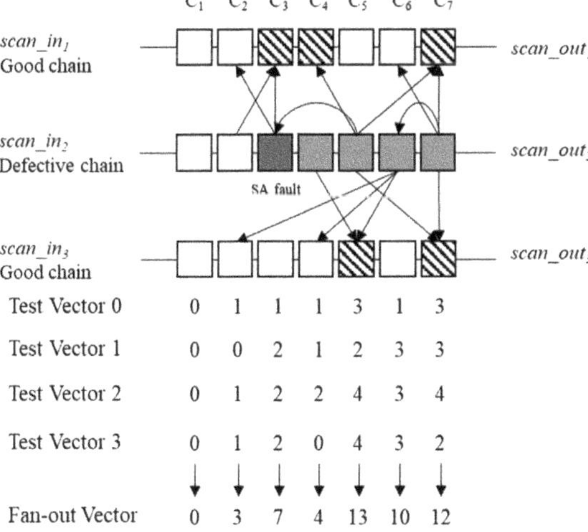

Figure 7. Example of the fan-out filter.

Table 2. Fan-out connected cells of example in Figure 7.

Cell	Fan-Out Connected Cells
S_2C_1	
S_2C_2	S_2C_3
S_2C_3	S_1C_2 and S_2C_3
S_2C_4	S_3C_5
S_2C_5	S_2C_3, S_1C_4, S_1C_7, and S_3C_7
S_2C_6	S_3C_2, S_3C_4, and S_3C_5
S_2C_7	S_1C_7, S_2C_7, and S_3C_7

3.2.3. Scan-Chain Diagnosis with Regressions

In this study, linear and logistic regressions were used for scan-chain diagnosis. The proposed scan-chain diagnosis requires only one trained model for the target chain and a target fault type, such as stuck-at 0, stuck-at 1, fast-to-rise, fast-to-fall, slow-to-rise, and slow-to-fall, to determine the accurate candidate of the scan-chain faults. With N_c scan chains and f fault types, $N_C \times f$ models are trained to support the proposed scan-chain diagnosis.

The input vector of the proposed methodology, which is called the failure feature vector, is formed by combining three vectors. The first vector is the fan-in vector from the fan-in filter, the second vector is the fan-out vector from the fan-out filter, and the last vector is the IFV. These failure feature vectors are generated from the failure log datasets using the fan-in and fan-out filters. Furthermore, the generated failure feature vectors are categorized into train and test sets for training and testing, respectively. An example of the training vector is illustrated in Figure 8.

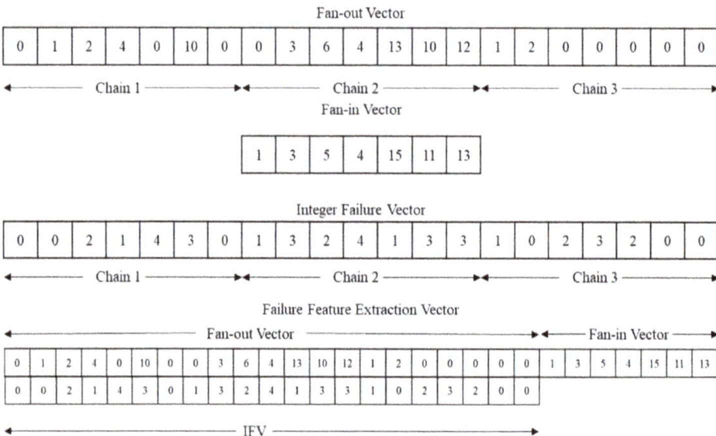

Figure 8. Example of failure feature vector.

Linear regression and logistic regression were used as the machine-learning models through scikit-learn [39]. Linear regression analysis involves quantitatively determining the relation between the D-dimensional vector-independent variable x and the corresponding scalar-dependent variable y.

The linear regression model involves obtaining a function $f(x)$ that outputs a value $\hat{y}$ that is closest to the corresponding dependent variable y for the independent variable x, Equation (1).

$$\hat{y} = f(x) \approx y \tag{1}$$

If the relation between x and y is the following linear function $f(x)$, it is known as a linear regression function Equation (2).

$$\hat{y} = w_0 + w_1x_1 + w_2x_2 + \cdots + w_Dx_D, \tag{2}$$

where $w_0, \dots, w_D$ are the coefficients of function $f(x)$ and the parameters of this linear regression model. We considered a linear regression for solving the single-fault problem. The failure log dataset of a scan-chain fault exhibits a failure distribution tendency, and a linear regression is a suitable algorithm for solving this problem.

The logistic regression is used to determine the probability of the existence of a certain class, such as pass/fail, win/lose, or healthy/sick. This scan-chain diagnosis problem can be categorized into a pass/fail class in the output element of each vector, which is the same as the logistic regression problem. This problem is difficult to solve because the output label is only 0/1 in the output vector, as shown in Figure 9a. Therefore, logistic regression is used to solve the pass/fail problem using logistic functions.

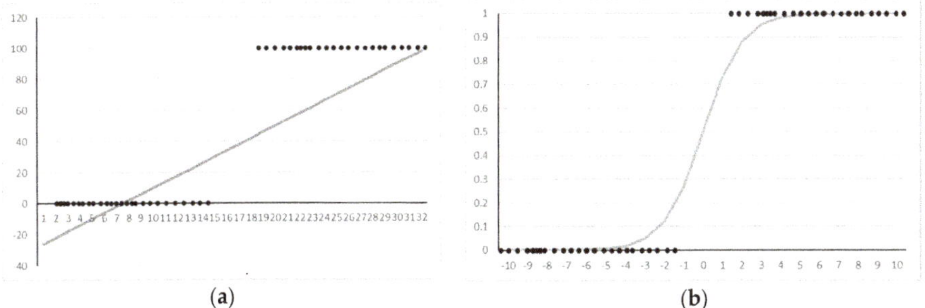

Figure 9. Pass/fail class with linear regression (**a**) and logistic regression (**b**).

In several natural and social phenomena, the probability value for a specific variable often follows the form of an S-curve rather than a linear one. The logistic function expresses this S-curve as a function. Logistic functions can assume any value as an x value, but the output is always between 0 and 1. In other words, this function satisfies the requirement of a probability distribution function. The formula (3) is given below:

$$y = \frac{1}{1 + e^{-x}} \tag{3}$$

If the logistic function is used as an activation function in neural networks, we can easily solve the pass/fail problems using the neural networks, as shown in Figure 9b.

For a future multiple-fault problem, logistic regression is performed to solve a single scan-chain fault problem. If multiple faults occur in a scan-chain diagnosis, an output label cannot be obtained. Alternatively, a binary output label having the same length as that of the target scan chain is obtained. For algorithm optimization, scikit-learn can decide to use various solvers for logistic regression. The solvers used for scan-chain diagnosis are "A library for large linear classification (liblinear)", "limited-memory Broyden–Fietcher–Goldfarb–Shanno (lbfgs)", "stochastic average gradient (sag)", and "saga". Appropriate solvers for each training case are listed in Table 3.

Table 3. Appropriate solvers for each case.

Case	Solver
L1 penalty	"liblinear" or "saga"
Multinomial Loss	"lbfgs," "sag," "saga," or "newton-cg"
Very Large dataset	"sag" or "saga"

The "liblinear" solver uses the coordinate descent algorithm. Hence, it successively solves the problem by performing an approximate minimization along the coordinate directions or hyperplanes. However, it may not be able to solve the nonstationary point or learn a multiclass model.

The "lbfgs" solver uses the Hessian matrix; however, it is approximated using updates specified by gradient evaluations. In addition, its limited memory stores merely a few vectors that represent the approximation implicitly. Therefore, if the training dataset is small, then "lbfgs" delivers the best performance, compared with the other methods. However, it may not converge if it is not safeguarded.

The "sag" solver uses optimization for the sum of a finite number of smooth convex functions. Therefore, its iteration cost is independent from the number of terms in the sum. However, it is faster than the other solvers for large datasets, as it incorporates the memory of previous gradient values when both the numbers of samples and features are large.

Additionally, the "saga" solver is a variant of the sag solver that supports the non-smooth option. This solver is for sparse multinomial logistic regression and is suitable for extremely large datasets, similar to the "sag" solver. An appropriate solver must be used to optimize the accuracy of scan-chain diagnosis. Therefore, the characteristics of the scan-chain failure data must be considered. As listed in Table 3, the size of the dataset is critical when selecting a solver. The size of the failure log datasets depends on the circuit size. Therefore, the solver for a circuit is selected based on the circuit size.

4. Experimental Results

The accuracy, training time, and diagnosis time of the proposed scan-chain diagnosis were evaluated through the experiments. The ITC'99 benchmark, which represents a wide range of industry-representative circuits and was used in this study, openrisc1200 (OR1200) and advanced encryption standard (AES) circuit are summarized in Table 4 [40].

Table 4 summarizes the characteristics of benchmark circuits. In the first and second columns, the circuit and gate count denote the circuit name and number of gates, respectively. The third and fourth columns show the numbers of scan cells and chains in the circuit, respectively. The fifth and sixth columns denote the maximum and minimum numbers of fan-in and fan-out cells in the circuit, respectively; the maximum number is shown on the left side of the slash (/), whereas the minimum number is shown on the right side of the slash (/).

The specifications of the computer used in the experiment are as follows:

CPU: Intel Core i7-770 @ 3.60 GHz

RAM: 64 GB

The scikit-learn solver does not use a graphics processing unit (GPU). Therefore, a GPU was not used. The datasets were generated with the probabilities of 10%, 20%, 30%, 40%, 50%, 60%, 70%, 80%, 90%, and 100%. The training and test datasets were generated through 10 iterations. The training and test data were obtained by splitting the datasets into 75% and 25%, respectively, using the train_test_split function in scikit-learn.

Table 4. Characteristics of benchmark circuits.

Circuit	Gate Count	Number of Scan Cells	Number of Scan Chains	Fan-In	Fan-Out
B12	1000	120	10	37/0	65/1
B14	3461	220	10	183/0	220/0
B15	6931	420	10	235/3	326/0
B17	21,191	1320	10	243/2	371/0
B20	7931	430	10	211/0	288/0
B22	12,128	620	10	227/0	285/0
AES	10,656	554	10	35/0	128/0
OR1200	16,637	3420	30	313/0	2905/0

4.1. Scan-Chain Diagnosis Accuracy

The accuracy of scan-chain diagnosis was evaluated using the accuracy classification score. The results of the test data using our trained regressions were compared with the actual injected fault location. If the prediction results of our models are the same as those of the injected fault location, they are treated as successful cases; otherwise, they are considered unsuccessful. The percentage of successful cases in the total test cases represents the accuracy of the scan-chain diagnosis.

The accuracy of the scan-chain diagnosis of the intermittent SA0 faults occurring in the first scan chain is reported. For the training and test data, we attempted to generate as many cases as possible. Hence, after targeting each fault location and setting the probability of faults, a failure was generated several times to create various failure cases.

For example, as shown in Figure 7, the fault location was C_3 in the defective scan chain. Hence, a failure may occur in the test stimuli at C_4, C_5, C_6, and C_7. These test stimulus vector elements and the test response vector elements where failure can occur are known as possible failure points. The failure generator generates a random number from 0 to 99 at each possible failure point. Subsequently, if the generated random number is higher than the probability value, such as 10, 20, … , 100, a failure is injected at that point; otherwise, a failure is not injected at that point. Therefore, each failure is injected at the test stimuli C_3, C_4, C_5, C_6, and C_7 and at the test responses C_1, C_2, and C_3 if each generated random number is higher than the probability value. Consequently, even if the probability and fault location are the same, different failure log datasets are generated in each iteration.

The results are presented in Table 5. The Tessent tool [31] indicated that the scan-chain diagnosis had a 72.43% accuracy. In addition, the scan-chain diagnosis using multistage ANNs [33] achieved an average accuracy of 86.98%, as shown in the labeled column [33]. The accuracy of the proposed scan-chain diagnosis was improved to 90.15% through logistic regression, which implies that its accuracy was increased by 3.17%, compared with that achieved in [33].

In addition, the open-source cryptography circuits, AES [41], and the open source core OR1200 [42] have been used to evaluate the performance of the proposed scan-chain diagnosis in the cryptographic circuits and the processor. "lbfgs" solver has been selected to optimize the logistic regression model of AES and OR1200. Subsequently, the scan-chain diagnosis accuracy of AES and OR1200 are 90.06% and 82.36%, respectively. However, it can be increased by changing the configuration of the logistic regression.

Table 5. Accuracy of scan-chain diagnosis for SA0 faults occurring in the first scan chain.

Circuit	Prediction Accuracy (%)		
	[31]	[33]	Proposed Method
B12	83.70%	79.00%	81.60%
B14	24.90%	88.40%	96.24%
B15	93.80%	80.30%	98.62%
B17	61.30%	85.20%	83.93%
B20	95.30%	97.50%	86.49%
B22	75.60%	91.50%	94.01%
AES	-	-	90.06%
OR1200	-	-	82.36%

4.2. Training Time and Diagnosis Time

Table 6 presents the construction times for the scan-chain diagnosis of a target chain. Compared with the previous scan-chain diagnosis [33], the proposed diagnosis additionally requires the filter generation time (Filter Gen. Time). However, as shown in Table 6, the filter generation time is shorter than the training time. Therefore, it does not affect the construction time. Let n be the number of training samples and d be the dimensionality of the features. The required training time is generally $O(nd^2)$. The dimensionalities of the features of the previous scan-chain diagnosis (D_M) [33] and the proposed scan-chain diagnosis (D_P) are defined as follows:

$$D_M = SC_{total} \times N_{test\ patterns} \tag{4}$$

$$D_P = (2 \times SC_{total}) + SC_{defective} \tag{5}$$

where SC_{total} is the number of scan cells, $N_{test\ patterns}$ is the number of test patterns, and $SC_{defective}$ is the number of scan cells in the defective chain. The proposed method reduces the dimensionality of the features from D_M to D_P.

Table 6. Model construction time.

Circuit	Model Construction Time (h)			
	[33]	Filter Gen. Time	Proposed Method	Reduction Ratio
B12	0.10	2.8×10^{-4}	0.08	20.00%
B14	2.70	9.2×10^{-2}	0.21	92.22%
B15	4.30	5.5×10^{-4}	0.76	82.32%
B17	71.60	2.8×10^{-2}	1.40	98.04%
B20	3.40	9.2×10^{-2}	0.38	88.82%
B22	25.90	1.5×10^{-1}	0.68	97.37%
AES	-	1.5×10^{-2}	0.18	-
OR1200	-	9.5×10^{-2}	9.05	-

Therefore, the training time required by the proposed scan-chain diagnosis is shorter than that required by the previous scan-chain diagnosis [33], as presented in Table 6. Moreover, with the increase in the circuit size, the previous scan-chain diagnosis becomes slower due to the requirement of multiple ANNs even for a single scan chain.

Therefore, the proposed scan-chain diagnosis reduced the construction time by more than 79%, compared with the previous scan-chain diagnosis. The diagnosis time, i.e., the inference time of all the test cases, is listed in Table 7. The proposed scan-chain diagnosis is significantly faster as it uses one model for one chain, whereas the previous scan-chain diagnosis [33] uses multiple ANNs for one chain. As the sizes of the circuit and layers increased, the diagnosis time of the proposed scan-chain diagnosis slightly increased. Moreover, the diagnosis time of the proposed scan-chain diagnosis was reduced by more than 99.9984%, compared with that of the previous diagnosis. The results of experiment in the cryptographic circuit, AES are also added in Tables 6 and 7. The construction time and diagnosis time of AES are 0.18 h and 3.9 microseconds, respectively.

In addition, the results in the processor, OR1200 are included in Tables 6 and 7. Because the number of scan cell of the circuit is increased to 3420, the construction time and diagnosis time are increased to 9.05 h and 45.77 microseconds, respectively.

Table 7. Diagnosis time.

Circuit	Diagnosis Time (ms)		
	[33]	Proposed Method	Reduction Ratio
B12	48,000	0.89	99.9981%
B14	181,500	2.00	99.9989%
B15	231,000	4.00	99.9983%
B17	3,663,000	48.85	99.9987%
B20	290,250	8.93	99.9969%
B22	1,224,500	7.98	99.9993%
AES	-	3.9	-
OR1200	-	45.77	-

The reduction in training and diagnosis times facilitates the application of scan-chain diagnosis in semiconductor industries. As the training and diagnosis times are critical to satisfy the time-to-market goal for industrial circuits, the proposed scan-chain diagnosis is suitable, as its training time is 50% of that of the previous diagnosis. Therefore, the preparation time for diagnosing industrial circuits reduces to 79.80% of that of the previous methods. In addition, the proposed scan-chain diagnosis enables the time-to-market goal to be achieved by improving the diagnosis speed in the semiconductor industry, where several cases require diagnosis.

The training time, diagnosis time, and accuracy of scan-chain diagnosis of logistic regression change depend on the solvers. Moreover, the regularization strength affects the accuracy of logistic regression. The regularization strength refers to a penalty to increase the magnitude of parameter values to reduce overfitting. Hence, we measured the accuracy of scan-chain diagnosis, training time, and diagnosis time of B12 while changing the solver and the inverse of the regularization strength of the logistic regression.

Figure 10 shows the accuracy of scan-chain diagnosis depending on the inverse of the regularization strength with the lbfgs solver. Moreover, the inverse of the regularization strength was changed to a logarithmic scale (i.e., $0.01, 0.1, 1, 10, 10^2 \dots$). The accuracy increases as the inverse of the regularization strength increases from $1 \times e^{-6}$ to $1 \times e^{-2}$. Subsequently, the accuracy gradually decreases as the inverse of the regularization strength increases. Therefore, the accuracy of scan-chain diagnosis is the highest when the inverse of regularization strength is $1 \times e^{-2}$. Figure 10a demonstrates the impact of the inverse of the regularization strength on the accuracy.

Additionally, we measured the accuracy of scan-chain diagnosis by changing the solvers with a fixed inverse of the regularization strength, $1 \times e^{-2}$. The accuracy varies slightly depending on the solver. However, this difference is small compared with the difference according to the inverse of the regularization strength; hence, if the appropriate regularization strength is determined, the solver does not significantly affect the scan-chain diagnosis.

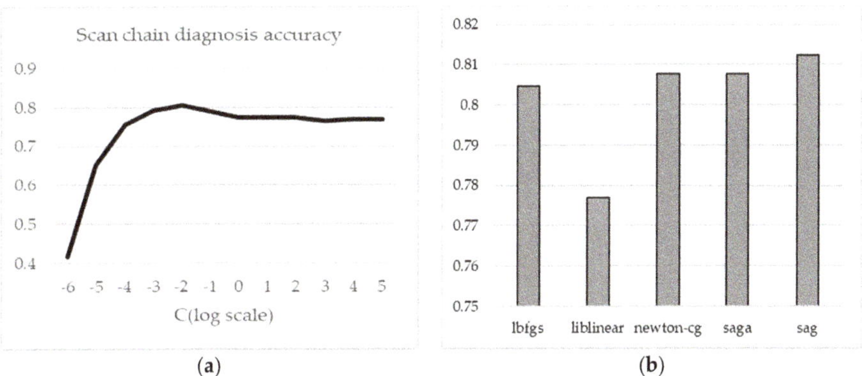

Figure 10. Accuracy of scan-chain diagnosis depending on regularization strength value and solvers on B12: (**a**) accuracy of scan-chain diagnosis depending on the regularization strength value (C); (**b**) accuracy of scan-chain diagnosis with various solvers at the inverse of regularization strength value e^{-2}.

4.3. Accuracy through Failure Log Dataset Property

The correlation between the circuit characteristics and accuracy was analyzed to improve the accuracy of the scan-chain diagnosis. For the analysis, B12 was used owing to its significantly lower training time than those of other circuits, with a comparable accuracy of scan-chain diagnosis.

First, the accuracy was analyzed according to the location of each cell in the scan chain by predicting the test stimulus. This model was trained to determine whether it can predict the test stimulus from the failure log dataset. Therefore, only logistic regression, specifically multiclass logistic regression (OneVsRest-Classification), was used. In this training, the input vector was a binary failure vector, and the output label was a test stimulus failure vector. The binary failure vectors were obtained from the failure log dataset. The total binary failure vector was divided among all test patterns, namely, test pattern 0, test pattern 1, ... test pattern N. The binary failure vector that matched each test pattern was the input vector. In addition, the test stimulus failure vectors were obtained during the failure generating sequence. Resembling the scan-chain diagnosis, the training data and test data were obtained from 75% and 25% of the dataset, respectively, using the *train_test_split* function in scikit-learn.

As shown in Figure 6, a fault occurred in C_3 in the target scan chain. Consequently, failures occurred at the test stimuli $C_3, C_4, C_5,$ and C_7. Hence, the test stimulus failure vector was <0 0 1 1 1 0 1>. Additionally, the binary failure vectors from this test pattern were <0 0 1 1 0 0 1>, <1 1 0 0 0 1 0>, and <0 0 0 0 1 0 1>. These models predicted the test stimulus failure vector from a cascaded set of binary failure vectors, known as a cascaded binary vector. Accordingly, the input vector was <0 0 1 1 0 0 1 1 1 0 0 0 1 0 0 0 0 0 1 0 1>, and the output label was <0 0 1 1 1 0 1>. In the prediction, the incorrect case was obtained by comparing the prediction and output labels. If the obtained prediction label is <0 1 1 1 1 0 1>, then the output of Cell 1 is incorrect. In this case, the accuracy of Cell 1 is reduced.

Table 8 lists the fan-in and fan-out characteristics and the accuracy of scan-chain diagnosis of each cell. The first and second columns show the cell number and the cell prediction accuracy of each scan cell, respectively. The third and fourth columns show the numbers of fan-in and fan-out cells of each cell, respectively; the number of cells in the other chains is shown on the left side of the slash (/), whereas the number of cells in the defective chain is shown on the right side of the slash (/). As presented in Table 8, the accuracy of stimulus failure prediction is high when a fault occurs in Cell 0. However, the accuracy decreases as a defective cell approaches the scan-out cell. Subsequently, it increases when the fault occurs at the scan-out cell. The correlation between the accuracy and the numbers of fan-in and fan-out cells is insignificant. However, if the number of fan-in cells is 0, the accuracy decreases rapidly, as shown by Cell 11. Therefore, if a cell has no fan-in cells, it must be located at the beginning of the chain. This reordering can increase the accuracy of scan-chain diagnosis.

Table 8. Accuracy of scan-chain diagnosis for SA0 faults occurring in the first scan-chain cells.

Cell	Cell Accuracy	Fan-Out Cells	Fan-In Cells
Cell 0	86.64%	65/1	7/1
Cell 1	87.62%	66/1	7/2
Cell 2	70.95%	66/1	7/2
Cell 3	64.34%	68/1	7/2
Cell 4	62.13%	66/1	7/2
Cell 5	56.50%	31/6	16/7
Cell 6	49.39%	31/6	16/6
Cell 7	44.24%	31/6	16/6
Cell 8	53.06%	31/6	16/6
Cell 9	49.75%	31/6	16/6
Cell 10	39.59%	31/6	16/6
Cell 11	11.76%	1/1	0/0
Cell 12	56.00%	1/1	0/2

Next, the correlation between the magnitude of the same fault effect between different faults in the chain, the magnitude of the failure effect of a cell in the chain, and the accuracy are analyzed. If the failure tendency of each cell in the same scan chain is similar, it is difficult to find the defective cell. Suppose there are two scan chains of three cells. The failure effect of the first cell on a chain is the first and second cells of another chain. The failure effect of the second cell on the chain is also the first and second cells of another chain. In this case, the exact defective scan cell cannot be clearly distinguished by any scan-chain diagnosis.

It is also difficult to predict the defective cell if the failure of a cell is not affected anywhere. Suppose the failure effect of the third cell does not affect any cell. The failure on the third cell of the chain does not make any failure on another chain. Therefore, the exact defective scan cell cannot be clearly distinguished by any scan-chain diagnosis.

To measure the magnitude of the same fault effect between different faults in the chain and the magnitude of the failure effect of a cell in the chain, the fan-out similarity is proposed, as well as fan-in similarity, average number of fan-out, and average number of fan-in. Fan-out and fan-in similarity are a measure of how many cells in the same chain share the same cells as the fan-out and fan-in cells of a specified cell. The similarity between Cell A and Cell B (SIM_{AB}) is defined as follows:

$$SIM_{AB} = \frac{N_{AB}}{N_A} \tag{6}$$

where N_A is the number of fan-out cells of Cell A and N_{AB} is the number of the same fan-out cells between Cell A and Cell B. Then, the similarity of a chain is the value that totalizes the similarity of all cells in the chain except the similarity of itself such as SIM_{AA}. Therefore, the similarity of the chain (SIM_{chain}) is defined as follows.

$$SIM_{chain} = \sum_{i=0}^{n} \sum_{j=0}^{n} SIM_{ij} - \sum_{i=0}^{n} SIM_{ii} \tag{7}$$

where n is the number of cells in the scan chain. Let us explain with an example. Suppose there are two scan chains of three cells. The fan-out cells of the first cell for the first chain are the first and second cells of another chain; the fan-out cells of the second cell are the second cells of another chain. Finally, the fan-out cell of the third cell is the first cell in another chain. In this case, the fan-out similarity between cells are measured to $SIM_{12} = \frac{1}{2}$; $SIM_{13} = \frac{1}{2}$; $SIM_{21} = 1/1$; $SIM_{23} = 0$; $SIM_{31} = 1/1$; $SIM_{32} = 0/1$. Thus, in this case, the fan-out similarity of the first chain is 3. Likewise, the fan-in similarity is also measured.

The average number of fan-out, the average number of fan-in is a measure of the magnitude of the failure effect of a cell in the chain. In the below example, the total number of fan-out cells of chain 1 is 4, and the number of scan cells is 3. Therefore, the average number of fan-out is 4/3. Likewise, the average number of fan-in is also measured.

Table 9 lists the fan-out and fan-in similarity, the numbers of fan-out and fan-in cells, and the accuracy of scan-chain diagnosis of each chain. The first and second columns show the chain number and accuracy of each scan chain, respectively. The third and fourth columns show the fan-out and fan-in similarity in the corresponding chain, respectively. The fifth and sixth columns show the average numbers of fan-in and fan-out cells in the corresponding chain, respectively.

Table 9. Accuracy of scan-chain diagnosis for SA0 faults with the fan-out similarity, fan-in similarity, average fan-out cells, and average fan-in cells.

Chain	Scan-Chain Diagnosis Accuracy	Fan-Out Similarity	Fan-In Similarity	Average Fan-Out Cells	Average Fan-In Cells
B12.Chain 0	77.74%	74.27	51.70	13.76923	43.61538
B12.Chain 1	87.12%	36.62	40.00	18.58333	27.25
B12.Chain 2	63.89%	51.77	34.15	14.66667	16.41667
B12.Chain 3	47.85%	97.11	30	9	2
B12.Chain 4	48.61%	97.11	30	9	2
B12.Chain 5	50.88%	98.11	30	8.916667	2
B12.Chain 6	48.23%	97.11	30	9	2
B12.Chain 7	50.51%	81.95	25	10.41667	1.833333
B12.Chain 8	71.97%	62.15	26.11	17.91667	16.08333
B12.Chain 9	80.30%	93.03	31.77	21.33333	9.25
B14.Chain 0	96.24%	582.68	457.12	229.52	20.28
B14.Chain 1	100%	191.59	504.86	71.2	45.76
B14.Chain 2	100%	458.33	546.60	111	39.56
B14.Chain 3	100%	89.29	524.10	51	39.44
B14.Chain 4	100%	40.03	434.03	33.84	63.36
B14.Chain 5	100%	415.03	499.50	54.72	54.72
B14.Chain 6	100%	452.94	486.34	93.36	118.96
B14.Chain 7	100%	404.59	473.91	82.52	128.84
B14.Chain 8	100%	393.20	478.06	69.28	139.08
B14.Chain 9	100%	394.05	446.78	74.12	130.64

The scan-chain diagnosis accuracy increased with the average number of fan-in and decreased with the fan-out similarity as shown in Table 9, the average numbers of fan-ins of Chains 8 and 9 in B12 are higher than that of Chain 1, but the fan-out similarities of the two chains are higher than that of Chain 1, therefore, the accuracies of Chains 8 and 9 are less than that of Chain 1. In addition, the lowest scan chain of B14 has the lowest average fan-in cells and the highest fan-out similarity. Therefore, Table 9 shows that the accuracy increases if the effect of the failure spreads more and decreases if the magnitude of the same fault effect between different faults is high.

The scan chains were reordered for increasing the number of fan-in cells in the other chains, to improve the accuracy of scan-chain diagnosis using machine learning. Subsequently, cells that had no fan-in cells were placed at the beginning of the chain. The scan-chain reordering method in these analyses can increase the accuracy of scan-chain diagnoses.

5. Conclusions

Machine learning has recently emerged as an important tool for integrated circuit testing; it possesses several features suitable for practical classification forecasting applications. Scan-chain diagnosis is a typical classification problem that can achieve high accuracy for big data in integrated circuit (IC) testing. However, training and testing are difficult to perform due to the extremely long failure log datasets.

Novel fan-in and fan-out filters that highlight failure features and remove unnecessary features in a failure log dataset were proposed herein. These filters can not only reduce the lengths of the input vectors and the size of the models, but also increase the accuracy of scan-chain diagnosis.

The experimental results indicated that the proposed scan-chain diagnosis can achieve a higher accuracy than the previous scan-chain diagnosis with ANN. Specifically, the proposed scan-chain diagnosis achieved an improvement of 3.17% and reduced the construction time by 79.80%, compared with the previous diagnosis. Furthermore, it reduced the diagnosis time by 99.9984%. Therefore, the proposed scan-chain diagnosis is applicable to industrial circuits to achieve the time-to-market goal by reducing the training and diagnosis times.

Henceforth, we will propose an appropriate scan-chain reordering method to obtain a higher accuracy of scan-chain diagnosis. We will also study the multiple scan-chain diagnosis with ML for multiple scan-chain faults in the manufacturing process.

Author Contributions: Conceptualization, S.K.; Formal analysis, H.L.; Investigation, M.C.; Methodology, H.L. and S.K.; Project administration, S.K.; Software, H.L.; Supervision, S.K.; Validation, M.C.; Visualization, M.C.; Writing original draft, H.L.; Writing review & editing, M.C. and S.K. All authors have read and agreed to the published version of the manuscript.

Funding: This research was supported by Multi-Ministry Collaborative R&D Program (R&D program for complex cognitive technology) through the National Research Foundation of Korea (NRF) funded by MOTIE (2018M3E3A1057248).

References

1. Xu, J.; Song, C. International Conference on Multimedia Technology and Enhanced Learning. In *Design and Improvement of Optimal Control Model for Wireless Sensor Network Nodes*; Springer: Berlin/Heidelberg, Germany, 2020; pp. 215–227.
2. Chen, Z.; Teng, G.; Zhou, X.; Chen, T. Passive-event-assisted approach for the localizability of large-scale randomly deployed wireless sensor network. *Tsinghua Sci. Technol.* **2018**, *24*, 134–146. [CrossRef]
3. Xie, H.; Yan, Z.; Yao, Z.; Atiquzzaman, M. Data Collection for Security Measurement in Wireless Sensor Networks. *IEEE Internet Things J.* **2019**, *6*, 2205–2224. [CrossRef]
4. Wang, J.; Gao, Y.; Liu, W.; Wu, W.; Lim, S.-J. An asynchronous clustering and mobile data gathering schema based on timer mechanism in wireless sensor networks. *Comput. Mater. Contin.* **2019**, *58*, 711–725. [CrossRef]
5. Karakaya, A.; Akleylek, S. A survey on Security Threats and Authentication Approaches in Wireless Sensor Networks. In Proceedings of the 2018 6th International Symposium on Digital Forensic and Security (ISDFS), Antalya, Turkey, 22–25 March 2018; pp. 1–4.
6. Hao, W.; Xiang, L.; Li, Y.; Yang, P.; Shen, X. Reversible natural language watermarking using synonym substitution and arithmetic coding. *Comput. Mater. Contin.* **2018**, *55*, 541–559.

7.	Guo, W.Z.; Chen, J.Y.; Chen, G.L.; Zheng, H.F. Trust dynamic task allocation algorithm with Nash equilibrium for heterogeneous wireless sensor network. *Secur. Commun. Netw.* **2015**, *8*, 1865–1877. [CrossRef]

8.	Mangard, S.; Aigner, M.; Dominikus, S. A highly regular and scalable AES hardware architecture. *IEEE Trans. Comput.* **2003**, *52*, 483–491. [CrossRef]

9.	Guo, R.; Venkataraman, S. A Technique for Fault Diagnosis of Defects in Scan Chains. In Proceedings of the International Test Conference 2001 (Cat. No. 01CH37260), Baltimore, MD, USA, 1 November 2001; pp. 268–277.

10.	Yang, J.-S.; Huang, S.-Y. Quick Scan Chain Diagnosis Using Signal Profiling. In Proceedings of the 2005 International Conference on Computer Design, San Jose, CA, USA, 2–5 October 2005; pp. 157–160.

11.	Wang, W.; Deng, Z.; Wang, J. Enhancing sensor network security with improved internal hardware design. *Sensors* **2019**, *19*, 1752. [CrossRef] [PubMed]

12.	Cui, A.; Luo, Y.; Chang, C.-H. Static and dynamic obfuscations of scan data against scan-based side-channel attacks. *IEEE Trans. Inf. Forensics Secur.* **2016**, *12*, 363–376. [CrossRef]

13.	De, K.; Gunda, A. Failure Analysis for Full-Scan Circuits. In Proceedings of the 1995 IEEE International Test Conference (ITC), Washington, DC, USA, 21–25 October 1995; pp. 636–645.

14.	Hirase, J.; Shindou, N.; Akahori, K. Scan Chain Diagnosis Using IDDQ Current Measurement. In Proceedings of the Eighth Asian Test Symposium (ATS'99), Shanghai, China, 18 November 1999; pp. 153–157.

15.	Song, P.; Stellari, F.; Xia, T.; Weger, A.J. A Novel Scan Chain Diagnostics Technique Based on Light Emission from Leakage Current. In Proceedings of the 2004 International Conferce on Test, Charlotte, NC, USA, 26–28 October 2004; pp. 140–147.

16.	Khusyari, K.; Ng, W.T.; Jaarsma, N.; Abraham, R.; Ng, P.W.; Ang, B.H.; Ong, C.H. Diagnosis of Voltage Dependent Scan Chain Failure Using Vbump Scan Debug Method. In Proceedings of the 2008 17th Asian Test Symposium, Sapporo, Japan, 24–27 November 2008; p. 271.

17.	Schafer, J.L.; Policastri, F.A.; McNulty, R.J. Partner SRLs for Improved Shift Register Diagnostics. In Proceedings of the 1992 IEEE VLSI Test Symposium, Atlantic City, NJ, USA, 7–9 April 1992; pp. 198–201.

18.	Narayanan, S.; Das, A. An Efficient Scheme to Diagnose Scan Chains. In Proceedings of the International Test Conference 1997, Washington, DC, USA, 6 November 1997; pp. 704–713.

19.	Wu, Y. Diagnosis of Scan Chain Failures. In Proceedings of the 1998 IEEE International Symposium on Defect and Fault Tolerance in VLSI Systems (Cat. No. 98EX223), Austin, TX, USA, 2–4 November 1998; pp. 217–222.

20.	Tekumulla, R.; Lee, D. On Identifying and Bypassing Faulty Scan Segments. In Proceedings of the 16th North Atlantic Test Workshop, Boxborough, MA, USA, 9–11 May 2007; pp. 134–143.

21.	Kao, Y.-L.; Chuang, W.-S.; Li, J.C.-M. Jump Simulation: A Technique for Fast and Precise Scan Chain Fault Diagnosis. In Proceedings of the 2006 IEEE International Test Conference, Santa Clara, CA, USA, 22–27 October 2006; pp. 1–9.

22.	Huang, Y. Dynamic Learning Based Scan Chain Diagnosis. In Proceedings of the 2007 Design, Automation & Test in Europe Conference & Exhibition, Nice, France, 16–20 April 2007; pp. 1–6.

23.	Guo, R.; Venkataraman, S. An algorithmic technique for diagnosis of faulty scan chains. *IEEE Trans. Comput.-Aided Des. Integr. Circuits Syst.* **2006**, *25*, 1861–1868.

24.	Li, J.C.-M. Diagnosis of single stuck-at faults and multiple timing faults in scan chains. *IEEE Trans. Very Large Scale Integr Syst.* **2005**, *13*, 708–718.

25.	Guo, R.; Huang, Y.; Cheng, W.-T. Fault Dictionary Based Scan Chain Failure Diagnosis. In Proceedings of the 16th Asian Test Symposium (ATS 2007), Beijing, China, 8–11 October 2007; pp. 45–52.

26.	Lei, Y.G.; Yang, B.; Jiang, X.W.; Jia, F.; Li, N.P.; Nandi, A.K. Applications of machine learning to machine fault diagnosis: A review and roadmap. *Mech. Syst. Signal Process.* **2020**, *138*, 106587. [CrossRef]

27.	Tian, Y.; Liu, X. A deep adaptive learning method for rolling bearing fault diagnosis using immunity. *Tsinghua Sci. Technol.* **2019**, *24*, 750–762. [CrossRef]

28.	Stratigopoulos, H.-G. Machine Learning Applications in IC Testing. In Proceedings of the 2018 IEEE 23rd European Test Symposium (ETS), Bremen, Germany, 28 May–1 June 2018; pp. 1–10.

29.	Zhong, J.-H.; Wong, P.K.; Yang, Z.-X. Simultaneous-fault diagnosis of gearboxes using probabilistic committee machine. *Sensors* **2016**, *16*, 185. [CrossRef]

30.	Yang, Z.-X.; Wang, X.-B.; Wong, P.K. Single and simultaneous fault diagnosis with application to a multistage gearbox: A versatile dual-ELM network approach. *IEEE Trans. Ind. Inform.* **2018**, *14*, 5245–5255. [CrossRef]

31. Huang, Y.; Benware, B.; Klingenberg, R.; Tang, H.; Dsouza, J.; Cheng, W.-T. Scan Chain Diagnosis Based on Unsupervised Machine Learning. In Proceedings of the 2017 IEEE 26th Asian Test Symposium (ATS), Taipei, Taiwan, 27–30 November 2017; pp. 225–230.
32. Chern, M.; Lee, S.-W.; Huang, S.-Y.; Huang, Y.; Veda, G.; Tsai, K.-H.; Cheng, W.-T. Improving Scan Chain Diagnostic Accuracy Using Multi-Stage Artificial Neural Networks. In Proceedings of the 24th Asia and South Pacific Design Automation Conference, Tokio, Japan, 21–24 January 2019; pp. 341–346.
33. Chern, M.; Lee, S.-W.; Huang, S.-Y.; Huang, Y.; Veda, G.; Tsai, K.-H.; Cheng, W.-T. Diagnosis of Intermittent Scan Chain Faults Through a Multi-Stage Neural Network Reasoning Process. *IEEE Trans. Comput. Aided Des. Integr. Circuits Syst.* **2019**. [CrossRef]
34. Huang, Y.; Cheng, W.-T.; Guo, R.; Tai, T.-P.; Kuo, F.-M.; Chen, Y.-S. Scan Chain Diagnosis by Adaptive Signal Profiling with Manufacturing ATPG Patterns. In Proceedings of the 2009 Asian Test Symposium, Taichung, Taiwan, 23–26 November 2009; pp. 35–40.
35. Huang, Y.; Cheng, W.-T.; Hsieh, C.-J.; Tseng, H.-Y.; Huang, A.; Hung, Y.-T. Intermittent scan chain fault diagnosis based on signal probability analysis. In Proceedings of the Design, Automation and Test in Europe Conference and Exhibition, Paris, France, 16–20 February 2004; pp. 1072–1077.
36. Lo, W.-H.; Hsieh, A.-C.; Lan, C.-M.; Lin, M.-H.; Hwang, T. Utilizing circuit structure for scan chain diagnosis. *IEEE Trans. Very Large Scale Integr. Syst.* **2014**, *22*, 2766–2778. [CrossRef]
37. Chen, H.; Qi, Z.; Wang, L.; Xu, C. A Scan Chain Optimization Method for Diagnosis. In Proceedings of the 2015 33rd IEEE International Conference on Computer Design (ICCD), New York, NY, USA, 18–21 October 2015; pp. 613–620.
38. Ye, J.; Huang, Y.; Hu, Y.; Cheng, W.-T.; Guo, R.; Lai, L.; Tai, T.-P.; Li, X.; Changchien, W.; Lee, D.-M. Diagnosis and layout aware (DLA) scan chain stitching. *IEEE Trans.Very Large Scale Integr. Syst.* **2014**, *23*, 466–479. [CrossRef]
39. Pedregosa, F.; Varoquaux, G.; Gramfort, A.; Michel, V.; Thirion, B.; Grisel, O.; Blondel, M.; Prettenhofer, P.; Weiss, R.; Dubourg, V. Scikit-learn: Machine learning in Python. *J. Mach. Learn. Res.* **2011**, *12*, 2825–2830.
40. Corno, F.; Reorda, M.S.; Squillero, G. RT-level ITC'99 benchmarks and first ATPG results. *IEEE Des. Test Comput.* **2000**, *17*, 44–53. [CrossRef]
41. OpenCores AES (Rijndael) IP Core. Available online: https://opencores.org/projects/aes_core (accessed on 23 August 2020).
42. OpenCores OpenRisc 1200 HP. Available online: https://opencores.org/projects/or1200_hp (accessed on 23 August 2020).

Article

HKF-SVR Optimized by Krill Herd Algorithm for Coaxial Bearings Performance Degradation Prediction

Fang Liu [1,2], Liubin Li [1], Yongbin Liu [1,2,*], Zheng Cao [1], Hui Yang [1] and Siliang Lu [1,2]

[1] College of Electrical Engineering and Automation, Anhui University, Hefei 230601, China; ufun@ahu.edu.cn (F.L.); 15056086408@139.com (L.L.); caozheng@ahu.edu.cn (Z.C.); 15086@ahu.edu.cn (H.Y.); lusliang@mail.ustc.edu.cn (S.L.)

[2] National Engineering Laboratory of Energy-Saving Motor and Control Technology, Anhui University, Hefei 230601, China

* Correspondence: lyb@ustc.edu.cn; Tel.: +86-55163861905

Received: 19 December 2019; Accepted: 20 January 2020; Published: 24 January 2020

Abstract: In real industrial applications, bearings in pairs or even more are often mounted on the same shaft. So the collected vibration signal is actually a mixed signal from multiple bearings. In this study, a method based on Hybrid Kernel Function-Support Vector Regression (HKF–SVR) whose parameters are optimized by Krill Herd (KH) algorithm was introduced for bearing performance degradation prediction in this situation. First, multi-domain statistical features are extracted from the bearing vibration signals and then fused into sensitive features using Kernel Joint Approximate Diagonalization of Eigen-matrices (KJADE) algorithm which is developed recently by our group. Due to the nonlinear mapping capability of the kernel method and the blind source separation ability of the JADE algorithm, the KJADE could extract latent source features that accurately reflecting the performance degradation from the mixed vibration signal. Then, the between-class and within-class scatters (SS) of the health-stage data sample and the current monitored data sample is calculated as the performance degradation index. Second, the parameters of the HKF–SVR are optimized by the KH (Krill Herd) algorithm to obtain the optimal performance degradation prediction model. Finally, the performance degradation trend of the bearing is predicted using the optimized HKF–SVR. Compared with the traditional methods of Back Propagation Neural Network (BPNN), Extreme Learning Machine (ELM) and traditional SVR, the results show that the proposed method has a better performance. The proposed method has a good application prospect in life prediction of coaxial bearings.

Keywords: rolling bearing; performance degradation; hybrid kernel function; krill herd algorithm; SVR

---

1. Introduction

Roller bearings are key components of rotating machinery and they are widely used in aerospace, railway and other industries [1]. Economic losses and major safety accidents can be avoided in industry through an accurate evaluation of the bearing degradation status of the equipment and a timely detection of bearing failures [2]. Two issues are key in the performance degradation evaluation of rolling bearings. One is to extract the performance degradation indicators [3] and the other is to establish effective prediction models [4]. Performance degradation index extraction is essential for bearing performance degradation assessment. In current studies, kurtosis, root mean square and peak indicators are used as indicators of bearing performance degradation [5]. However, completely reflecting the entire degradation process of the bearing using a single indicator parameter is difficult. Therefore, multi-domain features are extracted from the time domain and frequency domain. Then, these features are fused to remove the redundant features and used to characterize the bearing degradation process [6]. This process facilitates bearing degradation evaluation. Feature fusion techniques are generally divided into linear and nonlinear feature fusion methods [7]. Given that the vibration signal of the bearing is usually nonlinear, the nonlinear method has unique advantages in the fusion of bearing features [8]. For example, Zhang et al. used the Kernel Principal Component Analysis (KPCA) algorithm to fuse features [9].

Recently, a new algorithm named Kernel Joint Approximate Diagonalization of Eigen-matrices (KJADE) is invented by our group for feature fusion. This method is a combination of the kernel method and the traditional JADE algorithm [10]. Due to the nonlinear mapping capability of the kernel method and the blind source separation ability of the JADE algorithm. KJADE could extract latent source features that accurately reflecting the performance degradation from the mixed vibration signal.

On the other hand, an effective prediction model is critical to accurate performance evaluation [11]. In recent years, data-driven prediction models have been widely applied to performance degradation assessment [12]. Artificial neural network [13] and support vector regression are the two of most widely used prediction models for bearing performance degradation evaluation and residual life prediction [14]. These two models are based on statistical learning theory and data-driven model [15]. Liu et al. used neural network method to predict the performance degradation of rolling bearings [16]. Qian et al. used recurrence quantification analysis and auto-regression model for bearing degradation monitoring and state prediction [17]. Shen et al. used Support Vector Regression (SVR) and statistical parameters of wavelet packet paving to diagnose faults of rotating machinery [18]. Wang et al. used two novel mixed effects models to predict the performance of rolling element bearings [19]. Zhang et al. used SVR to achieve bearing remaining life prediction [20]. Ling et al. used Improved Empirical Wavelet Transform-Least Square Support Vector Machine (IEWT-LSSVM) and bird swarm algorithm to predict wind speed [21]. However, predicting bearing degradation is difficult owing to the non-linearity of bearing data. Due to the nonlinear mapping capability, the kernel methods have attracted the attention of many researchers in recent years. However, different kernel functions have different characteristics [22]. Choosing different kernel functions is crucial for dealing with different problems [23]. To deal with this method, in recent years, different forms of hybrid kernel functions have been studied. Zhou et al. used LSSVM with mixed kernel function build predictive model [24]; Cheng et al. used mixed kernel function support vector regression for global sensitivity analysis [23]. Wu et al. used mixed-kernel based weighted extreme learning machine for the influence of the imbalance datasets problem [25]. Although the hybrid kernel function is applied to many fields but the parameters of the kernel function have a great influence on the prediction results and these parameters are difficult to decide. In this study, Hybrid Kernel Function-Support Vector Regression (HKF–SVR) is proposed to predict bearing performance degradation. Taking into account the uncertainty of the model parameters, the krill herd (KH) algorithm is then used to optimize the parameters of the model.

The rest of this paper is arranged as follows: The second part provides a brief introduction of the KH algorithm and the HKF–SVR. The third part introduces HKF–SVR for the prediction of bearing performance degradation. The fourth part presents two case studies using the proposed method and other methods. The final part presents the conclusions and acknowledgments.

2. Theoretical Background

2.1. KH Algorithm

The KH algorithm is a bionic intelligent optimization algorithm for the simulation of krill foraging behavior [26]. The local optimal solution is found by attracting or repelling each adjacent krill, the model is simple and fast. At the same time, the krill herd algorithm has good robustness and faster convergence by using group search, compared with other algorithms. Due to using the Lagrange model, the performance of the algorithm is better than other bionic optimization algorithms [27]. Similar to most intelligent optimization algorithms, the KH algorithm generally uses real-coded methods to generate initial populations randomly. The evolution of particles is influenced by three motion components (neighbor induction, foraging movement and random diffusion). The population is increasingly diversified by crossing or mutating individuals until the set termination conditions are met. The KH algorithm is as follows:

Assume that the position of each krill at time t is $x(t)$ and the position after Δt time is $x(t + \Delta t)$. On the basis of the basic theory of the KH algorithm [28], the position update of each krill is affected by three speeds, namely, the speed of movement induced by the surrounding krill N_i, the foraging speed of the krill individual F_i and the random diffusion movement D_i. Here, N_i is expressed as follows:

$$N_i = N^{max}\alpha_i + w_n N_i^{old}, \tag{1}$$

where N^{max} is the maximum induction velocity, a_i is the induction direction, $\omega_n \in (0,1)$ is the inertial weight and N_i^{old} is the velocity vector of the last induced motion. a_i is affected by the surrounding krill and the current optimal particles, as shown in the following formula:

$$\begin{cases} \alpha_i = \alpha_i^{local} + \alpha_i^{target} \\ \alpha_i^{local} = \sum\limits_{j=1}^{NP} \hat{K}_{i,j}\hat{x}_{i,j} \\ \hat{K}_{i,j} = \frac{K_i - K_j}{K^{worst} - K^{best}} \\ \hat{x}_{i,j} = \frac{x_j - x_i}{\|x_j - x_i\| + \varepsilon} \end{cases} \tag{2}$$

where α_i^{local} is the direction of induction by the surrounding krill, α_i^{target} is the direction of induction by the current globally optimal individual, $\hat{K}_{i,j}$ is the force of the surrounding krill, $\hat{x}_{i,j}$ is the current particle's orientation to the neighbor and NP is the population. K_i and K_j are the fitness values of the current particle and the neighboring particle, respectively. K^{worst} and K^{best} are the fitness values of the worst individual and the optimal individual in the current population, respectively. The krill individual foraging speed F_i can be expressed by the following formula:

$$F_i = V_f\beta_i + w_f F_i^{old}, \tag{3}$$

where V_f is the maximum foraging speed, β_i is the foraging direction and $\omega_f \in (0,1)$ is the foraging inertial weight.

$$\begin{cases} \beta_i = \beta_i^{food} + \beta_i^{ibest} \\ \beta_i^{food} = 2\left(1 - \frac{I}{I_{max}}\right)\hat{K}_{i,food}\hat{X}_{i,food} \\ X_{food} = \frac{\sum\limits_{i=1}^{N} \frac{X_i}{K_i}}{\sum\limits_{i=1}^{N} \frac{1}{K_i}} \end{cases} \tag{4}$$

where β_i^{food} and β_i^{ibest} are the directions induced by the best individuals of food and particles themselves; X_{food} is the position of food in which $\hat{K}_{i,food}$ is the influence of food on current particles and $\hat{X}_{i,food}$ is the orientation of current food to particles. The random diffusion motion D_i is expressed as:

$$D_i = D_{max}\left(1 - \frac{t}{t_{max}}\right)\delta, \tag{5}$$

where D_{max} is the maximum random diffusion velocity and δ is the random diffusion direction.

The previous theory indicates that the position update of each krill individual is affected by the above three speeds, namely, the motion of the surrounding krill, the krill's foraging speed and the random diffusion motion. The speed and position update of the particles are expressed as follows:

$$\frac{dx_i}{dt} = N_i + F_i + D_i \tag{6}$$

$$x_i(t + \Delta t) = x_i(t) + \Delta t \frac{dx_i}{dt}, \tag{7}$$

where Δt represents the time interval.

2.2. HKF–SVR

SVR is a well-known method to solve the regression problem [29]. Assume the data set is $\{x_i, y_i\}$, where x_i is the input sample and y_i is the corresponding output value. Then, SVR can be represented by linear function $f(x) = \omega x + b$. The SVR function can be represent by introducing ε insensitive loss function:

$$\begin{aligned} y_i - \omega \cdot x_i - b \leq \varepsilon, \quad i = 1, 2, \cdots, n \\ -y_i + \omega \cdot x_i + b \leq \varepsilon \end{aligned}, \tag{8}$$

next we can get the Convex optimization problem by minimizing the $\frac{1}{2}\|\omega^2\|$.

$$\begin{aligned} &\min \tfrac{1}{2}\|\omega\|^2 + C \sum_{i=1}^{n} \xi_i + \xi_i^* \\ &\text{s.t} \begin{cases} y_i - \omega x_i - b \leq \varepsilon + \xi_i & i = 1, 2, \cdots, n \\ -y_i + \omega x_i + b \leq \varepsilon + \xi_i^* \end{cases} \end{aligned}, \tag{9}$$

where $C > 0$ is the regularization parameter controlling the punishment degree for the sample beyond the error. the relaxation variables $\xi_i \geq 0$ and $\xi_i^* \geq 0$. According to the optimization conditions, we can obtain the dual problem of the support vector regression machine [30] and satisfy the constraint conditions.

$$\begin{cases} \max \omega(a, a^*) = -\varepsilon \sum_{i=1}^{n} y_i(a_i^* - a_i) - \tfrac{1}{2} \sum_{i,j=1}^{n} (a_i^* - a_i)(a_j^* - a_j)(x_i.x_j) \\ \text{s.t.} \begin{cases} \sum_{i=1}^{n} (a_i - a_i^*) = 0 \\ 0 \leq a_i, a_i^* \leq C \end{cases} \end{cases} \tag{10}$$

where a_i and a_i^* are the Lagrange multipliers. Finally, the regression function [31] is obtained as follows:

$$f(x) = \sum_{i=1}^{n} (\alpha_i^* - \alpha_i)\langle x_i \cdot x \rangle + b^*. \tag{11}$$

Different kernel function have different effects on the predicted results [32]. Before the kernel function is constructed, the mapping of input space to feature space must be known. However, if we want to know the mapping of input space to map space, the distribution of data in the input space should be clarified. In most cases, the specific distribution of acquired data is consistently unknown. Thus, constructing a kernel function that fully conforms to the input space is generally difficult. Well-known kernel functions include linear, polynomial, radial basis and sigmoid kernel functions [33]. The linear kernel function is mainly used for linear problems and it has some merits, such as few parameters, fast calculation speed and improved effect for linear separable data. The polynomial kernel function is a global kernel function with many parameters. The radial basis kernel function is a locally strong kernel function that maps a sample into a high-dimensional space and the most widely used of all kernel functions; it has fewer parameters compared with the polynomial kernel function [34]. Thus, the radial basis kernel function is used in most cases.

In this study, a hybrid kernel function is proposed for support vector regression and the model parameters are optimized by the KH algorithm. For the rolling bearing performance degradation prediction, the polynomial and Gaussian kernel functions are selected to construct the hybrid kernel function for SVR and the parameters and the hybrid coefficient of the hybrid kernel function are optimized by the KH algorithm. The constructed HKF–SVR is shown as follows:

$$K_h = \lambda K_{poly} + (1 - \lambda) K_{rbf},\tag{12}$$

where K_{poly} is polynomial kernel function, $\lambda \in (0, 1)$ is a hybrid coefficient, K_{rbf} is the radial basis kernel function and K_h is a hybrid kernel function.

The parameters of HKF–SVR include the polynomial kernel function's highest degree d, a gamma1 parameter, the coef0 of the kernel function, the gamma2 parameter of the RBF kernel function, a penalty coefficient c and a hybrid coefficient λ. The error between the real and predicted values is used as the objective function. The specific parameters and ranges are shown in the Table 1 below. The optimization flow chart of the above parameters by KH algorithm is shown in Figure 1. The specific process is described as follows:

(1)   The number of iterations t, the number of krill and the maximum number of cycles are initialized.
(2)   The value range of parameters is set and the set of parameters is randomly generated as the initial position.
(3)   Particle motion and generalization error calculation are conducted.
(4)   If the error at a given moment meets the requirements or reaches the number of iterations, Step 7 is performed.
(5)   The number of iterations $t = t + 1$.
(6)   The current particle position and velocity are updated on the basis of Equations (6) and (7), new training parameters are found and then Step 3 is repeated.
(7)   The optimal parameters are obtained.

Table 1. Optimized parameters.

Description	Notation
Polynomial kernel function parameter	g1
Polynomial kernel function parameter	coef0
Polynomial kernel function parameter	d
Gaussian kernel function parameters	g2
SVR penalty coefficient	c
Kernel function hybrid coefficient	λ

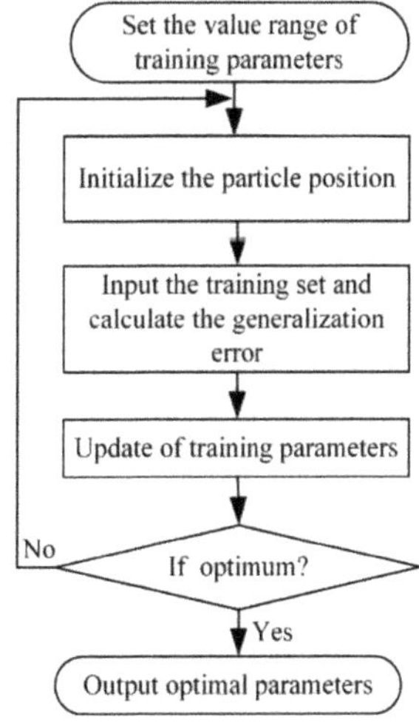

Figure 1. Flow chart of parameters optimization by Krill Herd (KH) algorithm.

3. HKF–SVR Method for Bearing Performance Degradation Prediction

In most real industrial applications, bearings in pairs or even more are mounted on the same shaft. For example, as shown in Case I, there are four bearings on one shaft. So, in this situation, the vibration signal acquired by the sensor mounted on each bearing block will be mixed by the signals propagated through the shaft from the other three bearings. In this study, our recently developed KJADE algorithm is studied on this issue. KJADE is a combination of kernel method and the traditional JADE algorithm. Through the kernel method the latent feature vector can be extracted in the high-dimensional feature space. On the other hand, due to the super blind source separation ability of the JADE algorithm, the correlative feature that could reflect the health status of the monitored bearing can be extracted. The integration evaluation factor of SS is then employed to further evaluate the performance degradation in real time. After that, considering the ability of mixed kernel function to deal with nonlinear problems and the regression prediction ability of SVR, the HKF–SVR is proposed to predict bearing performance degradation. Taking into account the uncertainty of the model parameters, the KH algorithm is then used to optimize the parameters of the model.

The steps of the whole method are shown in Figure 2 and described as follows:

(1) Multi-domain features extraction. The performance degradation process of bearings has a certain non-linearity. It is difficult for a single feature to accurately reflect the degradation process. In order to comprehensively reflect the bearing state, in this study, eight time-domain features and eight frequency-domain features $F1$-$F16$ shown in Tables 2 and 3 are extracted from the mixed bearing vibration signal. Where the $F1$-$F8$ stand for mean, root mean square value, square root amplitude, absolute mean, skewness, waveform indicators, pulse indicator and margin index, respectively. $F9$-$F12$ stand for mean frequency, standard deviation of frequency, center frequency, frequency RMS and $F13$-$F16$ stand for the degrees of dispersion or concentration of the spectrum where s_i is a spectrum for $i = 1, 2, \ldots , N$ (N is the number of spectrum lines) and f_i is the frequency value of the i-th spectrum line, which indicates the degree of dispersion or concentration of the spectrum and the change of the dominant frequency band. Assume that the sample of the healthy state is X, the sample of the current monitoring data sample is Y. The two samples are both divided into n_i segments $\{X_1, X_2, \ldots , X_{ni}\}$ and $\{Y_1, Y_2, \ldots , Y_{ni}\}$ and then the 16 features of each segment are calculated as the original feature vectors $\{F^x\} = \{F^x_{1,1}, F^x_{1,2}, \ldots , F^x_{1,ni}; F^x_{2,1}, F^x_{2,2}, \ldots , F^x_{2,ni}; \ldots ; F^x_{16,1}, F^x_{16,2}, \ldots , F^x_{16,ni}\}_{16 \times ni}$ and $\{F^y\} = \{F^y_{1,1}, F^y_{1,2}, \ldots , F^y_{1,ni}; F^y_{2,1}, F^y_{2,2}, \ldots , F^y_{2,ni}; \ldots ; F^y_{16,1}, F^y_{16,2}, \ldots , F^y_{16,ni}\}_{16 \times ni}$.

(2) Feature fusion using KJADE. In this step, KJADE is used to further extract latent sensitive source features that could accurately reflect the performance degradation of the monitored bearing from the features $\{F^x\}$ and $\{F^y\}$ extracted in the previous step. To facilitate visualization of results, the dimension of the latent sensitive source feature vector is set to be 3. So, after this step, the latent sensitive source features are transformed to be $\{F^x\} = \{F^x_{1,1}, F^x_{1,2}, \ldots , F^x_{1,ni}; F^x_{2,1}, F^x_{2,2}, \ldots , F^x_{2,ni}; F^x_{3,1}, F^x_{3,2}, \ldots , F^x_{3,ni}\}_{3 \times ni}$ and $\{F^y\} = \{F^y_{1,1}, F^y_{1,2}, \ldots , F^y_{1,ni}; F^y_{2,1}, F^y_{2,2}, \ldots , F^y_{2,ni}; F^y_{3,1}, F^y_{3,2}, \ldots , F^y_{3,ni}\}_{3 \times ni}$.

(3) Performance degradation index calculation. The integration evaluation factor of SS between the $\{F^x\}$ and $\{F^y\}$ obtained in the previous step is calculated as the comprehensive performance degradation index. First, the between-class scatter matrix is calculated as follows:

$$S_b = \sum_{i=1}^{C} p_i \|m_i - m\|^2. \tag{13}$$

Then, the inter-class scatter matrix is calculated as follows:

$$S_w = \sum_{i=1}^{C} p_i \frac{1}{n_i} \sum_{k=1}^{n_i} \|x_k^i - m_i\|^2, \tag{14}$$

where C is the number of categories, m_i is the feature mean in category i, m is the mean of the entire feature sample. Finally, the SS is calculated using the following equation:

$$SS = trace(S_b/S_w). \tag{15}$$

After this step, the SS that standing for the performance degradation index of the current monitored data sample can be obtained.

(4) Prediction model constructed through HKF-SVR. In practical engineering application, after continuous monitoring for a period of time, a continuous monitoring vibration data can be obtained. One SS value corresponding to each monitoring moment can be obtained by using steps 1–4 and the performance degradation prediction model can be conducted by using HKF-SVR.

(5) Performance degradation prediction using the constructed model. The performance degradation of the next moment can be predicted using the constructed model obtained in the previous step. Meanwhile, the model is updated in real time with the current and historical data.

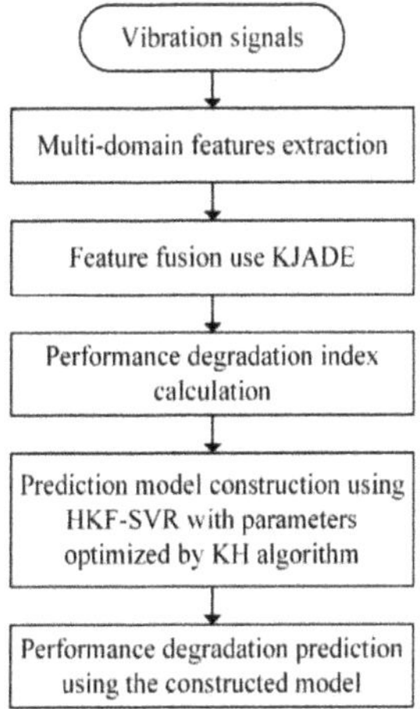

Figure 2. Flow chart of performance degradation prediction by Hybrid Kernel Function-Support Vector Regression (HKF–SVR).

Table 2. Time-domain features.

$F1 = \frac{1}{N}\sum_{i=1}^{N} x_i$	$F2 = \sqrt{\frac{1}{N}\sum_{i=1}^{N} x_i^2}$
$F3 = \left[\frac{1}{N}\sum_{i=1}^{N} \sqrt{|x_i|}\right]^2$	$F4 = \frac{1}{N}\sum_{i=1}^{N} |x_i|$
$F5 = \frac{1}{N}\sum_{i=1}^{N} x_i^3$	$F6 = \dfrac{\sqrt{\frac{1}{N}\sum_{i=1}^{N} x_i^2}}{F4}$
$F7 = \dfrac{max(x)}{\frac{1}{N}\sum_{i=1}^{N} |x_i|}$	$F8 = \dfrac{\frac{1}{N}\sum_{i=1}^{N} x_i^4}{\left(\sqrt{\frac{1}{N}\sum_{j=1}^{N} x_j^2}\right)^4}$

Table 3. Frequency-domain features.

$F9 = \frac{1}{N}\sum_{i=1}^{N} s_i$	$F10 = \frac{1}{N}\sum_{j=1}^{N} \left(s_j - \frac{1}{N}\sum_{i=1}^{N} s_i\right)^2$
$F11 = \dfrac{\sum_{i=1}^{N} f_i s_i}{\sum_{j=1}^{N} s_j}$	$F12 = \sqrt{\dfrac{\sum_{i=1}^{N} f_i^2 s_i}{\sum_{j=1}^{N} s_j}}$
$F13 = \dfrac{\frac{1}{N}\sum_{j=1}^{N} \left(s_j - \frac{1}{N}\sum_{i=1}^{N} s_i\right)^3}{\left(\sqrt{F10}\right)^3}$	$F14 = \sqrt{\frac{1}{N}\sum_{i=1}^{N} s_i(f_i - F12)^2}$
$F15 = \sqrt{\dfrac{\sum_{i=1}^{N} f_i^4 s_i}{\sum_{j=1}^{N} f_j^2 s_j}}$	$F16 = \dfrac{\sum_{i=1}^{N} f_i^2 s_i}{\sqrt{\sum_{j=1}^{N} s_j \sum_{k=1}^{N} f_k^4 s_k}}$

4. Case Studies

4.1. CASE I

In this case, the full-life-cycle bearing vibration signals provided by the Intelligent Maintenance System (IMS) Center of the University of Cincinnati are analyzed using the proposed method. The experimental platform is shown in Figure 3. The four Rexnord ZA-2115 bearings are mounted on the same shaft. The rotational speed of the experimental shaft was maintained at 2000 rpm, the radial load was 6000 lbs., the sampling frequency was 20 kHz and the data length was 20,480 points. The PCB 353B33 quartz sensor was mounted in the horizontal and vertical directions of each bearing and data were collected by the NI data acquisition card DAQ6062E. The acquisition interval between each signal was 10 min.

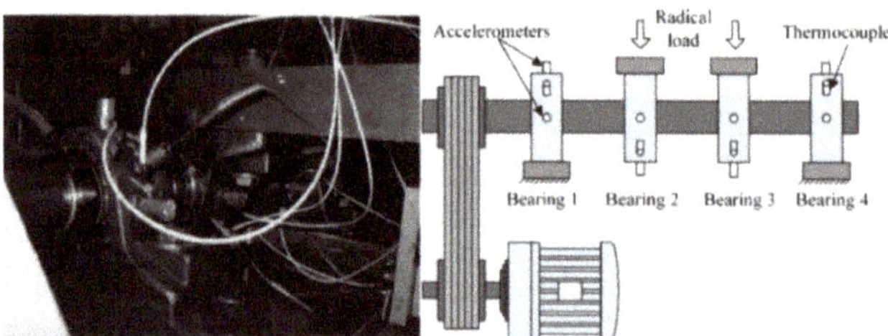

Figure 3. Experimental setup.

The entire experiment was completed in three groups. In the first group of experiments, 2156 documents were obtained intermittently. The inner ring of bearing 3 was damaged and the rolling elements of bearing 4 were damaged due to bearing disassembly. In the second set of tests, a total of 984 documents were collected and the outer part of bearing 1 was faulty. In the third set of tests, 4448 documents were obtained and the outer ring fault occurred in the third bearing. Bearings 3 and 4 in the first group of experiments and bearing 1 in the second group of experiments were analyzed. As shown in Figure 4, (**a**) is the full-life-cycle vibration signal of bearing 3, (**b**) is the full-life-cycle vibration signal of bearing 4 and (**c**) is the full-life-cycle vibration signal of bearing 1.

The lifetime data with inner-ring fault, roller fault and outer-ring fault are shown in Figure 4a–c, respectively. The first sample is taken as a healthy sample and the subsequent samples are analyzed as the current monitoring samples. During the analysis, 16,800 points are taken for each analysis sample and divided into 30 segments, so $n_i = 30$ in the step. The method in step one of the third part of this paper is used to extract the 16 dimensions features of the original signal, as shown in Tables 2 and 3 and then KJADE and SS are used to calculate performance degradation indicators. Then HKF-SVR was used to construct a prediction model to predict performance degradation at the next moment.

On the basis of the method proposed in Section 3, the hybrid kernel function of the support vector regression machine was constructed using polynomial and radial basis kernel functions. After the model was established, the parameters of the model were optimized by the KH algorithm. The initial parameters are shown as follows: an initial population of 20, five iterations, a maximum cycle number of 20, a maximum induction velocity $N^{max} = 0.01$, a maximum random diffusion velocity $D^{max} = 0.005$ and a maximum foraging speed $V_f = 0.02$. These parameters are the optimal parameters obtained through comparative testing. Finally, the HKF–SVR model was obtained to predict the degradation trend of bearing performance.

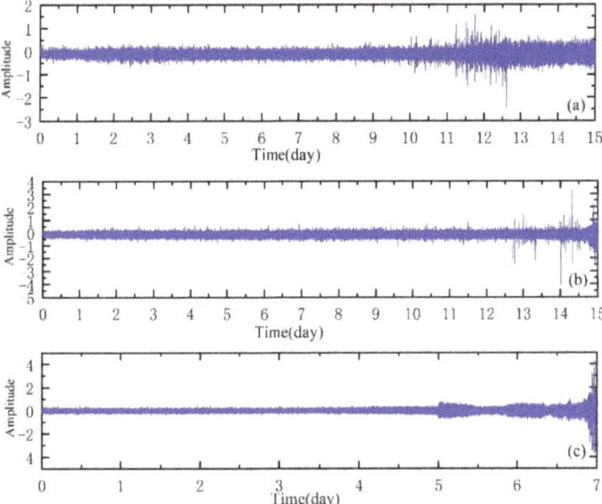

Figure 4. Whole-life vibration signals of bearings.

In this study, the root mean square error (RMSE) was used to evaluate the pros and cons of the method and the results were compared with those of the traditional support vector regression method. The calculation formula of the RMSE is as follows:

$$RMSE = \sqrt{\frac{1}{n}\sum_{i=1}^{n}(y_i - \hat{y}_i)^2}.$$

(16)

The proposed method was compared with the SVR and support vector regression optimized by the KH algorithm. The results showed that the method could track the performance degradation trend of a bearing; a good prediction result was also obtained. Figures 5–7 show the performance degradation prediction graphs of bearing 1 in the second experiment group and bearings 3 and 4 in the first experiment group, respectively.

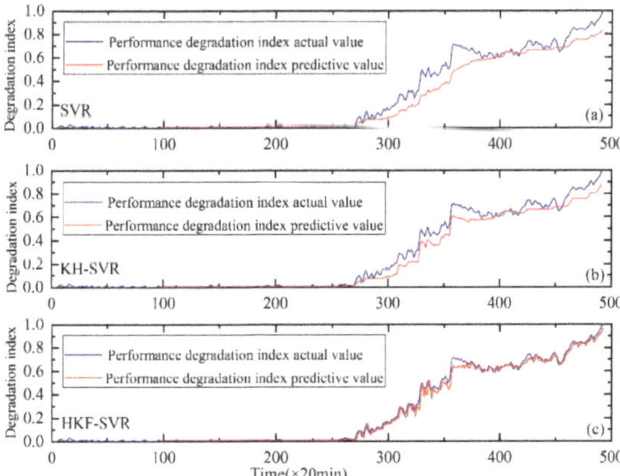

Figure 5. Performance degradation prediction of bearing 1. (**a**) is the result of SVR, (**b**) is the result of SVR with parameters optimized by the KH algorithm and (**c**) is the result of HKF–SVR with parameters optimized by the KH algorithm.

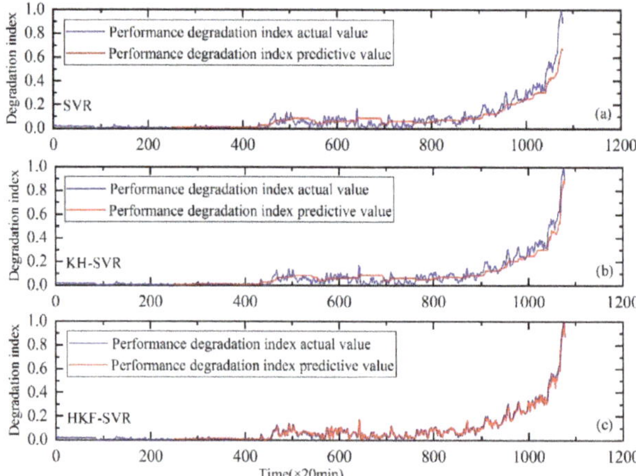

Figure 6. Performance degradation prediction of bearing 3. (**a**) is the result of SVR, (**b**) is the result of SVR with parameters optimized by the KH algorithm and (**c**) is the result of HKF–SVR with parameters optimized by the KH algorithm.

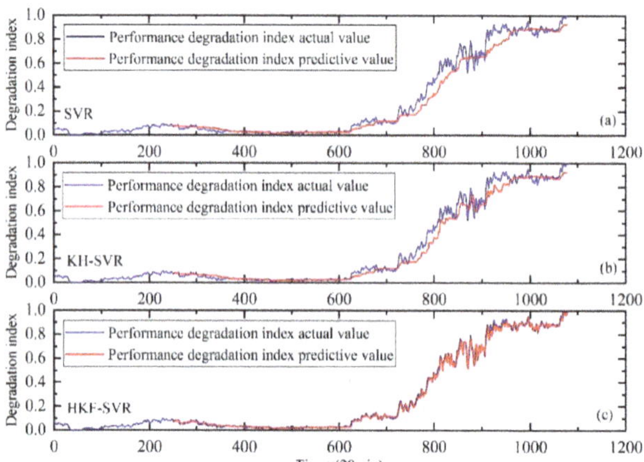

Figure 7. Performance degradation prediction of bearing 4. (**a**) is the result of SVR, (**b**) is the result of SVR with parameters optimized by the KH algorithm and (**c**) is the result of HKF–SVR with parameters optimized by the KH algorithm.

The results indicate that the method proposed in this study can effectively predict the performance degradation trend for bearings 1, 3 and 4. Meanwhile, the RMSE values shown in Table 4 indicate that the prediction error of the proposed prediction method is smaller than the other two methods. Thus, this method has advantages over the other two methods.

Table 4. Prediction errors comparison.

Method	Bearing 1	Bearing 4	Bearing 3
SVR	0.104	0.082	0.062
KH–SVR	0.079	0.069	0.047
HKF–SVR	0.026	0.027	0.022

In order to further prove the validity of the proposed method, the comparison with the Back Propagation Neural Network (BPNN) and Extreme Learning Machine (ELM) are carried out. The results of the comparison are shown in Figures 8–10. It can be seen that the prediction results of the HKF-SVR method are more accurate than the others. The RMSE values shown in Table 5 show that the proposed method achieved the minimum prediction error.

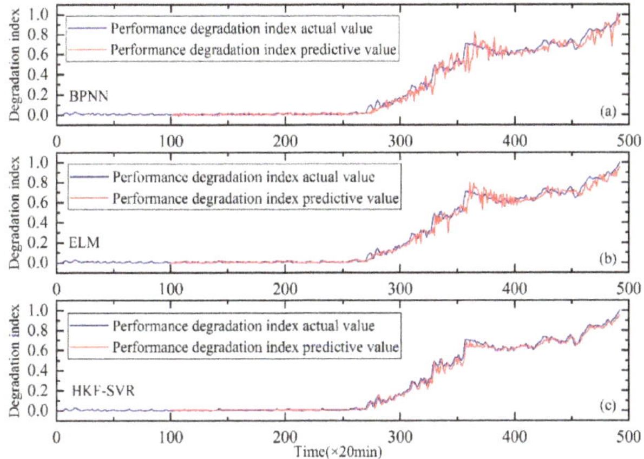

Figure 8. Performance degradation prediction of bearing 1. (**a**) is the result of BPNN, (**b**) is the result of ELM and (**c**) is the result of HKF–SVR with parameters optimized by the KH algorithm.

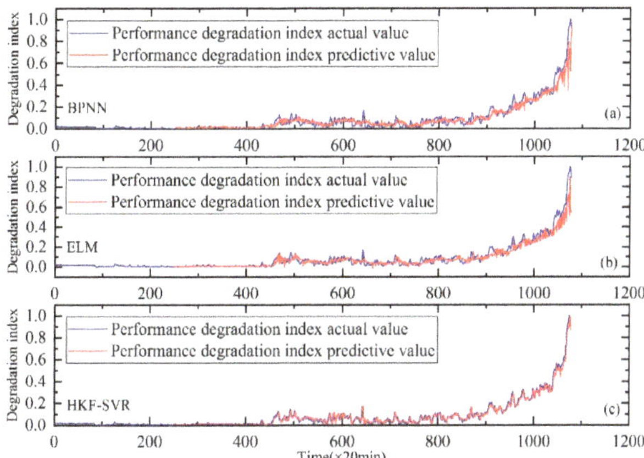

Figure 9. Performance degradation prediction of bearing 3. (**a**) is the result of BPNN, (**b**) is the result of ELM and (**c**) is the result of HKF–SVR with parameters optimized by the KH algorithm.

Table 5. Prediction errors comparison.

Method	Bearing1	Bearing4	Bearing3
BPNN	0.051	0.04	0.047
ELM	0.042	0.055	0.05
HKF-SVR	0.026	0.027	0.022

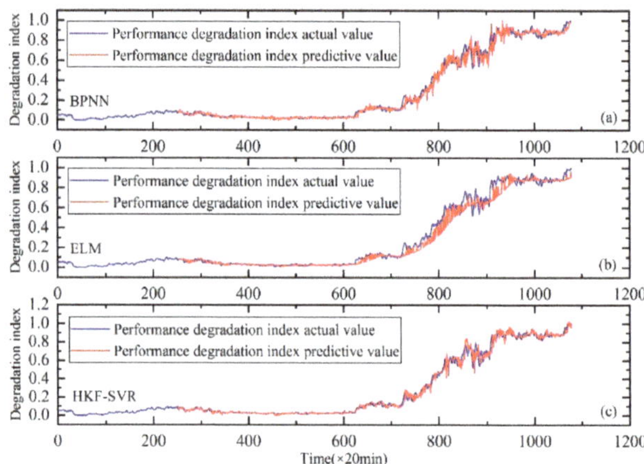

Figure 10. Performance degradation prediction of bearing 4. (**a**) is the result of BPNN, (**b**) is the result of ELM and (**c**) is the result of HKF–SVR with parameters optimized by the KH algorithm.

In order to verify the superiority of the KH method, the classic Genetic Algorithm (GA) method is used to optimize the parameters. The comparative results are shown in Figures 11–13. RMSE is used to further compare the results of the two methods, as shown in the Table 6. Through the above comparative analysis, it can be seen that the prediction result based on the KH method is more accurate.

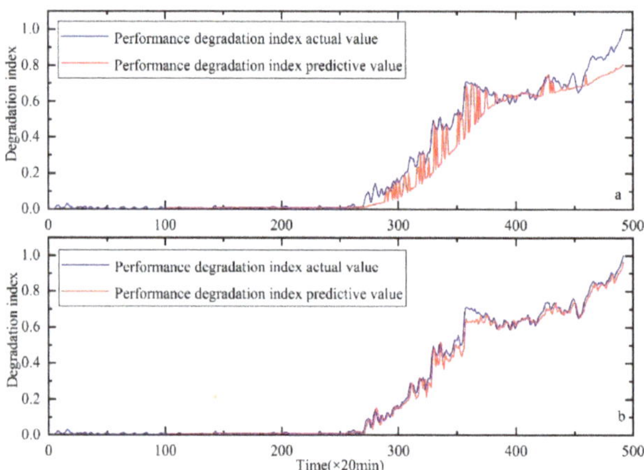

Figure 11. Performance degradation prediction of bearing 1, (**a**) is the result of HKF–SVR with parameters optimized by the GA and (**b**) is the result of HKF–SVR with parameters optimized by KH.

Table 6. Prediction errors comparison.

Method	Bearing1	Bearing4	Bearing3
GA-HKFSVR	0.078	0.052	0.054
KH-HKFSVR	0.026	0.027	0.022

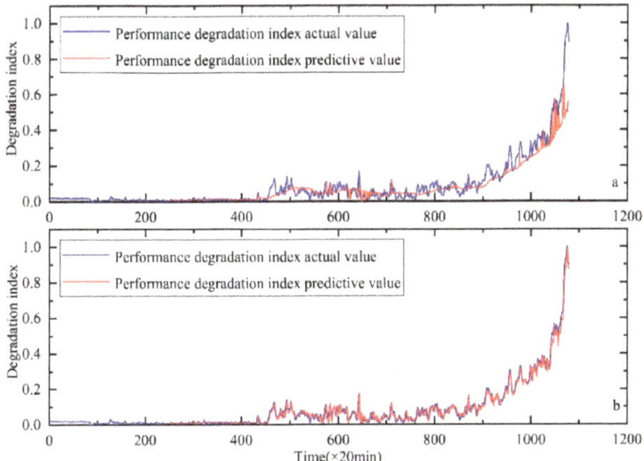

Figure 12. Performance degradation prediction of bearing 3, (**a**) is the result of HKF–SVR with parameters optimized by the GA and (**b**) is the result of HKF–SVR with parameters optimized by KH.

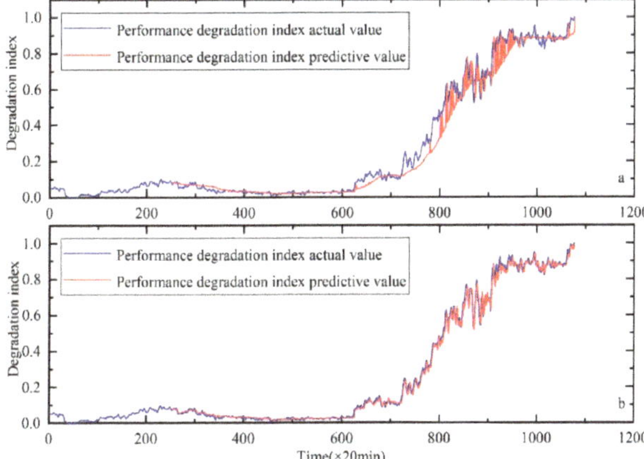

Figure 13. Performance degradation prediction of bearing 4, (**a**) is the result of HKF–SVR with parameters optimized by the GA and (**b**) is the result of HKF–SVR with parameters optimized by KH.

4.2. CASE II

The bearing test platform is shown in Figure 14, which includes the ABLT tester and the signal acquisition system based on LabVIEW and NI PXI platforms. The ABLT test machine was produced by the Hangzhou Bearing Test Center, which consists of three systems, namely, control and drive, loading and lubrication systems. The control and drive system enable real-time monitoring of the temperature and vibration signals of the bearings. Four HRB6305 bearing models, which were fixed on the same shaft and driven by an AC motor and connected by a belt, were used in this experiment. The failure of the bearing was accelerated by loading 750 kg in the radial direction of each bearing. After some fatigue tests, three types of faults for the inner ring, outer ring and rolling elements were obtained. Full-life vibration signals were acquired every 5 min by the NI PXI acquisition system. All data were collected at a frequency of 20 kHz and a bearing speed of 3000 rpm.

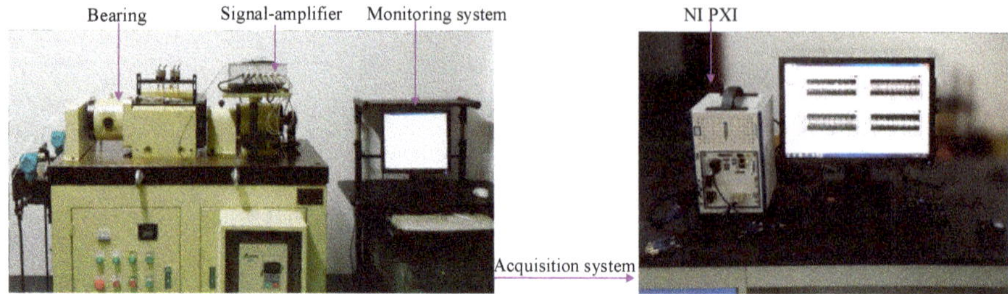

Figure 14. Experimental setup.

The full-life original vibration signal of the rolling element is shown in Figure 15.

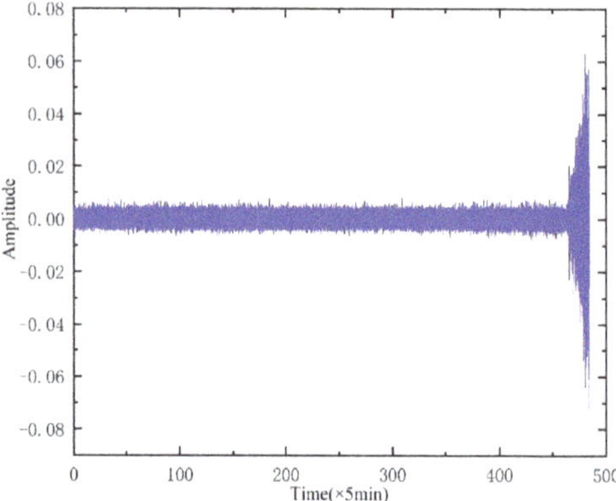

Figure 15. Whole-life vibration signal of bearing.

Similar to Case I, the features of the time and frequency domains were extracted first. Then, KJADE was used to fuse the original features and the performance degradation index was calculated from inter and between class distances. Finally, the performance degradation of the rolling bearings was predicted using the method proposed in the second part. The results are shown in Figure 16.

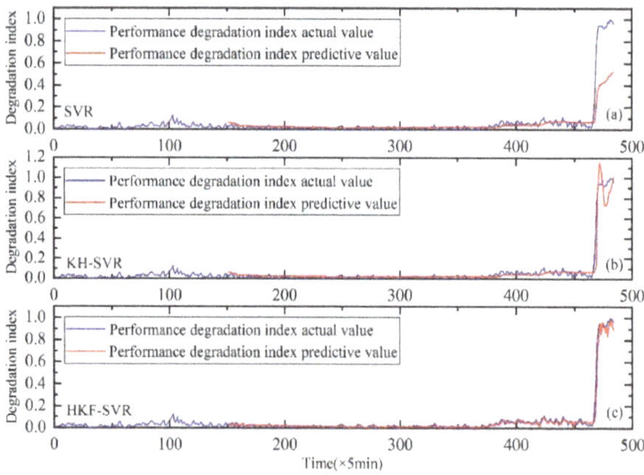

Figure 16. Performance degradation prediction of the rolling bearings, (**a**) is the result of SVR, (**b**) is the result of SVR with parameters optimized by the KH algorithm and (**c**) is the result of HKF–SVR with parameters optimized by the KH algorithm.

The prediction results of SVR, SVR with parameters optimized by KH algorithm and HKF–SVR with parameters optimized by KH algorithm are shown in Table 7, indicating that the method proposed in this study is more effective than the other two methods.

Table 7. Prediction errors comparison.

	SVR	KH–SVR	HKF–SVR
RMSE	0.225	0.077	0.035

Similar to Case I, the method proposed in this paper is compared with BPNN and ELM. As show in Figure 17, the results also prove the effectiveness of the proposed method.

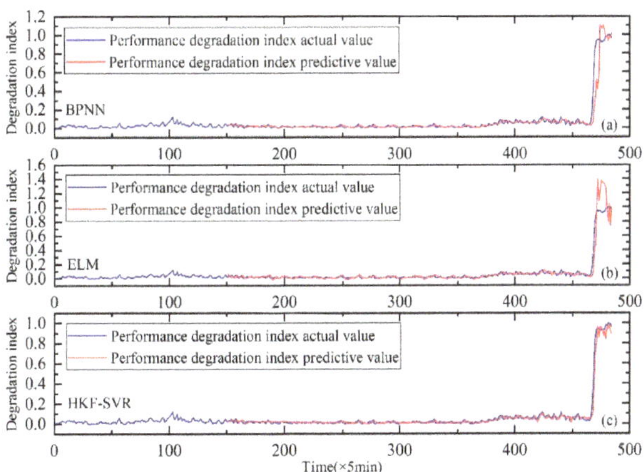

Figure 17. Performance degradation prediction of the rolling bearings. (**a**) is the result of BPNN, (**b**) is the result of ELM and (**c**) is the result of HKF–SVR with parameters optimized by the KH algorithm.

The RMSE values of the BPNN, ELM and HKF-SVR methods are shown in Table 8.

Table 8. Prediction errors comparison.

	BPNN	ELM	HKF-SVR
RMSE	0.067	0.065	0.035

As shown in Figure 18. Meanwhile, RMSE values are shown in Table 9. From the results, we can also see the advantage of the proposed method.

Table 9. Prediction errors comparison.

	GA-HKF-SVR	KH-HKF-SVR
RMSE	0.163	0.035

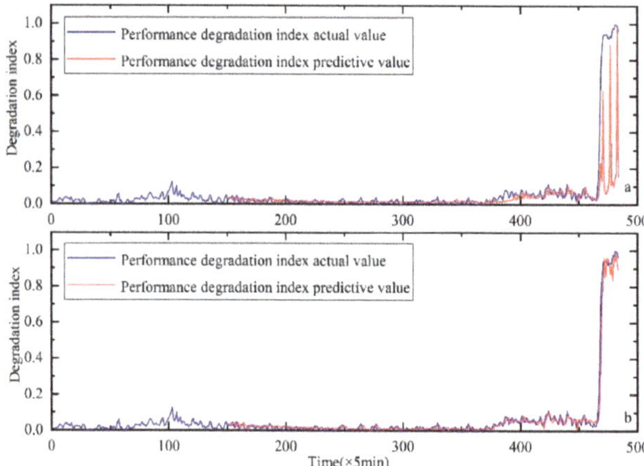

Figure 18. Performance degradation prediction of the rolling bearings, (**a**) is the result of HKF–SVR with parameters optimized by the GA and (**b**) is the result of HKF–SVR with parameters optimized by the KH.

5. Conclusions

In this study, the HKF–SVR optimized by the KH algorithm is proposed to predict the degradation of rolling bearing performance for coaxial bearings. It can effectively solve the problem of parameter selection of prediction model. On the other hand, our recently developed KJADE algorithm and SS is studied on performance degradation features extraction for coaxial bearing signals. The proposed method is compared with the SVR, SVR optimized by the KH algorithm, ELM and BPNN. Results have verified the effectiveness and advantage of the proposed method. The proposed method has a good application prospect in life prediction of coaxial bearings.

Author Contributions: In this article, the author's contributions are shown below: Methodology, F.L.; Writing-original draft preparation, L.L.; Project administration, Y.L.; Data curation, Z.C.; Investigation, H.Y.; Resources, S.L. All authors have read and agreed to the published version of the manuscript.

Funding: This work is supported by the National Natural Science Foundation of China (51675001, 51875001) the State Key Program of National Natural Science of China (51637001) and the Key Research and Development Plan of Anhui Province (201904A05020034).

Acknowledgments: The authors thank the IMS Center of the University of Cincinnati for providing free downloads of the rolling element bearing fault data sets.

Conflicts of Interest: The authors declare that there is no conflict of interest in the publication of this article. Again, the author states that there is no commercial or corporate interest in this submission.

References

1. Zheng, G.; Zhao, H.; Wu, D.; Li, X. Study on a Novel Fault Diagnosis Method of Rolling Bearing in Motor. *Recent Pat. Mech. Eng.* **2016**, *9*, 144–152.
2. Nohál, L.; Vaculka, M. Experimental and computational evaluation of rolling bearing steel durability. In *IOP Conference Series: Materials Science and Engineering*; IOP Publishing: Bristol, UK, 2017.
3. Dong, W.; Tsui, K.L.; Qiang, M. Prognostics and Health Management: A Review of Vibration based Bearing and Gear Health Indicators. *IEEE Access* **2017**, *6*, 665–676.
4. Liao, W.Z.; Dan, L. An improved prediction model for equipment performance degradation based on Fuzzy-Markov Chain. In Proceedings of the IEEE International Conference on Industrial Engineering & Engineering Management, Singapore, 6–9 December 2015.
5. Liu, P.; Hongru, L.I.; Baohua, X.U. A performance degradation feature extraction method and its application based on mathematical morphological gradient spectrum entropy. *J. Vib. Shock* **2016**, *35*, 86–90.

6. Duan, L.; Zhang, J.; Ning, W.; Wang, J.; Fei, Z. An Integrated Cumulative Transformation and Feature Fusion Approach for Bearing Degradation Prognostics. *Shock Vib.* **2018**, *2018*, 9067184. [CrossRef]

7. Liu, W.B.; Zou, Z.Y.; Xing, W.W. Feature Fusion Methods in Pattern Classification. *J. Beijing Univ. Posts Telecommun.* **2017**, *40*, 1–8.

8. Liu, P.; Li, H.; Ye, P. A Method for Rolling Bearing Fault Diagnosis Based on Sensitive Feature Selection and Nonlinear Feature Fusion. In Proceedings of the International Conference on Intelligent Computation Technology and Automation, Nanchang, China, 14–15 June 2015; pp. 30–35.

9. Zhang, M.; Shan, X.; Yang, Y.U.; Na, M.I.; Yan, G.; Guo, Y. Research of individual dairy cattle recognition based on wavelet transform and improved KPCA. *Acta Agric. Zhejiangensis.* **2017**, *29*, 2000–2008.

10. Liu, Y.; He, B.; Liu, F.; Lu, S.; Zhao, Y. Feature fusion using kernel joint approximate diagonalization of eigen-matrices for rolling bearing fault identification. *J. Sound Vib.* **2016**, *385*, 389–401. [CrossRef]

11. Han, T.; Jiang, D.; Zhao, Q.; Wang, L.; Yin, K. Comparison of random forest, artificial neural networks and support vector machine for intelligent diagnosis of rotating machinery. *Trans. Inst. Meas. Control* **2018**, *40*, 2681–2693. [CrossRef]

12. Shao, H.; Cheng, J.; Jiang, H.; Yang, Y.; Wu, Z. Enhanced deep gated recurrent unit and complex wavelet packet energy moment entropy for early fault prognosis of bearing. *Knowl. Based Syst.* **2019**. [CrossRef]

13. Qi, Y.; Shen, C.; Dong, W.; Shi, J.; Zhu, Z. Stacked Sparse Autoencoder-Based Deep Network for Fault Diagnosis of Rotating Machinery. *IEEE Access* **2017**, *5*, 15066–15079. [CrossRef]

14. Fan, G.F.; Peng, L.L.; Hong, W.C.; Fan, S. Electric load forecasting by the SVR model with differential empirical mode decomposition and auto regression. *Neurocomputing* **2016**, *173*, 958–970. [CrossRef]

15. Bahmani, S.; Romberg, J. Phase Retrieval Meets Statistical Learning Theory: A Flexible Convex Relaxation. *arXiv* **2016**, arXiv:1610.04210.

16. Liu, Z.; Yu, G. A neural network approach for prediction of bearing performance degradation tendency. In Proceedings of the 2017 9th International Conference on Modelling, Identification and Control (ICMIC), Kunming, China, 10–12 July 2017.

17. Qian, Y.; Hu, S.; Yan, R. Bearing performance degradation evaluation using recurrence quantification analysis and auto-regression model. In Proceedings of the 2013 IEEE International Instrumentation and Measurement Technology Conference (I2MTC), Minneapolis, MN, USA, 6–9 May 2013.

18. Shen, C.; Dong, W.; Kong, F.; Tse, P.W. Fault diagnosis of rotating machinery based on the statistical parameters of wavelet packet paving and a generic support vector regressive classifier. *Measurement* **2013**, *46*, 1551–1564. [CrossRef]

19. Dong, W.; Tsui, K.L. Two novel mixed effects models for prognostics of rolling element bearings. *Mech. Syst. Signal Process.* **2018**, *99*, 1–13.

20. Wang, X.L.; Han, G.; Li, X.; Hu, C.; Hui, G. A SVR-Based Remaining Life Prediction for Rolling Element Bearings. *J. Fail. Anal. Prev.* **2015**, *15*, 548–554. [CrossRef]

21. Xiang, L.; Deng, Z.; Hu, A. Forecasting Short-Term Wind Speed Based on IEWT-LSSVM model Optimized by Bird Swarm Algorithm. *IEEE Access* **2019**, *7*, 59333–59345. [CrossRef]

22. Ma, X.; Zhang, Y.; Wang, Y. Performance evaluation of kernel functions based on grid search for support vector regression. In Proceedings of the IEEE International Conference on Cybernetics & Intelligent Systems, Siem Reap, Cambodia, 15–17 July 2015.

23. Kai, C.; Lu, Z.; Wei, Y.; Yan, S.; Zhou, Y. Mixed kernel function support vector regression for global sensitivity analysis. *Mech. Syst. Signal Process.* **2017**, *96*, 201–214.

24. Zhou, J.M.; Wang, C.; He, B. Forecasting model via LSSVM with mixed kernel and FOA. *Comput. Eng. Appl.* **2013**, *33*, 964–966.

25. Wu, D.; Wang, Z.; Ye, C.; Zhao, H. Mixed-kernel based weighted extreme learning machine for inertial sensor based human activity recognition with imbalanced dataset. *Neurocomputing* **2016**, *190*, 35–49. [CrossRef]

26. Madamanchi, D. Evaluation of a New Bio-Inspired Algorithm: Krill Herd. Master's Thesis, North Dakota State University, Fargo, ND, USA, 2014.

27. Ayala, H.V.H.; Segundo, E.V.; Coelho, L.D.S.; Mariani, V.C. Multiobjective Krill Herd Algorithm for Electromagnetic Optimization. *IEEE Trans. Magn.* **2015**, *52*, 1–4. [CrossRef]

28. Ren, Y.T.; Qi, H.; Huang, X.; Wang, W.; Ruan, L.M.; Tan, H.P. Application of improved krill herd algorithms to inverse radiation problems. *Int. J. Therm. Sci.* **2016**, *103*, 24–34. [CrossRef]

29. Yan, T.; Pang, B.; Hua, W.; Gao, X. Research on the Optimized Algorithms for Support Vector Regression Model of Slewing Bearing's Residual life Prediction. In Proceedings of the 2017 International Conference on Artificial Intelligence, Automation and Control Technologies, Wuhan, China, 7–9 April 2017.

30. A Gentle Introduction To Support Vector Machines in Biomedicine: Volume 1: Theory and Methods. Available online: https://www.researchgate.net/publication/311233724_A_gentle_introduction_to_support_vector_machines_in_biomedicine_Volume_1_Theory_and_methods (accessed on 24 January 2020).

31. Lu, S.; Tao, C.; Tian, S.; Lim, J.H.; Tan, C.L. Scene text extraction based on edges and support vector regression. *Int. J. Doc. Anal. Recognit.* **2015**, *18*, 125–135. [CrossRef]

32. Wei, C.; Pourghasemi, H.R.; Naghibi, S.A. A comparative study of landslide susceptibility maps produced using support vector machine with different kernel functions and entropy data mining models in China. *Bull. Eng. Geol. Environ.* **2017**, *77*, 647–664.

33. Ouyang, T.; Zha, X.; Qin, L.; Xiong, Y.; Xia, T.; Huang, H. Short-term wind power prediction based on kernel function switching. *Electr. Power Autom. Equip.* **2016**, *9*, 12.

34. Chen, D.; Yuan, Z.; Wang, J.; Chen, B.; Gang, H.; Zheng, N. Exemplar-Guided Similarity Learning on Polynomial Kernel Feature Map for Person Re-identification. *Int. J. Comput. Vis.* **2017**, *123*, 392–414. [CrossRef]

Article

The Enhancement of Leak Detection Performance for Water Pipelines through the Renovation of Training Data

Tu T.N. Luong [1] and Jong-Myon Kim [2,*]

[1] Department of Computer Engineering, University of Ulsan, Ulsan 44610, Korea; ngtu.mta@gmail.com
[2] School of IT Convergence, University of Ulsan, Ulsan 44610, Korea
* Correspondence: jmkim07@ulsan.ac.kr; Tel.: +82-52-259-2217

Received: 3 April 2020; Accepted: 28 April 2020; Published: 29 April 2020

Abstract: Leakage detection is a fundamental problem in water management. Its importance is expressed not only in avoiding resource wastage, but also in protecting the environment and the safety of water resources. Therefore, early leak detection is increasingly urged. This paper used an intelligent leak detection method based on a model using statistical parameters extracted from acoustic emission (AE) signals. Since leak signals depend on many operation conditions, the training data in real-life situations usually has a small size. To solve the problem of a small sample size, a data improving method based on enhancing the generalization ability of the data was proposed. To evaluate the effectiveness of the proposed method, this study used the datasets obtained from two artificial leak cases which were generated by pinholes with diameters of 0.3 mm and 0.2 mm. Experimental results show that the employment of the additional data improving block in the leak detection scheme enhances the quality of leak detection in both terms of accuracy and stability.

Keywords: intelligent leak detection; acoustic emission signals; statistical parameters; support vector machine; wavelet denoising; Shannon entropy

---

1. Introduction

Leakage detection is a primary problem in water management [1,2]. About 20–30% of the water has been lost in water supply system every year. Especially, the loss of water can be up to 50% in some systems [2]. The growing demand for water inspires reconsideration of the management and supply of pipeline systems. Complications in exploiting new water bodies can be beaten by decreasing water losses [3]. Furthermore, the present attention of environmental protection and issues related to water quality encourages a growing interest in leakage detection. The community concerned with water resources has been concentrated more on the natural environment. However, the guardianship of water against incursions in pipes and the protection of the environment from the arrival of a transported contaminant are as significant as the protection of aquifers and well-fields from contaminant discharge [4]. Subsequently, methodologies for early leak detection are strongly urged. Additionally, they should not induce the interruption of piping actions, and they should be simple enough to actualize in practice.

Many studies on leak detection for water supply systems have been conducted and published. The avenues may be passive or active [5], and hardware-based or software-based [6]. Passive methods require direct visual investigation or supervision of sites, while active methods include a signal analysis. Signals used in active methods can be acoustic, vibration, flow rate, or pressure. Besides, hardware-based methods are classified depending on the type of special sensing devices such as acoustic monitoring, vibration analysis, cable sensor, etc. On the other hand, software-based methods are categorized based on the type of software programs and techniques used for leak detection such as support vector machine, harmonic wavelet analysis, genetic algorithm, etc. Among those avenues, acoustic emission (AE)-based methods, which are passive and hardware-based, are auspicious, since AE sensors can quickly recognize small leaks, offering high sensitivity in relation to fault buildup in a piping system. Accordingly, AE-based methods for pipeline diagnostics have been exploited [7–12].

In recent years, defect diagnosis methods based on modelling have been extensively used to improve the availability and reliability of mechanical systems subject to defects [13–15]. These avenues use high-dimensional signature vectors to prevent the hazards of dropping likely essential information. Nevertheless, some defect signatures are repetitious or inapplicable to the predicting models (namely unsupervised and supervised learning). As a result, these defect signatures can be a fundamental source of diagnostic efficiency deterioration. To address this issue, discriminative defect signature selection has turned into an indispensable part of trustworthy diagnosis. Basically, the subsequent two steps are carried out in the signature selection procedure, namely a configuration step of signature subsets and an assessment step of signature subset quality. Specifically, a number of signature subsets are first assembled and then assessed. Based on the assessment step, signature selection strategies are fundamentally assorted into wrappers or filters. Filter avenues use an assessment strategy that is separated from any classification strategy, while wrapper strategies employ accuracy estimates for particular classifiers during the evaluation of signature subset quality [16]. As a result, wrapper methodologies give better diagnostic efficiency for predetermined classifiers than filter strategies, theoretically. Nevertheless, filter avenues are computationally profitable because they bypass the accuracy estimation step for a specific classifier.

Practically, various conditions influence leakage signals, such as pipe diameter, surrounding environment, pipeline material, flow rate, and pressure [17,18]. Therefore, the data collected for training classifiers may not be large enough and extracted features from it may not be smooth enough to cover the whole probability space of features. As a result, the accuracy of feature evaluation and selection based on these data may be reduced. Furthermore, leak detection is a real application, and thus the techniques should be simple, effective, and easy to implement by hardware. Recently, Tu et al. offered an effective multivariable signature assessment coefficient (MSAC) to simultaneously evaluate the interclass separability and intraclass compactness depending on predicting the signature space from a restricted data point number [12]. Based on this coefficient, the diagnostic performance, in case the training data is not broad enough, is considerably improved. Nevertheless, the quality of leak detection is also affected by outsiders. These effects are regarded as noise data points with a low probability distribution, and they are far from the central data point in the same class. The accuracy and stability of a model greatly depends on the training dataset. If the training dataset is less generalized, the diagnostic model built on it will have reduced reliability and stability of performance. To deal with this problem, a data renovation method was introduced in this study. Particularly, the MSAC was first used to evaluate signatures as a filter method, and then the most discriminative signature subset was produced. Based on the selected signature subset, detecting and removing outsiders from the known dataset before training a diagnostic model is a key issue.

Once the discriminatory feature subset is determined and the known dataset is renovated, they are further employed to train a Support Vector Machine (SVM) classifier, which is a supervised model with higher accuracy than unsupervised models such as k-NN classifier, and with faster processing speed and lower hardware requirements than deep learning. In this work, the offered method was used to detect artificial leaks created in a laboratory with hole diameters of 0.3 mm and 2.0 mm.

The organization of this paper is as follows. The offered method is presented in Section 2. The data collecting method for leak detection is illustrated in Section 3. The efficacy of the proposed method is validated in Section 4, and the final section shows the conclusions.

2. The Offered Method

The overall flow diagram of leak detection is illustrated in Figure 1. First, the acquired AE signals were denoised by a Wavelet algorithm based on normalized Shannon entropy, which was also adopted in some recent studies of leak detection [12,19,20]. After that, the denoised signals were divided into separate analysis and evaluation datasets. The isolation of the evaluation dataset from the analysis dataset was to ensure the reliability of the performance evaluation results. Based on the analysis dataset, a defect signature pool was configured and the most discriminative signature subset, which was also applied on the evaluation dataset, was determined. Subsequently, based on selected features, the analysis dataset was renovated by detecting and removing outsiders before it was used to train SVM classifiers. Finally, the efficacy verification of the proposed method was carried out on the evaluation dataset. Each specific part is described in detail as follows.

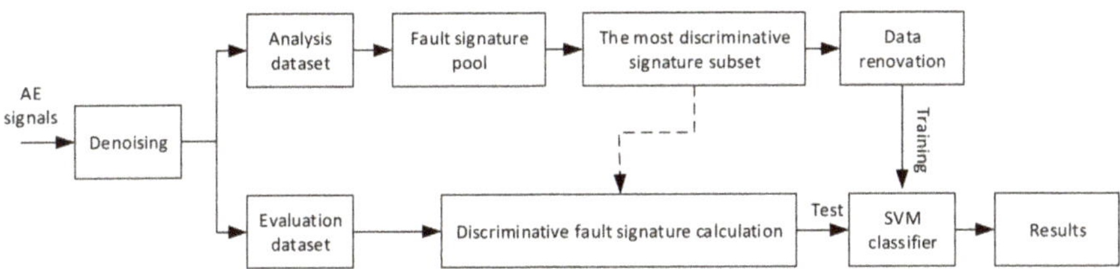

Figure 1. The overall flow diagram of the leak detection model.

2.1. Noise Reduction Using a Wavelet Transform and Shannon Entropy

Due to the nature of the AE mechanism, leakage noise is commonly nonstationary [21,22]. Time-frequency analysis methods, which are powerful tools to analyze the time-varying nonstationary signals, are recommended to study a signal in both the time and frequency domain simultaneously. Many studies have adopted the wavelet transform to detect the leak by reason of its multiresolution capability [23–25].

A form of wavelet transform which allows multiresolution investigation is known as a Wavelet packet transform (WPT) [26]. Signals can be decomposed into both wavelet coefficients and the scaling values through the WPT technique. Based on this technique, the complete decomposition hierarchy is provided. As a result, because of uniform frequency secondary groups, the decomposition becomes extremely adoptable [27].

A signal $\psi(t)$ with a fixed energy, which is expressed as a mother wavelet, is a consecutive vacillating function of intensely short duration as indicated in Equation (1):

$$\psi_{s,\tau}(t) = \frac{1}{\sqrt{s}}\psi(\frac{t-\tau}{s}), s > 0; -\infty < \tau < \infty, \tag{1}$$

where $\psi_{s,\tau}(t)$ consists of the total standardized expressions (expansions) in time t designated by $s > 0$ (scale factor) and translation in time t is designated by $-\infty < \tau < \infty$. Equation (2) expresses a cross correlation of $x(t)$ with $\psi_{s,\tau}(t)$ which depicts the Wavelet transformation of a signal $x(t)$ [24,27–29]. Mathematically, the similarity between two signals can be identified by cross-correlation analysis. Given two sets of signals x_i and y_i, where $i = 0, 1, 2, \ldots, N-1$, Equation (3) describes the function of normalized cross correlation with zero time-lag. The normalized cross correlation is a numerical quantity between 0 and 1, which predicts the closeness in characterization between two signals. Two signals which have identical characterizations generate a normalized cross correlation coefficient of 1.0 [30]:

$$WT_x(s,\tau) \triangleq \int x(t)\psi_{s,\tau}(t)dt, \tag{2}$$

$$R = \frac{\sum x_i y_i}{(\sum x_i^2)^{\frac{1}{2}}(\sum y_i^2)^{\frac{1}{2}}}, \tag{3}$$

The determination coefficient is made by executing the WPT with filter banks through recursive schemes. Low-frequency components (approximations) and high-frequency components (details) at each resolution level are obtained by transmitting the signal *x(t)* to a two-channel filter. Compared to the wavelet transform technique, which decomposes only the approximations, the WPT technique decomposes both details and approximations at every resolution level.

The most indispensable challenge in wavelet analysis is the selection of the mother wavelet function as well as the decomposition level of signal. Among orthogonal wavelets, Daubechies (DB) wavelets have been widely implemented, as they match the transient components in acoustic and vibration signals [31]. The order of the mother wavelet function and the level of decomposition were often determined by trial-and-error methods based on intrinsic characteristics of the data [31,32]. In this study, the selected mother function is DB15, and the number of levels was experimentally determined by Equation (3). Figure 2 illustrates the binary hierarchical tree of discrete wavelet packet transform (DWPT) coefficients. Each node of this tree was considered as a sub-band and numbered according to its level and its ordinal in level. Here, hierarchical levels and ordinals are numbered from 1.

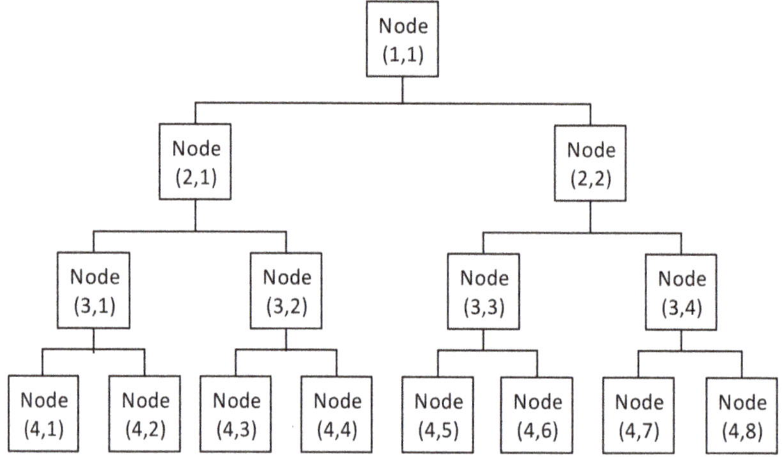

Figure 2. The binary tree of the organization and numbering of sub-bands in discrete wavelet packet transform (DWPT).

An algorithm based on informative entropy was utilized to detect the unnecessary signatures in an AE signal acquired during a test, where the informative entropy was considered a cost function. In this method, only the sub-bands which focus the major information carried by the signal are intended to be picked. Generally, the following equations denote the Shannon entropy $H(X_j)$ if $X_j = (x_{j,k})$ is a cluster of coefficients of a specified sub-band of the WPT tree at stage of resolution *j*:

$$H(X_j) = -\sum_k P_k \ln(P_k),$$ (4)

$$P_k = \frac{|x_{j,k}|^2}{\|X_j\|^2},$$ (5)

Here, $\|X_j\|^2 = \sum_k x_{j,k}^2$ signifies a norm of X_j [26]. A large value of $H(X_j)$ means that the signal is in higher disorder and carries less information. As a result, the corresponding sub-band and its subordinates are discarded. This implies that the entropy computes a correlation of energy among the sub-bands. At this moment, the aim is to select the WPT branch which transports the minor disorders and has minimum conceivable energy. If the informative entropy of the current resolved sub-band is smaller than that of the subsequent resolved sub-band, then the total data is conserved. Otherwise, a lesser energy level of resolution is essential. In other words, the selected sub-band should have the lowest entropy value and the highest resolution level. After that, the preferred sub-bands are used to reconstruct the AE signal such that the most significant part of the signal is saved, and the complementary component which is known to be noise is removed.

2.2. Fault Signature Pool Configuration

According to the authors of [11,12], intelligent leak detection schemes are well corroborated with statistical parameters from the time and frequency domains. Thus, this study used them as fault signatures for the identification of leaks. Statistical parameters for the given one-second AE data, $x(n)$, are defined in Tables 1 and 2. These parameters were computed in the frequency and time domain, and involved the peak (sp_1), the root-mean-square (sp_2), kurtosis (sp_3), crest-factor (sp_4), impulse factor (sp_5), shape factor 1 (sp_6), skewness (sp_7), the square-mean-root (sp_8), margin factor (sp_9), peak-to-peak (sp_{10}), kurtosis factor (sp_{11}), energy (sp_{12}), clearance factor (sp_{13}), shape factor 2 (sp_{14}), the fifth normalized moment (sp_{15}), the sixth normalized moment (sp_{16}), entropy (sp_{17}), spectral centroid (sp_{18}), the root-mean-square of frequency (sp_{19}), root variance of frequency (sp_{20}), and the frequency spectrum energy (sp_{21}).

Table 1. General statistical parameters in time-space of acoustic emission (AE) signals.

$sp_1 = \max\{\|\mathbf{x}\|\}$	$sp_2 = \sqrt{\frac{1}{N}\sum_{i=1}^{N} x_i^2}$	$sp_3 = \frac{1}{N}\sum_{i=1}^{N}\left(\frac{x_i - mean(\mathbf{x})}{std(\mathbf{x})}\right)^4$
$sp_4 = \frac{\max\{\|\mathbf{x}\|\}}{rms(\mathbf{x})}$	$sp_5 = \frac{peak(\mathbf{x})}{\frac{1}{N}\sum_{i=1}^{N}\|x_i\|}$	$sp_6 = \frac{rms(\mathbf{x})}{\frac{1}{N}\sum_{i=1}^{N}\|x_i\|}$
$sp_7 = \frac{1}{N}\sum_{i=1}^{N}\left(\frac{x_i - mean(\mathbf{x})}{std(\mathbf{x})}\right)^3$	$sp_8 = \left(\frac{1}{N}\sum_{i=1}^{N}\sqrt{\|x_i\|}\right)^2$	$sp_9 = \frac{peak(\mathbf{x})}{smr(\mathbf{x})}$
$sp_{10} = \max\{\|\mathbf{x}\|\} - \min\{\|\mathbf{x}\|\}$	$sp_{11} = \frac{1}{rms(\mathbf{x})^3}$	$sp_{12} = \sum_{i=1}^{N} x_i^2$
$sp_{13} = \frac{peak(\mathbf{x})}{smr(\mathbf{x})}$	$sp_{14} = \frac{pp(\mathbf{x})}{\frac{1}{N}\sum_{i=1}^{N}\|x_i\|}$	$sp_{15} = \frac{1}{N}\sum_{i=1}^{N}\left(\frac{x_i - mean(\mathbf{x})}{std(\mathbf{x})}\right)^5$
$sp_{16} = \frac{1}{N}\sum_{i=1}^{N}\left(\frac{x_i - mean(\mathbf{x})}{std(\mathbf{x})}\right)^6$	$sp_{17} = -\sum_{i=1}^{N} p_i \log_2 p_i$	

$\mathbf{x} = [x_1, x_2, \dots, x_N]$ denotes a signal in time-space, $mean(\mathbf{x}) = \frac{1}{N}\sum_{i=1}^{N} x_i$, $std(\mathbf{x}) = \sqrt{\frac{1}{N}\sum_{i=1}^{N}(x_i - mean(\mathbf{x}))^2}$, $p_i = \frac{x_i^2}{\sum_{i=1}^{N} x_i^2}$, $rms = sp_2$, $smr = sp_8$.

Table 2. General statistical parameters in frequency-space of AE signals.

$sp_{18} = \frac{\mathbf{X} \times \mathbf{f}^T}{\sum_{i=1}^{N} X_i}$	$sp_{19} = \sqrt{\frac{1}{N}\sum_{i=1}^{N} X_i^2}$
$sp_{20} = \sqrt{\frac{1}{N}\sum_{i=1}^{N}(X_i - mean(\mathbf{X}))^2}$	$sp_{21} = \sum_{i=1}^{N} X_i^2$

$\mathbf{X} = [X_1, X_2, \dots, X_N]$ expresses a signal in frequency-space corresponding to the frequency vector $\mathbf{f} = [f_1, f_2, \dots, f_N]$.

In summary, the dimensionality of the fault-signature pool used in the feature selection process is $N_{dp} \times N_{sp} \times N_{cl}$, where N_{dp}, N_{sp}, N_{cl} are the number of data points per leak condition class in the analysis dataset, the number of statistical parameters, and the number of classes to be discriminated in this study, respectively. Figure 3 illustrates an example of a data point configuration used to yield the most discriminatory feature subset. The set of elements in the fault-signature pool is denoted by $X = \{x(dp, sp, cl)\}$, with $dp = 1, \dots, N_{dp}$, $sp = 1, \dots, N_{sp}$, and $cl = 1, \dots, N_{cl}$. Variables dp, sp, cl represent coordinates of data point x in the dataset X.

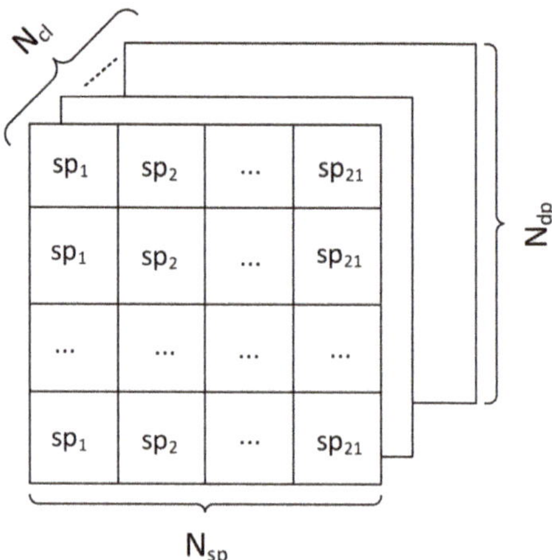

Figure 3. The illustration of data point configuration.

2.3. The Generation of the Discriminative Signature Subset

In order to achieve fairness, statistical parameters need to be standardized before evaluating and grading. This study used a simple scaling method with the following formula:

$$\widetilde{x}_i = \frac{x_i - min(X_k)}{max(X_k) - min(X_k)},\tag{6}$$

Here, $X_k = \{x_i | sp = k\}$, $\widetilde{X}_k = \{\widetilde{x}_i | sp = k\}$ denote original and standardized sets of values of the kth signature (i.e., k^{th} statistical parameter) respectively. After standardization, values of different signatures were all in the range [0,1].

To solve the small dataset problem, Tu et al. recently introduced an MSAC to evaluate the discrimination of fault signatures between two different classes [12]. The MSAC method estimates the potential value range of kth dimension of the signature sub-space of class i by interval $\left[mean\left(X_k^i\right) - 3std\left(X_k^i\right),\ mean\left(X_k^i\right) + 3std\left(X_k^i\right)\right]$, where X_k^i denotes the set of values of kth signature of data points in class i. Therefore, the crossing level between two signature sub-spaces of classes i, j at dimension k which is denoted by $MSAC_k^{i,j}$ is determined by Equation (7):

$$MSAC_k^{i,j} = 1 - \frac{3\left(std\left(X_k^i\right) + std\left(X_k^j\right)\right)}{\left|mean\left(X_k^i\right) - mean\left(X_k^j\right)\right|},\ X_k^i = \{x | sp = k, cl = i\}, X_k^j = \{x | sp = k, cl = j\},\tag{7}$$

Values of $std\left(X_k^i\right), std\left(X_k^j\right)$ represent the intraclass compactness of classes, and $\left|mean\left(X_k^i\right) - mean\left(X_k^j\right)\right|$ represents the interclass separability. According to the authors of [12], the bigger the MSAC, the better discrimination. Thus, MSAC expresses the distinguishable ability of signatures for each pair of classes. Although it is simple and has low computing cost, it is still effective and suitable for real applications such as leak detection. In this paper, the MSAC was used to rank signatures from top to bottom and the discriminatory signature subset was created by picking the signatures on top.

2.4. Data Renovation

To build a classification model, the correctness and generalization of the training dataset are extremely important. If the dataset is inaccurate or not generalized, then the accuracy, reliability, and stability of the trained model may be reduced. Related studies have most focused on big data [33–35].

Meanwhile, the problem with leak detection using a smart fault diagnostic model is related to the small data problem, because leakage signals are affected by many external factors. Therefore, it is necessary to revamp the dataset. In machine learning, the quality of samples is more important than their quantity, especially when the quantity is not large. The higher quality the samples, the greater the generalization ability and the better the accuracy. In a class, points that are far from the center and have a low probability distribution are known as outsiders. They are less significant than the rest and may be noise points. Consequently, they should be detected and removed.

This study focuses on improving the quality of data before training the classification model with a simple and effective technique. This technique includes three processes of detecting, eliminating outsiders, and updating dataset alternately until there no longer exist outsiders in the renovated dataset. In this study, we assumed that the statistical parameter values were Gaussian random variables. In term of statistics, the probability that each statistical parameter value in a specific class lies in the interval $[mean - 3std, mean + 3std]$ is equal to 99.73% [36], where *mean* is the mean and *std* is the standard deviation of their values in that class. This study used such range as the limit for outsider detection to ensure that outsiders were both far from central points and had a low probability distribution. Outsiders were defined as data points that were outside the confident interval (CI), which was determined through the central coordinate (CC) (i.e., the central point) and the standard deviation of each dimension (i.e., each statistical parameter or signature) of the signature space. Figure 4 illustrates an example about how to identify the central point, inner points, and outsider points in a signature space having two dimensions.

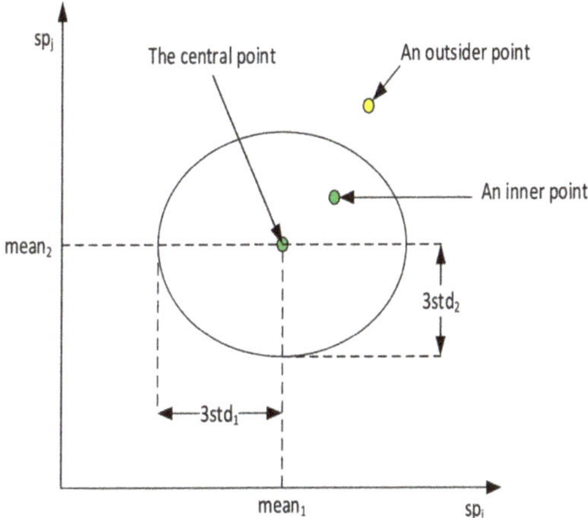

Figure 4. The illustration of how to classify data points.

Denote $X_k^i = \{x | sp = k, cl = i\}$ as the set of values of kth statistical parameter of all data points in class *i*. The CC of class *i*, CC_i, is defined in Equation (8). The CI of value of kth statistical parameter (i.e., k^{th} dimension) of data points in the signature space of class *i* is given in Equation (9). A data point is considered as an outsider if any dimension of that data point is outside the CI of such dimension. The process of improving the dataset was implemented separately for each class and illustrated in Figure 5. Whenever outsiders are detected and eliminated, the dataset needs to be updated. After that, values of CC and CIs also needs to be updated, and as a result, new outsiders can be detected and eliminated. This process ends when no outsider is detected in the updated dataset:

$$CC_i = \left(mean\left(X_1^i\right), \ mean\left(X_2^i\right), \ldots, \ mean\left(X_{N_{sp}}^i\right)\right), \tag{8}$$

$$CI_k^i = \left[mean\left(X_k^i\right) - 3std\left(X_k^i\right), \ mean\left(X_k^i\right) + 3std\left(X_k^i\right) \right], \tag{9}$$

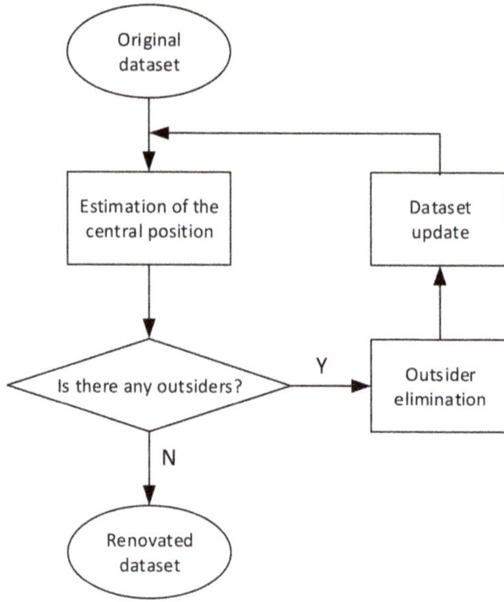

Figure 5. The process of improving the training dataset.

2.5. Classification

This study used a two-class SVM classifier whose theory is based on the idea of structural hazard minimization [37]. In the SVM method, the generalization error is minimized and the geometric margin between two classes is maximized. This method is also known as the maximum margin classifier. In this study, the kernel function was used to map the input data into a high-dimensional signature space and detect the best hyper plane to discriminate between the two classes of input data. The margin between two classes in the feature space was maximized by the best hyper plane. This quadratic optimization problem was worked out using Lagrange multipliers. The term "support vectors" is used to refer to the points which are nearest to the optimal hyper plane for each class [38]. Support vectors are selected along the surface of a kernel function which can be chosen among different functions such as polynomial, linear, radial-based, and sigmoid for the SVM during the training phase [39]. Based on a set of predetermined support vectors that are members of the set of training inputs, SVM distributes data with two class labels.

Kernel function parameter selection is one of the significant details of SVM modeling. In this paper, we used the radial based function (RBF), which is a common kernel function that can be employed to any sample distribution through parameter selection. The RBF has been used more and more in the nonlinear mapping of SVMs. The RBF kernel function expression is:

$$K(x_i, x_j) = \exp\left(-\gamma \lVert x_i - x_j \rVert^2\right), \tag{10}$$

The corresponding minimization problem of an SVM is expressed below:

$$\min_{\alpha_i} \frac{1}{2}\sum_{i=1}^{n}\sum_{j=1}^{n} y_i y_j \alpha_i \alpha_j \exp\left(-\gamma \lVert x_i - x_j \rVert^2\right) - \sum_{i=1}^{n} \alpha_i, \ \sum_{i=1}^{n} y_i \alpha_i = 0, \ 0 \leq \alpha_i \leq C, \tag{11}$$

The minimum value of Equation (11) depends on the choice of parameters (C, γ). In this study, the grid search method was used to get the final optimal parameters (C, γ) [40]. This method respectively takes m values in C and takes n values in γ, for the $m \times n$ combinations of (C, γ), trains different SVM respectively, then estimates the learning precision. We can obtain the highest study accuracy of the best combination as the optimal parameters in the $m \times n$ combinations of (C, γ).

3. Experiment Setup

Figure 6 shows the setup of the AE signal acquisition from a water pipeline system. The pipe, which was made of stainless steel 304, had an outside diameter of 34 mm and a wall thickness of 3.38 mm. A pump was employed to keep the water flow constant at a pressure of 3 bar. The experiments were executed under a balanced temperature of approximately 29 °C. AE sensors were mounted on both sides of the testing pipe fragment. The distance from sensors to the leak position was 1000 mm. In this study, wideband differential-auto sensor test (WDI-AST) sensors were used to provide high sensitivity and a wide frequency band. The characteristics of the sensors are recapped in Table 3.

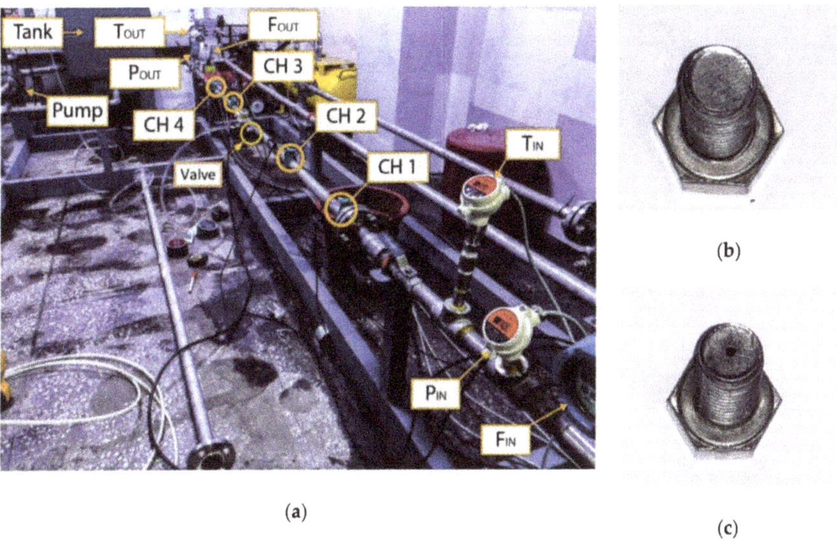

Figure 6. AE signal acquisition from a water pipeline system: (**a**) The test pipeline and component arrangement; (**b**) the pinhole size of 0.3 mm; (**c**) the pinhole size of 2.0 mm.

Table 3. Specifications of the wideband differential-auto sensor test (WDI-AST) sensors.

No	Parameters	Values
1	Peak sensitivity	96 dB
2	Operating frequency range	200–900 kHz
3	Directionality	+/−1.5 dB
4	Temperature range	−35 to 75 °C

The experiments were based on two leak cases with different pinhole diameters, i.e., 0.3 mm and 2.0 mm, which were considered dataset 1 and 2, respectively. The AE signals were collected in one-second time lengths and sampled at a frequency of 1 MHz. Details of the datasets which were used to assess the offered method are described in Table 4. In this table, "normal" means the no leakage case. Since different datasets were acquired in different dates, and operating conditions such as temperature, pressure, flow rate, etc. have impacts on AE signals, each "normal" data is taken accordingly with the leak data to have coherence with background condition.

Table 4. Details of the datasets employed to assess the offered method.

The Rate of Sampling = 1 MHz	Dataset 1		Dataset 2	
Signal Length = 1 s	Normal	0.3 mm	Normal	2.0 mm
Size of the training data	48	56	72	80
Size of the test data	12	14	18	20
Total	60	70	90	100

4. Results and Discussion

Figures 7 and 8 illustrate one obtained AE signal sample of each case for each dataset over the time and its fast Fourier transform in frequency domains. It is clear that these original signals contained noise, and that there was not much difference between signals at healthy and unhealthy states.

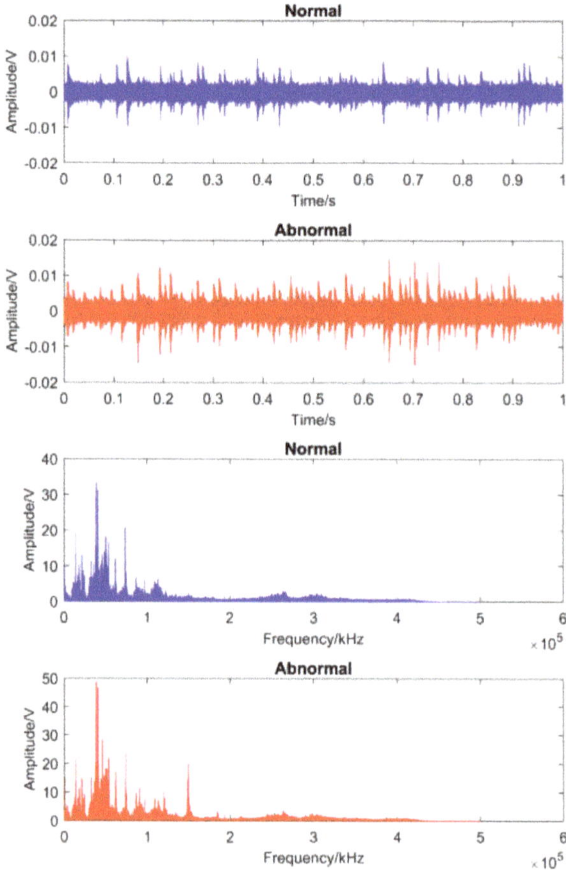

Figure 7. The illustration of the obtained AE signals in dataset 1 over the time and frequency domains.

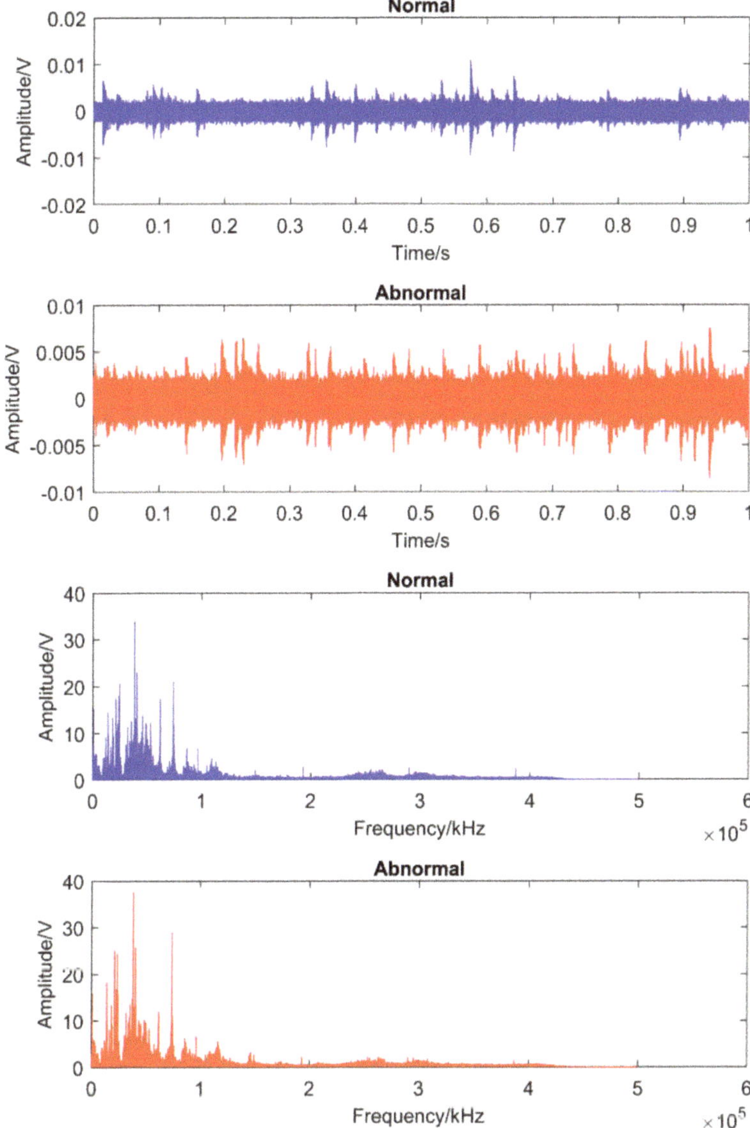

Figure 8. The illustration of the obtained AE signals in dataset 2 over the time and frequency domains.

To extract the most informative part of the signals, sub-bands were first produced by implementing the DWPT on each raw AE signal. Then, the optimal sub-band was selected depending on the minimum wavelet entropy before being employed to restore the AE signal. Figure 9 shows the difference between signals before and after denoising in both the time and frequency domains.

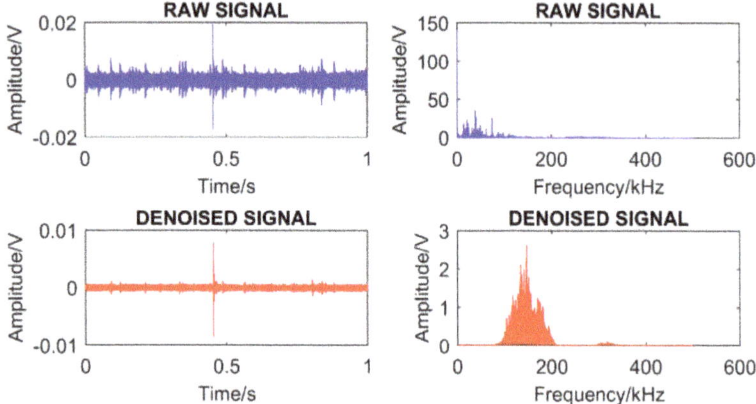

Figure 9. Differences between the signals before and after filtering noise in both the time and frequency domains.

In the next step, the fault signature pool was created from reconstructed AE signals in the analysis dataset. Then, the MSAC was used to evaluate the signatures. Table 5 lists signatures in order of best to worst in terms of leak detection, together with their MSACs corresponding to each case.

Table 5. Lists of signatures in order of best to worst together with their multivariable signature assessment coefficients (MSACs).

Grade	Dataset 1		Dataset 2	
	Signature	MSAC	Signature	MSAC
1	sp_8	0.4070	sp_{18}	0.1780
2	sp_{18}	0.1316	sp_8	−1.9828
3	sp_{20}	−0.2410	sp_{11}	−3.3668
4	sp_{19}	−0.4302	sp_{19}	−3.6743
5	sp_2	−0.4302	sp_2	−3.6743
6	sp_{11}	−0.7475	sp_{21}	−3.8037
7	sp_{21}	−0.8396	sp_{13}	−3.8037
8	sp_{13}	−0.8396	sp_{20}	−4.8188
9	sp_7	−3.4940	sp_{15}	−9.7089
10	sp_1	−5.1438	sp_{16}	−13.4133
11	sp_{10}	−5.4023	sp_7	−20.9734
12	sp_{15}	−5.8335	sp_6	−29.5249
13	sp_6	−6.5642	sp_1	−30.3707
14	sp_{14}	−8.6542	sp_{10}	−34.4620
15	sp_5	−8.6890	sp_9	−52.0857
16	sp_9	−8.7830	sp_3	−63.6919
17	sp_3	−8.8678	sp_{14}	−69.6629
18	sp_4	−8.8681	sp_5	−75.3590
19	sp_{17}	−9.4356	sp_4	−83.9089
20	sp_{12}	−12.4321	sp_{17}	−108.4910
21	sp_{16}	−12.5609	sp_{12}	−109.8144

After that, the two best signatures on top were selected as a discriminatory feature subset. In such a manner, the feature sub-set that was most discriminative for both cases included one parameter on the time domain, namely the square-mean-root, and one parameter on the frequency domain, namely the spectral centroid. Figure 10 illustrates the distribution of data points according to the selected features corresponding to each leak case. It can be seen that data points in the same class, in the case of a lower-level leak (pinhole size of 0.3 mm), had a higher concentration than in the case of a higher-level leak (pinhole size of 2.0 mm), while the separation between classes in the first case was lower than the other. The reason for this may be that instability of the AE signal increased belong with the leakage level. It follows that the leak detection method of using statistical parameters of AE signals was limited by leak level in both directions. Specifically, the greater the leakage level, the lower the concentration level in the same class and the greater the interclass separability.

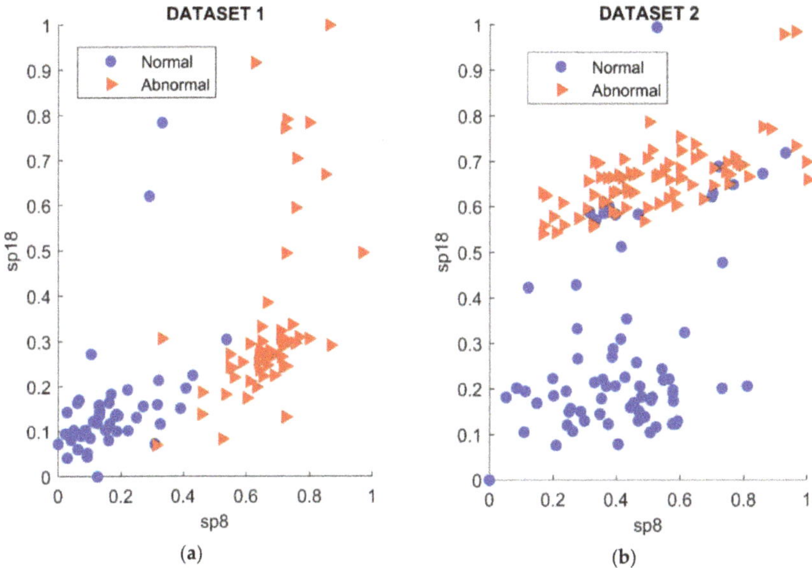

Figure 10. The distribution of data points according to the selected features corresponding to each leak case: (**a**) Leak size of 0.3 mm; (**b**) leak size of 2.0 mm.

To enhance the stability and quality of the SVM classifiers, the training dataset needs to be improved by detecting and removing outsiders, which may be noise data points because of their low probability distribution and weak generalization. Based on the renovated analysis dataset, the SVM classifiers were trained before being used to detect leaks in the evaluation dataset. To evaluate the proposed method, this study used a 10-fold cross validation to compare classification accuracies (CAs). The CA given in Equation (12) is the ratio between the number of correctly classified data points (i.e., true points), N_{TP}, and the total of data points, N_{total}. The results of CAs of three methods are shown in Table 6. Here, "All" represents the conventional method, which uses all of 21 fault signatures without signature selection and data renovation, whereas the conventional method [12] uses a signature selection with MSAC without data renovation. In general, the proposed method, which added the data enhancement block, outperformed the method in [12], which had the same signature subset. Specifically, the former had no worse results than the latter in 10 total assessments of both cases. In addition, the former surpassed the latter by four times in dataset 1 and three times in dataset 2. It follows that the former enhanced the average CA of 4.61% and 1.58% compared to the latter when datasets 1 and 2 were used, respectively. Therefore, it is proven that the proposed method is both more accurate and more stable than the previous method.

$$CA = \frac{N_{TP}}{N_{total}} \times 100\%$$ (12)

Table 6. The classification accuracy results (%) for 10-fold cross validation.

No	Dataset 1			Dataset 2		
	All (%)	[12] (%)	The Proposed Method (%)	All (%)	[12] (%)	The Proposed Method (%)
1	100	92.31	100	89.47	84.21	89.47
2	100	76.92	84.62	78.95	84.21	89.47
3	100	100	100	84.21	84.21	84.21
4	92.31	92.31	92.31	100	100	100
5	100	92.31	92.31	78.95	84.21	84.21
6	100	69.23	92.31	100	100	100
7	100	100	100	94.74	94.74	100
8	100	92.31	92.31	100	100	100
9	100	100	100	94.74	94.74	94.74
10	100	92.31	100	100	100	100
Average	99.23	90.77	95.38	92.11	92.63	94.21

Compared to the method of using all the signatures in terms of CA, the proposed method was better in dataset 1, but worse in dataset 2. However, the proposed method, which used only two features, significantly reduced the number of dimensions of the fault signature vector compared to the non-signature-selection method, which used 21 features. This means that it is possible to mitigate the computational responsibility in the configuration of signature vectors in real applications. Moreover, low-dimensional signature vectors can assist in reduction of consumed time to train classifiers. Table 7 shows computational time comparison between the proposed and conventional method which employed all 21 signatures. Compared to the conventional method, the processing speed of the proposed method was improved by 31.17% in training, 76.77% in test, and 40.14% in total for the dataset 1. Similarly, those improvements for the dataset 2 were 41.63%, 76.80%, and 48.63% respectively. All experiments were implemented with MATLAB R2018b on an Intel Core i7-7700 CPU operating at 3.60 GHz.

Table 7. Computational time comparison between the proposed and the conventional methods.

	Average Computational Time [Seconds]					
	All 21 Signatures Used for Leak Detection			The Proposed Method		
	Classifier Construction	Test	Total	Classifier Construction	Test	Total
Dataset 1	33.2924	8.1491	41.4415	22.9162	1.8927	24.8089
Dataset 2	48.2951	12.0008	60.2959	28.1876	2.7836	30.9712

5. Conclusions

In this paper, an intelligent leak detection method based on a model using statistical parameters extracted from AE signals was used for early leak detection. Since leak signals depend on many operation conditions, the training data in real-life situations usually has a small size. To solve the problem of a small dataset, a data improving method based on enhancing the generalization ability of the data was proposed. To evaluate the effectiveness of the proposed method, this study used the datasets obtained from two artificial leak cases which were generated by pinholes with diameters of 0.3 mm and 0.2 mm. Experimental results showed that the employment of the additional data improving block in the leak detection scheme enhances the quality of leak detection in both terms of accuracy and stability.

Author Contributions: Conceptualization, T.T.N.L. and J.-M.K.; Data curation, T.T.N.L.; Formal analysis, T.T.N.L.; Funding acquisition, J.-M.K.; Methodology, T.T.N.L. and J.-M.K.; Software, T.T.N.L.; Supervision, J.-M.K.; Validation J.-M.K.; Visualization, T.T.N.L.; Writing—original draft, T.T.N.L.; Writing—review & editing, J.-M.K. All authors have read and agreed to the published version of the manuscript.

Funding: This work was supported by the Korea Institute of Energy Technology Evaluation and Planning (KETEP) and the Ministry of Trade, Industry & Energy (MOTIE) of the Republic of Korea (No. 20192510102510).

Conflicts of Interest: The authors declare no conflict of interest.

References

1. Ferrante, M.; Brunone, B. Pipe system diagnosis and leak detection by unsteady-state tests. 1. Harmonic analysis. *Adv. Water Resour.* **2003**, *26*, 95–105. [CrossRef]
2. Frauendorfer, R.; Liemberger, R. *The Issues and Challenges of Reducing Non-Revenue Water*; Asian Development Bank: Mandaluyong, Philippines, 2010.
3. DJB. Theme introduction: Leak detection: A money-saving expense. *J. Am. Water Works Assoc.* **1979**, 51.
4. Funk, J.; VanVuuren, S.; Wood, D.; LeChevallier, M.; Friedman, M. Pathogen intrusion into water distribution systems due to transients. In Proceedings of the ASME/JSME Joint Fluids Engineering Conference, San Francisco, CA, USA, 18–22 July 1999.
5. Chan, T.; Chin, C.S.; Zhong, X. Review of current technologies and proposed intelligent methodologies for water distributed network leakage detection. *IEEE Access* **2018**, *6*, 78846–78867. [CrossRef]
6. Datta, S.; Sarkar, S. A review on different pipeline fault detection methods. *J. Loss Prev. Process Ind.* **2016**, *41*, 97–106. [CrossRef]
7. Martini, A.; Troncossi, M.; Rivola, A. Vibroacoustic measurements for detecting water leaks in buried small-diameter plastic pipes. *J. Pipeline Syst. Eng. Pract.* **2017**, *8*, 04017022. [CrossRef]
8. Li, S.; Song, Y.; Zhou, G. Leak detection of water distribution pipeline subject to failure of socket joint based on acoustic emission and pattern recognition. *Measurement* **2018**, *115*, 39–44. [CrossRef]
9. Mandal, S.K.; Chan, F.T.; Tiwari, M. Leak detection of pipeline: An integrated approach of rough set theory and artificial bee colony trained SVM. *Expert Syst. Appl.* **2012**, *39*, 3071–3080. [CrossRef]
10. Xiao, Q.; Li, J.; Bai, Z.; Sun, J.; Zhou, N.; Zeng, Z. A Small leak detection method based on VMD adaptive de-noising and ambiguity correlation classification intended for natural gas pipelines. *Sensors* **2016**, *16*, 2116. [CrossRef]
11. Luong, T.N.T.; Kim, J.-M. Leakage classification based on improved kullback-leibler separation in water pipelines. In Proceedings of the International Conference on Future Data and Security Engineering, Nha Trang City, Vietnam, 27–29 November 2019; pp. 56–67.
12. Tu, L.T.N.; Kim, J.-M. Discriminative feature analysis based on the crossing level for leakage classification in water pipelines. *J. Acoust. Soc. Am.* **2019**, *145*, EL611–EL617. [CrossRef]
13. Seshadrinath, J.; Singh, B.; Panigrahi, B.K. Vibration analysis based interturn fault diagnosis in induction machines. *IEEE Trans. Ind. Inform.* **2013**, *10*, 340–350. [CrossRef]
14. Gritli, Y.; Zarri, L.; Rossi, C.; Filippetti, F.; Capolino, G.-A.; Casadei, D. Advanced diagnosis of electrical faults in wound-rotor induction machines. *IEEE Trans. Ind. Electron.* **2012**, *60*, 4012–4024. [CrossRef]
15. Huang, S.; Tan, K.K.; Lee, T.H. Fault diagnosis and fault-tolerant control in linear drives using the Kalman filter. *IEEE Trans. Ind. Electron.* **2012**, *59*, 4285–4292. [CrossRef]
16. Li, B.; Zhang, P.-L.; Tian, H.; Mi, S.-S.; Liu, D.-S.; Ren, G.-Q. A new feature extraction and selection scheme for hybrid fault diagnosis of gearbox. *Expert Syst. Appl.* **2011**, *38*, 10000–10009. [CrossRef]
17. Hunaidi, O.; Chu, W.T. Acoustical characteristics of leak signals in plastic water distribution pipes. *Appl. Acoust.* **1999**, *58*, 235–254. [CrossRef]
18. Moore, S. A review of noise and vibration in fluid-filled pipe systems. In Proceedings of the Acoustics, Brisbane, Australia, 9–11 November 2016; pp. 9–11.
19. Duong, B.P.; Kim, J.-M. Pipeline fault diagnosis using wavelet entropy and ensemble deep neural technique. In Proceedings of the International Conference on Image and Signal Processing, Cherbourg, France, 2–4 July 2018; pp. 292–300.
20. Xiao, R.; Hu, Q.; Li, J. Leak detection of gas pipelines using acoustic signals based on wavelet transform and Support Vector Machine. *Measurement* **2019**, *146*, 479–489. [CrossRef]

21. Ghazali, M.; Beck, S.; Shucksmith, J.; Boxall, J.; Staszewski, W. Comparative study of instantaneous frequency based methods for leak detection in pipeline networks. *Mech. Syst. Signal Process.* **2012**, *29*, 187–200. [CrossRef]

22. Xiao, Q.; Li, J.; Sun, J.; Feng, H.; Jin, S. Natural-gas pipeline leak location using variational mode decomposition analysis and cross-time–frequency spectrum. *Measurement* **2018**, *124*, 163–172. [CrossRef]

23. Ferrante, M.; Brunone, B. Pipe system diagnosis and leak detection by unsteady-state tests. 2. Wavelet analysis. *Adv. Water Resour.* **2003**, *26*, 107–116. [CrossRef]

24. Ahadi, M.; Bakhtiar, M.S. Leak detection in water-filled plastic pipes through the application of tuned wavelet transforms to acoustic emission signals. *Appl. Acoust.* **2010**, *71*, 634–639. [CrossRef]

25. Zhang, Y.; Chen, S.; Li, J.; Jin, S. Leak detection monitoring system of long distance oil pipeline based on dynamic pressure transmitter. *Measurement* **2014**, *49*, 382–389. [CrossRef]

26. Bianchi, D.; Mayrhofer, E.; Gröschl, M.; Betz, G.; Vernes, A. Wavelet packet transform for detection of single events in acoustic emission signals. *Mech. Syst. Signal Process.* **2015**, *64*, 441–451. [CrossRef]

27. Morsi, W.G.; El-Hawary, M. A new reactive, distortion and non-active power measurement method for nonstationary waveforms using wavelet packet transform. *Electr. Power Syst. Res.* **2009**, *79*, 1408–1415. [CrossRef]

28. Saeidi, F.; Shevchik, S.; Wasmer, K. Automatic detection of scuffing using acoustic emission. *Tribol. Int.* **2016**, *94*, 112–117. [CrossRef]

29. Wang, X.; Zhu, C.; Mao, H.; Huang, Z. Wavelet packet analysis for the propagation of acoustic emission signals across turbine runners. *Ndt E Int.* **2009**, *42*, 42–46. [CrossRef]

30. Wren, T.A.; Do, K.P.; Rethlefsen, S.A.; Healy, B. Cross-correlation as a method for comparing dynamic electromyography signals during gait. *J. Biomech.* **2006**, *39*, 2714–2718. [CrossRef] [PubMed]

31. Rafiee, J.; Tse, P.; Harifi, A.; Sadeghi, M. A novel technique for selecting mother wavelet function using an intelli gent fault diagnosis system. *Expert Syst. Appl.* **2009**, *36*, 4862–4875. [CrossRef]

32. Kar, C.; Mohanty, A. Monitoring gear vibrations through motor current signature analysis and wavelet transform. *Mech. Syst. Signal Process.* **2006**, *20*, 158–187. [CrossRef]

33. Bosu, M.F.; MacDonell, S.G. Data quality in empirical software engineering: A targeted review. In Proceedings of the 17th International Conference on Evaluation and Assessment in Software Engineering, Porto de Galinhas, Brazil, 14–16 April 2013; pp. 171–176.

34. Shirai, Y.; Nichols, W.; Kasunic, M. Initial evaluation of data quality in a TSP software engineering project data repository. In Proceedings of the 2014 International Conference on Software and System Process, Nanjing, China, 26–28 May 2014; pp. 25–29.

35. Shepperd, M. Data quality: Cinderella at the software metrics ball? In Proceedings of the 2nd International Workshop on Emerging Trends in Software Metrics, Honolulu, HI, USA, 24 May 2011; pp. 1–4.

36. Ribeiro, M.I. Gaussian probability density functions: Properties and error characterization. *Inst. Syst. Robot. Lisboa Portugal* **2004**, 3–12.

37. Vapnik, V. *The Nature of Statistical Learning Theory*; Springer: Berlin/Heidelberg, Germany, 2013.

38. Kecman, V. *Learning and Soft Computing: Support Vector Machines, Neural Networks, and Fuzzy Logic Models*; MIT Press: Cambridge, MA, USA, 2001.

39. Huang, C.-L.; Wang, C.-J. A GA-based feature selection and parameters optimizationfor support vector machines. *Expert Syst. Appl.* **2006**, *31*, 231–240. [CrossRef]

40. Han, S.; Qubo, C.; Meng, H. Parameter selection in SVM with RBF kernel function. In Proceedings of the World Automation Congress 2012, Puerto Vallarta, Mexico, 24–28 June 2012; pp. 1–4.

Article

A Reliable Fault Diagnosis Method for a Gearbox System with Varying Rotational Speeds

Cong Dai Nguyen, Alexander Prosvirin and Jong-Myon Kim *

School of Electrical, Electronics and Computer Engineering, University of Ulsan, Ulsan 44610, Korea;
daimtavn@gmail.com (C.D.N.); a.prosvirin@hotmail.com (A.P.)
* Correspondence: jmkim07@ulsan.ac.kr; Tel.: +82-52-259-2217

Received: 9 April 2020; Accepted: 29 May 2020; Published: 31 May 2020

Abstract: The vibration signals of gearbox gear fault signatures are informative components that can be used for gearbox fault diagnosis and early fault detection. However, the vibration signals are normally non-linear and non-stationary, and they contain background noise caused by data acquisition systems and the interference of other machine elements. Especially in conditions with varying rotational speeds, the informative components are blended with complex, unwanted components inside the vibration signal. Thus, to use the informative components from a vibration signal for gearbox fault diagnosis, the noise needs to be properly distilled from the informational signal as much as possible before analysis. This paper proposes a novel gearbox fault diagnosis method based on an adaptive noise reducer–based Gaussian reference signal (ANR-GRS) technique that can significantly reduce noise and improve classification from a one-against-one, multiclass support vector machine (OAOMCSVM) for the fault types of a gearbox. The ANR-GRS processes the shaft rotation speed to access and remove noise components in the narrowbands between two consecutive sideband frequencies along the frequency spectrum of a vibration signal, enabling the removal of enormous noise components with minimal distortion to the informative signal. The optimal output signal from the ANR-GRS is then extracted into many signal feature vectors to generate a qualified classification dataset. Finally, the OAOMCSVM classifies the health states of an experimental gearbox using the dataset of extracted features. The signal processing and classification paths are generated using the experimental testbed. The results indicate that the proposed method is reliable for fault diagnosis in a varying rotational speed gearbox system.

Keywords: adaptive noise reducer; gaussian reference signal; gearbox fault diagnosis; one against on multiclass support vector machine; varying rotational speed

1. Introduction

Gearboxes are widely used in industrial applications, usually in harsh and continuous conditions, making them susceptible to a variety of failures. Defects can cause the gearbox system to break down and potentially damage complex mechatronic equipment or even cause a serious threat to safety, property, or customer satisfaction. Therefore, it is essential to diagnose gearbox faults regularly to ensure their early detection. The vibrations of gearbox systems have been studied since the 1980s, and previous researchers have found that gearbox vibrations have a keynote meshing frequency [1,2] with complex sidebands around it and its harmonics [3,4]. Therefore, the sideband frequencies and the meshing frequency and its harmonics are the informative components for identifying gear faults. Signal analysis is a backbone procedure for rotational-machine fault diagnosis research and applications. It works by decomposing the related fault features that are the groundwork for identifying fault patterns. The vibration characteristics of gearbox systems produce two major signals that can be analyzed for fault detection: acoustic signals and vibration signals [5]. Vibration signals are the most popularly used ones for gearbox fault monitoring because acquiring vibration data is easy [6]. However, vibration signals contain many types of noise from sources such as measurement systems (data acquisition systems), the environment, shafts, gears, and other related components and their impingement [7,8]. All that noise, which exceeds the signal, fills the frequency spectrum of the vibration signal, and eclipses it.

Many signal processing methods using advanced techniques have been presented by many researchers: frequency analysis focusing on Fourier transform [9], Wigner distribution [10], rank-order morphological filter [11], cyclostationary signals for mechanical applications [12], and the envelope analysis [13], which is the most well-known for rotational-machine fault diagnosis applications such as bearing-fault diagnosis. It detects the repeating shock amplitudes that appear as faulty teeth traverse each rotation cycle. Using this method, the vibration signal is first processed by a bandpass filter to achieve a high signal-to-noise ratio, and second, the Hilbert transform is used to achieve the envelope. If the sideband frequencies in a gearbox vibration signal appear in the envelope, the presence of faulty teeth in the gearbox can be deduced [14,15]. However, when the vibration signal is submerged in noise, it is difficult to recognize the informative components for fault diagnosis in the envelope.

Time-frequency analyses were developed to process non-stationary signals using a frequency transformation process divided based on windows across the time axis to capture informative events. The basic time-frequency analysis method is a short-time Fourier transform (STHT) or a spectrogram, such as a limited time window–width Fourier spectral analysis [16,17]. The challenges of the STHT method, such as a failure of the assumption that the pieces of a non-stationary signal are stationary, difficulty adapting the observation window size to the size of a real stationary piece of signal, and the conflict between frequency resolution and time resolution (which is related to the Heisenberg uncertainty principle) limit its usability. To resolve the disadvantages of the STHT method, the wavelet approach was developed as an adjustable window frequency spectral analysis method. The basic wavelet function can be modified to meet special needs, so the wavelet transform produces outputs with good resolution in the low-frequency range and good time resolution in the high-frequency range [18–20]. In the region relevant for rotational-machine fault diagnosis, wavelet-based decomposition has been widely used to apprehend the useful components of a vibration signal in a non-stationary condition (in this context, *non-stationary* is the notion that the sideband frequency information of a vibration signal is time-variant). Wavelet transform decomposes a vibration signal into many sub-bands that express the time-frequency distribution through the dilation and transition of the mother wavelet. The sub-bands that contain fault-related intrinsic features can then be used in the fault diagnosis process [21,22]. Nevertheless, the efficiency of the wavelet-based method correlates with the basic wavelet function, so informative components that do not correlate as well with the applied wavelet could be missed or lost in the transformed outcome. In addition, the white noise that is frequently parasitic in a vibration signal and appears across the whole range of the frequency spectrum gives correlated oscillations with a high potential to appear as excitation. In this paper, the effect of noise on the wavelet applied to the wavelet transform method is compared with the proposed method for processing the signal path in the experimental results.

The Hilbert-Huang transform (HHT) was introduced as a better methodology for analyzing non-linear and non-stationary signals [23]. This technique is now often used for rotational-machine fault diagnosis [24–27]. The HHT method uses a time adaptive operation known as empirical mode decomposition (EMD) to decompose the signal into a group of complete and orthogonal components, denoted as intrinsic mode functions (IMFs), that represent the intrinsic oscillation modes of the fault-related components of a vibration signal. The HHT method was shown to outperform wavelet transform in rotational-machine fault diagnosis in [28–30]. To capture the advantages of the HHT, several fault detection tools combine EMD with other methods, such as envelope analysis and the wavelet-based technique. EMD and the envelope analysis combine in series: the vibration signal is first decomposed by EMD to determine the number of IMFs; the envelope analysis then processes the IMFs to monitor for fault-related components. Compared with previous methodologies, this combined technique had better results [28,31,32]. Combining wavelet transform and EMD for time-frequency analysis is another currently used combination method. It takes advantage of the strong points of the two techniques and minimizes their limitations, particularly aliasing in the high-frequency band (wavelet transform) and difficulties in isolating the signals within the second harmonic (EMD) [33,34].

However, the EMD technique is sensitive to noise, so noise-related IMFs, which are not useful, can be generated by the EMD. As illustrated by Van M. et al. [35], EMD performs well in processing low-noise vibration signals and poorly in processing high-noise signals. In other words, even EMD combination techniques are unreliable in noisy environments. Therefore, to effectively apply enhanced signal analysis techniques to non-stationary vibration signals, a proper pre-processing method to reduce noise is required, such as narrowband demodulation [36] or discrete wavelet transform (DWT) [37,38]. Applying those de-noising methods effectively reduces the measurement noise, but the original informative signal is also distorted by the attenuation of a narrow bandpass filter in narrowband demodulation or the threshold in DWT-based de-noising. In other words, using one of those noise reduction methods can degrade the performance of a fault diagnosis system. Therefore, we have developed a new de-noising technique to reduce the noise from an original vibration signal by optimizing a process for filtering the weights and parameters of the reference input signals (adaptive) and considering the noise characteristics and rotational speed. We call our new technique the Adaptive Noise Reducer–based Gaussian Reference Signal (ANR-GRS).

The adaptive noise-controlling technique reduces noise by means of destructive interference. It consists of an adaptive filter and reference signals. The adaptive noise filter is a digital filter with an adaptive algorithm that adjusts the filtering coefficients (or tap weights) so the filter can be flexibly and optimally operated in unknown conditions with non-stationary signals to effectively remove low-level noise [39]. The typical performance criterion for adjusting the filtering weights (convergence condition) is based on the error signal, which is the difference between the output of the filter and the input reference signal as determined using the recursive least-squares or least mean square (LMS) algorithm. Between them, the LMS is more widely used because of its robustness and simplicity [40]. The ANR-GRS technique has three main function blocks: Gaussian reference signal (GRS) generation, adaptive noise filtering using the LMS algorithm, and optimal output sub-band selection. The generated GRS is a special signal consisting of a white-noise reference signal and a Gaussian reference signal with adjustable parameters (mean value and standard deviation) to identify noise components that are independent of the informative components in the frequency domains of the vibration signal from a varying speed gearbox. The adaptive noise filter consists of an M-tap digital Finite impulse response (FIR) filter and the LMS adaptive algorithm; it has two inputs: a reference input for the GRS signal with specific parameters and the desired input for a vibration signal. The noise-reduced sub-band is achieved as the output of the adaptive noise filter. The optimal output sub-band selection adjusts the parameter of the GRS signals to receive the set of noise-filtered sub-bands output by the adaptive filters and then selects the sub-band with the minimum mean square as the optimal sub-band, which is the final output of the ANR-GRS module. That output becomes the input for the feature pool configuration process used to extract the statistical features in the time and frequency domains of the vibration signal as feature vectors [41] to be classified.

The heterogeneous feature pool improves the efficiency of gearbox fault expression for fault diagnosis process; however, the high dimensionality of the feature vectors can be a challenge for various machine learning techniques that can be used for decision making. In comparison with the other artificial intelligence algorithms, the classification performance of support vector machines (SVM) classifier is not much sensitive to the dimensionality of the feature vectors, in other words, this algorithm is not affected by the problem called 'curse of dimensionality'. Furthermore, SVM demonstrates excellent generalization performance, so this technique is capable of achieving high accuracy while classifying mechanical faults in rotation machinery [42]. Also, with an appropriate kernel function, SVM can accurately separate the non-linear datasets by hyperplanes in high-dimensional feature space using the non-linear mapping [43]. SVMs are widely used for fault diagnosis in many real-world applications [44]. They were originally designed for binary classification and then improved for multiclass classification using the one against one, one against all, or hierarchical strategy. Among them, the one against one strategy is the most reliable for our purposes [45–47]. Therefore, a one-against-one multiclass SVM (OAOMCSVM) is used in this proposed methodology.

The new hybrid technique employs the ANR-GRS, which produces an optimal sub-band, and then uses a machine-learning classification of fault types based on the OAOMCSVM on features extracted from that optimal subband to identify faults in a gearbox system. The experimental results show that the proposed method outperforms the aforementioned denoising methods, which verifies that the "clean" input can be used to produce correct output from the signal processing and classification paths.

The rest of this paper is organized as follows: Section 2 provides the characteristics of a gearbox vibration signal and the experimental test setup used in this study. The proposed methodology is explained in detail, from theory to the construction of the ANR-GRS, feature pool configuration, and OAOMCSVM classification, in Section 3. Section 4 demonstrates our experimental results in signal processing and classification. Finally, Section 5 concludes the paper.

2. The Characteristics of a Gearbox Vibration Signal and Experimental Testbed Setup

2.1. The Characteristics of a Gearbox Vibration Signal

The defects of a gearbox can be classified into three major categories: manufacturing defects (tooth profile error, eccentricity of the wheel, etc.), installation defects (parallelism), and operational defects (tooth wear, case wear, tooth spalling, tooth cracks). This research considers operational faults. A one-stage transmission gearbox, which consists of two rigid blocks with a pinion (on the drive side) and a gear (on the non-drive side), is illustrated in Figure 1. A healthy gear in normal condition working smoothly and periodically generates a linear and periodic vibration signal [3]. The vibration signal, $s_h[\text{mV/(m/s}^2)]$, of a fault-free normal pair of gears meshing under a constant load speed can be formulated as [48]:

$$s_h(t) = \sum_{i=0}^{N} S_i \cos(2\pi i f_M t + \varphi_i), \tag{1}$$

where S_i and φ_i are the amplitude and phase of the i-*th* meshing frequency harmonics; f_M is the meshing frequency (for the pinion: P is the number of teeth in the pinion wheel, f_P is the pinion rotation frequency, $f_M = P.f_P$; or $f_M = G.f_G$, G is the number of teeth in the gear wheel, f_G is the gear rotation frequency), and N is the total number of f_M harmonics in the frequency range of a vibration signal. Figure 2a shows the spectrum of the output vibration signal of a fault-free gearbox; it is filled with the frequency tones of the meshing frequency and its harmonics.

In a fault case, when the motion transferred from the drive shaft to the non-drive shaft by the rotation between the pinion wheel and the gear wheel traverses a defective tooth (chipped, worn, or missing), an abnormal movement occurs that changes the impulses in the vibration signals. The vibration signal contains amplitude and phase modulations of the carrier frequency as the meshing frequency; its frequency spectrum includes sidebands, frequency components on two sides of the meshing frequency and its harmonics, as given in Equation (1). Thus, when the gear wheel has a faulty tooth, the velocity of the gear angle changes impulsively within the rotating functionality and generates a non-linear vibration signal in which issues such as speed variation, amplitude, and phase modulation prevail [4]. The vibration signal is formulated [3] as given by Equation (2), and an example of its spectrum is shown in Figure 2b:

$$s_f(t) = \sum_{k=0}^{N} S_k (1 + a_k(t)) \cos(2\pi k f_M t + \varphi_k + p_k(t)) \tag{2}$$

Here, $a_k(t) = \sum_{j=0}^{M} A_{kj} \cos(2\pi j f_G t + \mu_{kj})$ and $p_k(t) = \sum_{j=0}^{M} P_{kj} \cos(2\pi j f_G t + \xi_{kj})$.

A_{kj}, P_{kj} are amplitudes and μ_{kj}, ξ_{kj} are phases of the j-*th* sideband in the amplitude and phase modulation signals, respectively, around k meshing harmonics.

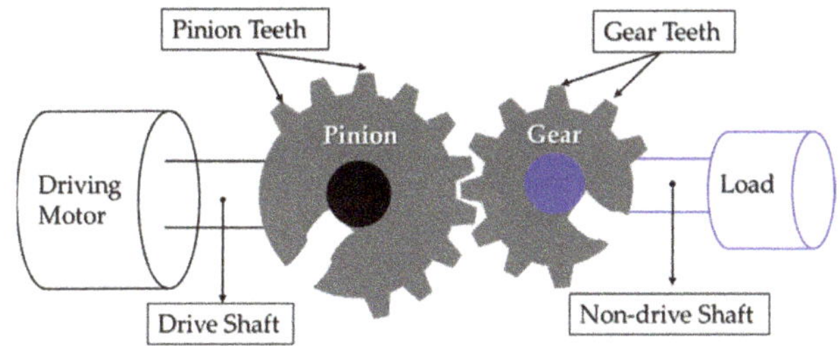

Figure 1. The spur gearbox model.

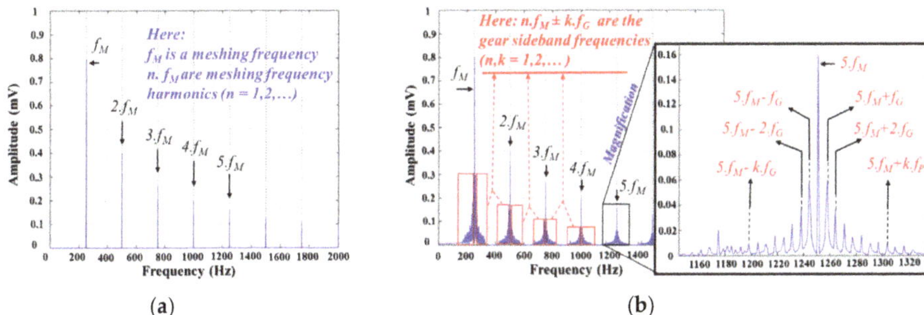

Figure 2. The frequency spectrum of the gearbox vibration signal: (**a**) a healthy gearbox and (**b**) a faulty gearbox.

2.2. The Experimental Testbed Setup

The experimental testbed is illustrated in Figure 3. The pinion wheel is fixed to a three-phase AC induction motor by a drive shaft (DS). The motion (torque) is transmitted from the AC motor to the load as adjustable blades, which are mounted on the end of the non-drive shaft by the engaged teeth of a pinion wheel and a gear wheel (a gearbox with a gear reduction ratio of 1:1.52).

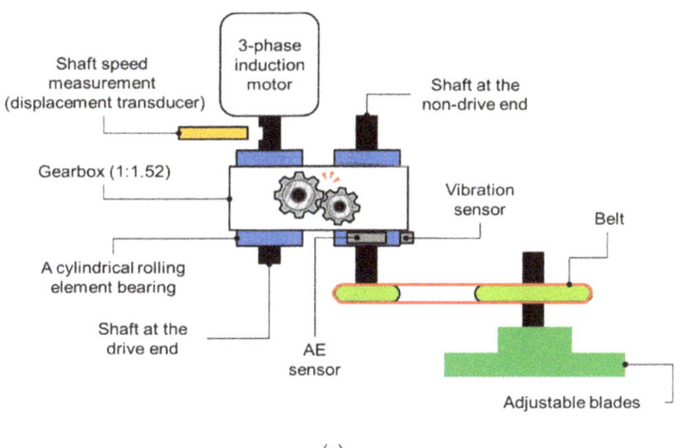

Figure 3. *Cont.*

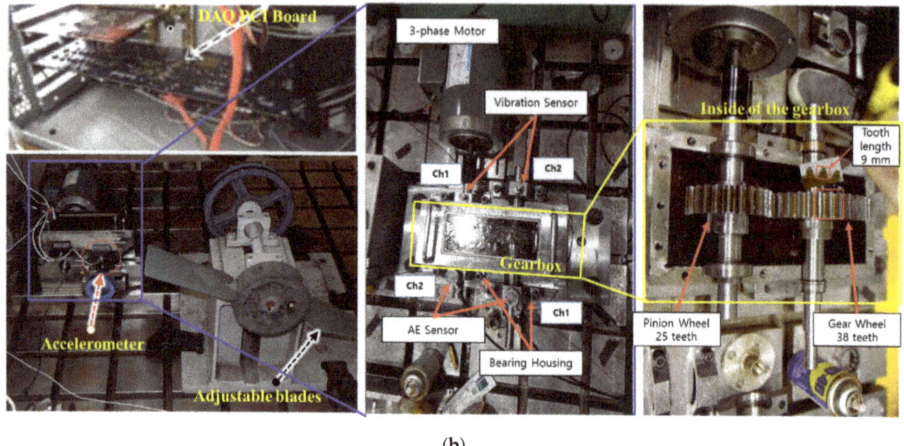

(**b**)

Figure 3. Experimental testbed setup: (**a**) function block diagram; (**b**) actual experimental assembly.

The number of teeth on the pinion wheel is 25 (P = 25), the gear wheel has 38 (G = 38), and the length of each tooth is 9 mm. Figure 4 depicts the seeded tooth failures on the gear wheel: a perfect or healthy gear (H), tooth cut 10% (F1), tooth cut 30% (F2), and tooth cut 50% (F3). To measure the speed of the shaft rotation, a displacement transducer is placed to track the hole in the DS once per rotation. The vibration signals from the gear wheel in the normal condition and three levels of tooth cut defects (shown in Figure 4) were continuously acquired from the vibration sensor (accelerometer 622B01 made by the IMI Sensor Company) mounted on the end of the DS, 72.5 mm from the pinion gear. The analog vibration was digitized using PCI-2 data acquisition (the specifications of the data acquisition system are provided in Table 1). The sample datasets for each health condition (H, F1, F2, F3) of the gearbox under four shaft speeds are provided in Table 2.

Table 1. Specification of the sensors and data acquisition system.

Device	Specification
Vibration sensor (Accelerometer 622B01)	Sensitivity (V/g): 10.2 mV/(m/s^2)
	Operational frequency range: 0.42 to 10 kHz
	Resonant frequency: 30 kHz
	Measurement range: ± 490 m/s^2
4- Channel DAQ PCI Board	18-bit 40MHz AD conversion, a sampling frequency of 65.536 kHz is used for each of two channels simultaneously
Displacement transducer	Distance from the head of a transducer to a hole: 1.0 mm
	Diameter of a hole: 12.80 mm
	Sensitivity: 0 to −3dB
	Frequency response: 0–10 kHz

Each health state is sampled by sampling frequency 65536 Hz in continuous 1 s (1-s sample) repeating by 300 times to receive 300 1-s samples for each shaft speed. Hence, the number of samples for each health state is 1200 vibration samples in four different shaft speeds, the total number of samples in this experimental testbed is 4800 of 1-s samples.

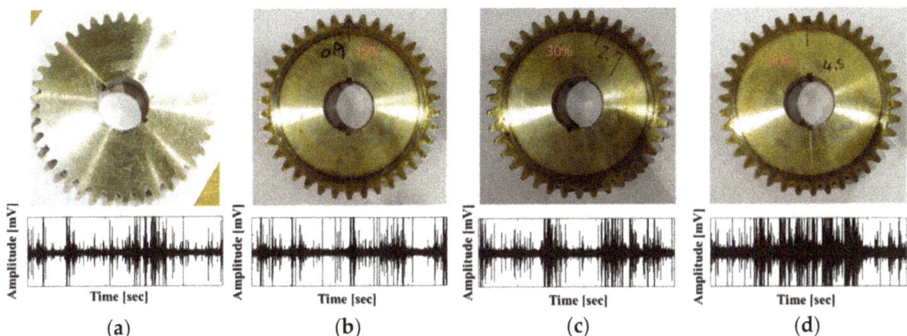

Figure 4. The health states of the gear wheel and examples of vibration signals at the rotation speed of 300 RPM: (**a**) no seeded fault, healthy gear, (**b**) tooth cut 10% (0.9 mm), (**c**) tooth cut 30% (2.7 mm), (**d**) tooth cut 50% (4.5 mm).

Table 2. A detailed description of the fault types and dataset.

Gearbox Health State	Description	Number of 1-s Data Samples Acquired for each Rotation Speed				Sampling Frequency (Hz)
		300 RPM	**600 RPM**	**900 RPM**	**1200 RPM**	
Healthy (H)	No seeded fault in the teeth of a gearbox	300	300	300	300	65536
Fault type 1 (F1)	Pinion tooth cut 10% (0.9 mm)	300	300	300	300	65536
Fault type 2 (F2)	Pinion tooth cut 30% (2.7 mm)	300	300	300	300	65536
Fault type 3 (F3)	Pinion tooth cut 50% (4.5 mm)	300	300	300	300	65536

3. The Gearbox Fault Diagnosis Methodology

A function block diagram of the method for gearbox fault diagnosis proposed in this study is provided in Figure 5. We use four main processing blocks: the sensor and DAQ block, ANR-GRS, feature pool configuration, and multiclass SVM-based classification. To acquire discrete samples of each captured signal event containing information about the defective gearbox in the acquisition dataset, the raw vibration signal was sampled at a high frequency of 65536 Hz to acquire rich digitized vibration sample data under different shaft rotation speeds (300, 600, 900 and 1200 RPM), and the adjustable load was non-stationary. Though the operation frequency range of a vibration sensor in this study is from 0.42 Hz to 10 kHz (this is presented in Table 1), thus fault-related components in the frequency domain of vibration signals mostly exist in the lowest segment 0–10 kHz of their frequency spectrums. Therefore, the sampling frequency of the raw vibration signal (all vibration signals acquired in this paper) was reduced by a factor of three using a down-sampling technique. However, implementing decimation involves aliasing, so a low-pass Chebyshev Type I Finite Impulse Response filter (the filter with the order of 35 and a cut-off frequency of 10 kHz) was used for antialiasing [49]. The output sub-band signal from the lowpass filter (lpf(n), n is denoted as discrete-time) which have frequency spectrums in the range from 0–10 kHz, were then optimized by the ANR-GRS to achieve the optimal sub-band, opt(n), from which twenty-one features were extracted through a feature pool configuration, F(k), for classification by the OAOMCSVM.

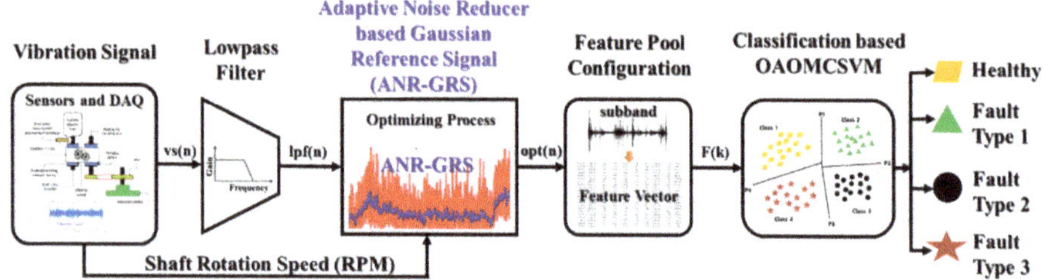

Figure 5. Function block diagram of the proposed methodology.

3.1. Adaptive Noise Reducer–based Gaussian Reference Signal

3.1.1. Adaptive Noise Filtering Technique

The Digital Filter

An adaptive filter combines the operation of a digital filter and an adaptive algorithm. The adaptive algorithm optimizes the coefficient (or weight) of a digital filter by using the feedback signal from the output (error signal) according to the signal condition or performance criteria [38]. Figure 6 illustrates the function of an adaptive filter constructed using a FIR filter and an adaptive algorithm. The output of the FIR filter is calculated as given in Equation (3):

$$g(n) = \sum_{m=1}^{M} c_m(n)r(n-m) = c^T(n)r(n) \tag{3}$$

where, c_m, m = 0,1, … , M−1 (M is the digital filter length) are the adjustable weights (coefficients) of the filter, which do not depend on the sample time. The weight vector (M × 1) is formed as:

$$c(n) \equiv [c_0, c_1, \ldots, c_{M-1}]^T, \tag{4}$$

and r (n−m), m = 0, 1, … , M−1 are samples of an input signal composed of the vector M × 1:

$$r(n) \equiv [r(n), r(n-1), \ldots, r(n-M+1)]^T. \tag{5}$$

T denotes the transpose operation of the matrix. Then, the error signal e(n) is the difference between the FIR filter response, y(n), and desired signal, d(n), which can be calculated as:

$$e(n) = d(n) - g(n) = d(n) - c^T(n)r(n). \tag{6}$$

A common criterion for tuning the convergence of the weight vector, c(n), is the minimization of the mean-square error (MSE):

$$\begin{aligned} J &\equiv E\{e^2(n)\} = E\{[d(n) - c^T(n)r(n)]^2\} \\ J &= c^T(n)Rc(n) - 2Pc^T(n) + E\{d^2(n)\}, \end{aligned} \tag{7}$$

where, $R \equiv E\{r(n)r^T(n)\}$ is the input autocorrelation matrix, and $P \equiv E\{r(n)d(n)\}$ is the cross-correlation vector between the input signal and the desired signal vector.

Equation (7) indicates that the MSE is a quadratic function of the filter weights (c), and its performance surface guarantees that it has a single global minimum MSE corresponding to the optimal vector c_o. The optimal vector c_o can be found by taking the first derivative of Equation (7) and setting it to zero, the result achieved by Wiener-Hopf equation (assuming that R has an inverse matrix):

$$c_o = R^{-1}P, \tag{8}$$

so that the minimum MSE is:

$$J_{min} = E\{\mathbf{d}^2(n)\} - P^T\mathbf{c}_o. \tag{9}$$

Adaptive Algorithm

The adaptive algorithm is a recursive function to automatically adjust the coefficient vector, $c(n)$, to minimize MSE (J_{min}) so that the weight vector converges to the optimum solution, $\mathbf{c}_o$, after iteration loops. Both the LMS and recursive least-squares algorithms can be used to fetch the optimal solution [39], but the LMS is the most broadly used. To calculate the updated weight vector in the recursive loop, the LMS algorithm is based on the steepest-descent procedure using a negative gradient of the instant square error, which was devised by Widrow and Stearns [50] as follows:

$$c(n+1) = c(n) + \mu r(n)e(n), \tag{10}$$

where μ is the step size (or convergence factor) that determines the stability and convergence rate of the LMS algorithm. The algorithm adapts the weight vector to the optimal Wiener-Hopf solution ($\mathbf{c}_o$) given in Eqaution (9) by an iterative process with the convergence factor. The step size is selected in the range [40]:

$$0 < \mu < \frac{2}{MS_u}, \tag{11}$$

where S_u is the average power of the input signal $\mathbf{r}(n)$.

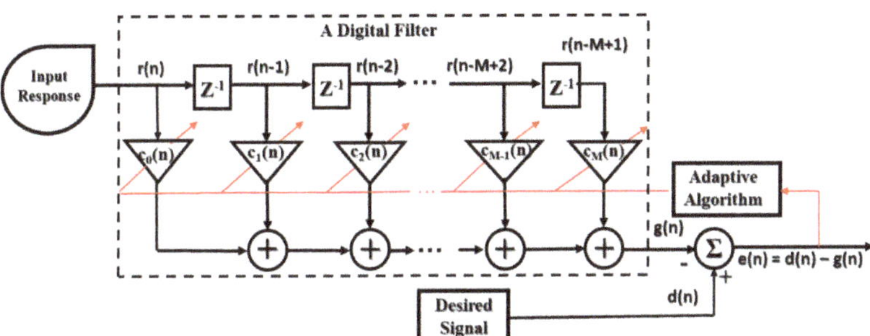

Figure 6. Function block diagram of an adaptive filter.

Adaptive Noise Filtering Technique Applied to a Vibration Signal

To construct the adaptive noise filter, the noise reference signal and observed signal are applied as the input signal of an adaptive filter (the input response in Figure 6) and the desired signal (d(n) in Figure 6), respectively. The observed signal is the vibration signal acquired from the accelerometer sensor and digitized by the DAQ block reflecting gearbox behavior as expressed by the informative signals (s(n)) and the noise (w(n)), as shown in Figure 7. As explained in Sections 1 and 2, the informative signals and noise are formed by different sources: the informative signal comes from the vibration of the gear and pinion teeth, whereas the noise comes from the measurement system, unrelated gearbox components, and mechanical resonances. Therefore, the informative signal and noise represent independent processes ($E\{\mathbf{s}(n)\mathbf{w}(n)\} = 0$). To implement an effective adaptive noise filtering system, the generated noise reference signal, r(n), should meet two conditions (A and B):

(A) The generated noise reference, r(n), and informative signal, s(n), are uncorrelated and independent ($E\{r(n)s(n)\} = 0$)

(B) The characteristics of the generated noise reference, r(n), and noise, w(n), are homologous as much as possible.

When those conditions are met, the MSE of the adaptive noise filter can be calculated as follows:

$$E\{e^2(n)\} = E\{[s^2(n) + (w(n) - r_o(n))]^2\}, \tag{12}$$

where $r_o(n) = c^T(n)r(n)$, and the adaptive filter uses an FIR filter,

$$E\{e^2(n)\} = E\{s^2(n)\} + E\{[w(n) - c^T(n)r(n)]^2\}. \tag{13}$$

The informative signal is independent of both the noise ($E\{s(n)w(n)\} = 0$) and the generated noise reference ($E\{r(n)s(n)\} = 0$). By implementing the LMS adaptive algorithm to adapt the filter coefficient vector, $c(n)$, to the optimal vector, c_o, the mean square of the output signal (error signal) approaches the single minimum of the performance surface. From Equation (13), the minimum MSE is taken to be:

$$\min_{c(n)} E\{e^2(n)\} = E\{s^2(n)\} + \min_{c(n)} E\{[w(n) - c^T(n)r(n)]^2\} \tag{14}$$

Therefore, the output signal of the adaptive noise filtering system carries the complete informative part of the gearbox vibration signal throughout the whole process of algorithm implementation. In addition, the noise integrated into the vibration signal is reduced; in the ideal case, the noise is removed ($\min_{c(n)} E\{[w(n) - c^T(n)r(n)]^2\} = 0$). Therefore, for adaptive noise control, we implement the ANR-GRS.

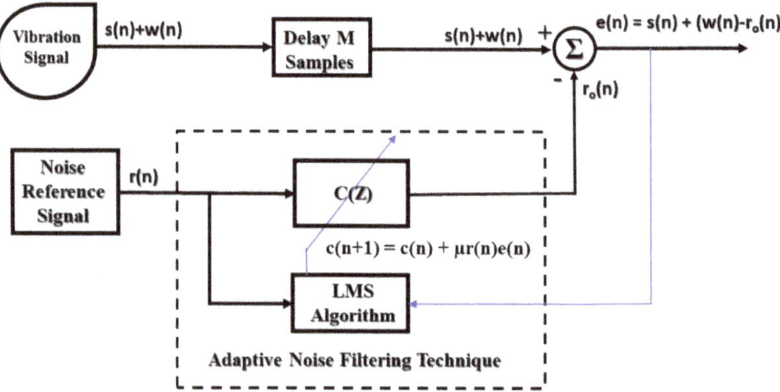

Figure 7. Function block diagram of an adaptive noise filtering technique.

3.1.2. ANR-GRS

In this paper, the noise ($w(n)$) in the gearbox vibration signal is divided into two types: white noise ($u(n)$) and band noise ($b(n)$). The white noise arises from the measurement system: the amplifier, detector, DC power supply, thermal vibration of the semiconductor atoms, etc. In the frequency domain, the power of the white noise is spread across the whole frequency spectrum of the vibration signal (theoretically, the power of white noise is spread from -∞ to ∞ in the frequency axis) [8]. Band noise, on the other hand, represents noise caused by unrelated components [7]. The frequency harmonics of the band noise are distributed around the informative components of the gear sideband frequency, meshing frequency, and their harmonics. Therefore, the informative signal inside the vibration signal is separately independent of both types of noise. The ANR-GRS module is built using the adaptive noise filtering technique and reference noise-related generation signals, as illustrated in Figure 8. To reduce the white noise, we apply a generated white noise signal with a uniform, random distribution function ($v(n)$). The oscillation form of the generated white noise is thus analogous to the white noise integrated into the vibration signal. Because its frequency spectrum is within the observed frequency range, the maximum level of the power spectrum average (PSA) of the reference white noise is reduced to less than 10% (10% in this study) of the PSA of the vibration signal to ensure that the informative signal can be eligible for conditions A or B. The GRS ($g(n)$) is created to adapt to the band noise inside the vibration signal. To make the proposed methodology as an invariant model, the GRS generation module uses the shaft rotation speed (RPM) information from the displacement transducer and the vibration signal as the input parameters. Then, the mean frequency (F_{Center}) and the standard deviation of the GRS are calculated based on the frequency of the defective wheel, which is a function of RPM (the gear frequency in this paper). The GRS window is confined entirely within the frequency space between two consecutive sideband frequencies (a sideband segment), pictorially described in Figure 9, and computed as follows:

$$W_{GRS}(k) = \sum_{k=1}^{N_b} e^{-\frac{(k-F_{Center})^2}{2\Delta}}, \tag{15}$$

where $\Delta = \sigma^2$ is the variance, σ is the standard deviation of the GRS window, and F_{Center} is the mean frequency of the GRS window. They function as the frequency of a faulty wheel (the gear frequency, $f_G = P.RPM/G$ in this research).

$$F_{Center} = \alpha.f_{FW}. \tag{16}$$

By linearization of the Gaussian function, the standard deviation (the characteristic of a Gaussian distribution) can be approximately calculated as:

$$\sigma = 0.318.F_{Center} = 0.318.\alpha.f_{FW}. \tag{17}$$

N_b is the number of frequency bins in a sideband segment and defined as follows:

$$N_b = \frac{2N_s}{F_s}.f_{FW}, \tag{18}$$

where N_s is the number of samples of the vibration signal, F_s is the sampling frequency of the vibration signal, and f_{FW} is the frequency of the faulty wheel (the gear frequency, f_G, in this paper).

To qualify condition B, the frequency components of a Gaussian window are separated from the informative frequencies (sideband frequencies). A Gaussian window is placed completely inside the space between two continual sideband frequencies in the visualization. Thus, the adaptation process for a band-noise reduction is to preserve the original informative frequency component (significantly reducing the noise components and causing negligible attenuation of the informative components). From Equations (15)–(17) and Figure 9, the coefficient α is selected in the range from 0.25 to 0.75, and the qualified Gaussian window signals are generated using the parameters in the following ranges:

The range of the mean value:

$$0.25. f_{FW} \leq F_{Center} \leq 0.75.f_{FW}. \tag{19}$$

The range of the standard deviations of the Gaussian reference signal:

$$\sigma = \begin{cases} 0.318.\alpha.f_{FW} & \text{when } 0.25 \leq \alpha \leq 0.5 \\ 0.318.(1-\alpha).f_{FW} & \text{when } 0.5 < \alpha \leq 0.75 \end{cases} \tag{20}$$

Therefore, the implementation of a stepping adjustment in the coefficient α drives a change in the mean value and standard deviation (the position and shape) of the Gaussian window, which defines the condition for fetching the optimal Gaussian window, as illustrated in Figure 9.

3.1.3. The Process for Calculating the Optimized Subband

First, the ANR-GRS algorithm germinates the initial parameters for the Gaussian signal generation module: starting value of $\alpha = 0.25$ in this paper, adaptive filter (M-tap, M = 40 in this study), coefficient vector $c(n) = [0, 0, \ldots , 0]$, and step size μ ($\mu = 0.01$). The parameter α is scanned in the range [0.25 0.75] in steps of 0.01 in company with the input rotation speed (RPM) to compute the F_{Center} (mean value) and standard deviation (σ) using Equations (16) and (17). To generate the specific GRS needed for the reference input of the adaptive filter $r(n)$, the output of the adaptive filter is connected to the minus port of the summation module, $r_o(n)$. The vibration signal, which contains both the informative component and noise, is entered as the desired input and delayed for M sampling time steps to be compatible with the delayed processing of the FIR digital filter. The LMS algorithm adjusts the coefficient vector to receive the LMS of the error, which is the output of the summation module. The output error signal, which has LMS (and to which the optimal coefficient vector is set), is pushed into the set of proposed optimized sub-bands.

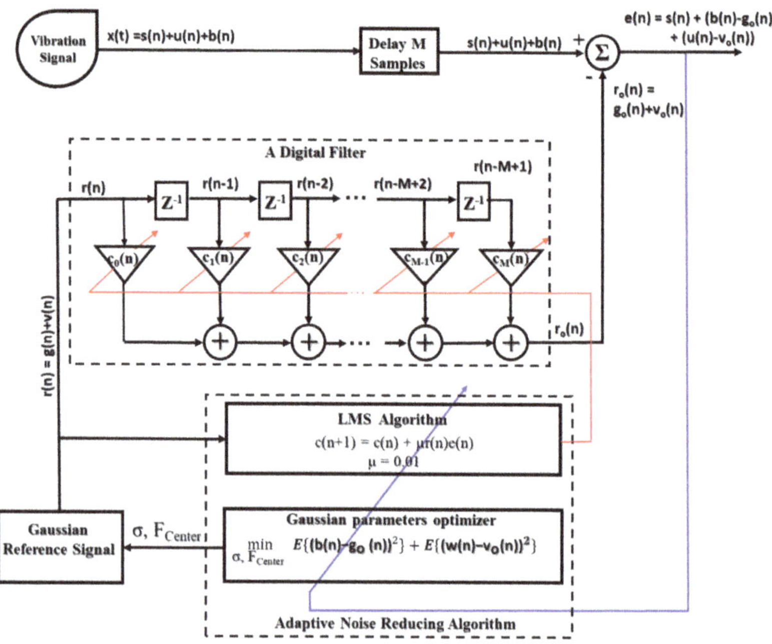

Figure 8. A function block diagram of the ANR-GRS module and parameter adjustments.

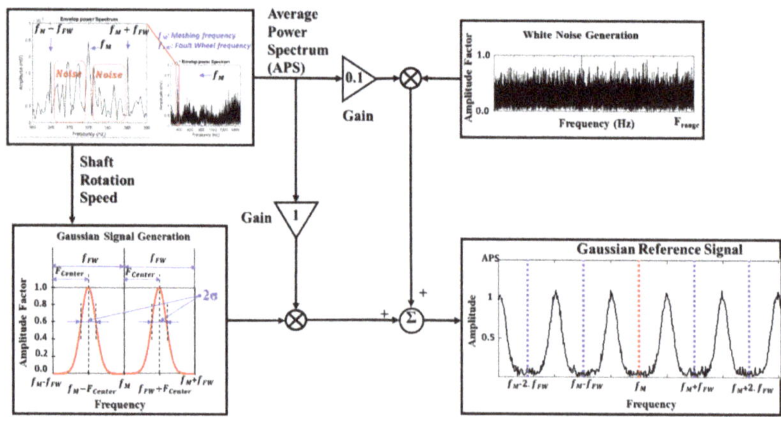

Figure 9. The overall flow chart of GRS signal generation for the ANR-GRS module.

Finally, the algorithm calculates the mean square value of each sub-band in that set and then selects the sub-band with the minimum value as the optimized sub-band and output of the AND-GRS module (Figure 10).

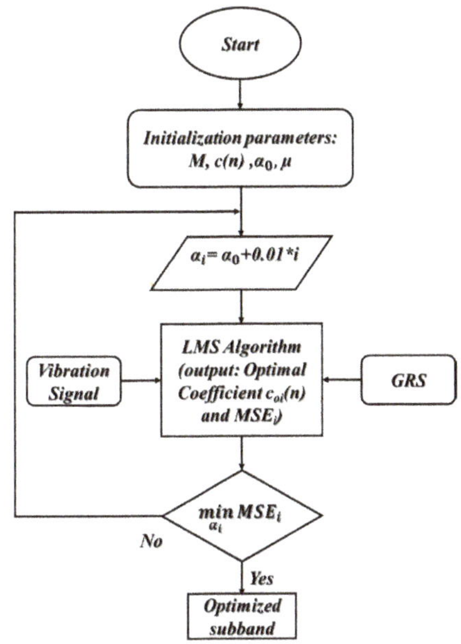

Figure 10. The algorithm flow chart of the ANR-GRS module.

3.2. Feature Pool Configuration

We found the ANR-GRS methodology to be highly effective in reducing most of the noise components from a 1-s raw vibration signal while leaving the information about gearbox faults intact. The optimized sub-band output from the ANR-GRS, i.e., the "clean" signal presenting the characteristics of the gearbox component vibration with trivial noise effects, carries the intrinsic fault symptoms of the cut tooth defects. We then use those optimal sub-bands, rather than the raw 1-s vibration signals, to extract features. According to Caesarendra et al. [51], the statistical parameters from the time and frequency domains of the signal are congruent and subservient for fault classification using machine learning. Table 3 displays twenty-one features, eighteen time-domain features (e.g., root means square, square mean root, kurtosis, skewness, margin, impulse, and peak-to-peak value) and three frequency-domain features (root mean square frequency, frequency center, and root variance frequency) for each optimal sub-band. The feature pool dimensionality is $N_{HS} \times N_{1\text{-SEC}} \times N_F$, where N_{HS} is the number of gearbox health states (number of classes) that need to be classified (4 classes in this study: healthy, pinion tooth cut 10%, pinion tooth cut 30%, and pinion tooth cut 50%), $N_{1\text{-SEC}}$ is the number of 1-s samples of each class (300 in this study), and N_F defines the number of features (21 in this study). Therefore, groups of 21-feature vectors were considered as the validating input dataset for our proposed intelligent fault-detection method based on a multiclass SVM.

3.3. Gearbox Fault Classification Using a Multiclass SVM Classifier

The principle operation of an SVM is based on the statistical learning theory of Vapnik [45] and quadratic programming [46]. It was actually designed to classify binary datasets by finding the optimal plane, generally called the *hyperplane*, with the largest margin-gap separating it from both binary classes. Let $\{(x_m, y_m), m=1, 2, \dots, M\}$ be the given training dataset with M samples, where each sample data $x_m \in \mathbb{R}^D$, $\mathbb{R}^D$ is a D-dimensional feature vector, and y_m ($y_m \in \{-1, +1\}$) are the class labels. The SVM is used to find a set of linearly separable hyperplanes between two classes and maintain the maximum distance (called the *margin*) from both of them.

Table 3. Definition of statistical features in the time and frequency domains.

Features	Equations	Features	Equations	Features	Equations
Peak	$\text{Max}(|s|)$	Shape factor	$\dfrac{s_{rms}}{\frac{1}{N}\sum_{n=1}^{N}|s_n|}$	Mean ($\bar{s}$)	$\dfrac{1}{N}\sum_{n=1}^{N} s_n$
Root mean square (s_{rms})	$\sqrt{\dfrac{1}{N}\sum_{n=1}^{N} s_n^2}$	Entropy	$-\sum_{n=1}^{N} p_n.log_2(p_n)$	Shape factor square mean root	$\dfrac{s_{srm}}{\frac{1}{N}\sum_{n=1}^{N}|s_n|}$
Kurtosis	$\dfrac{1}{N}\sum_{n=1}^{N}\left(\dfrac{s_n-\bar{s}}{\sigma}\right)$	Skewness	$\dfrac{1}{N}\sum_{n=1}^{N}\left(\dfrac{s_n-\bar{s}}{\sigma}\right)^3$	Margin factor	$\dfrac{\max(s)}{s_{smr}}$
Crest factor	$\dfrac{\text{Max}(|s|)}{s_{rms}}$	Square mean root (s_{smr})	$\left(\dfrac{1}{N}\sum_{n=1}^{N}\sqrt{|s_n|}\right)^2$	Peak to peak	$\max(s)-\min(s)$
Clearance factor	$\dfrac{\text{Max}(|s|)}{s_{smr}}$	5th normalized moment	$\dfrac{1}{N}\sum_{n=1}^{N}\left(\dfrac{s_n-\bar{s}}{\sigma}\right)^5$	Kurtosis factor	$\dfrac{Kurtoris}{s_{rms}^4}$
Impulse factor	$\dfrac{\text{Max}(|s|)}{\frac{1}{N}\sum_{n=1}^{N}|s_n|}$	6th normalized moment	$\dfrac{1}{N}\sum_{n=1}^{N}\left(\dfrac{s_n-\bar{s}}{\sigma}\right)^6$	Energy of signal	$\sum_{n=1}^{N} s_n^2$
Frequency center (FC)	$\dfrac{1}{N_f}\sum_{f}^{N_f} S(f)$	Root mean square frequency	$\sqrt{\dfrac{1}{N_f}\sum_{f}^{N_f} S(f)^2}$	Root variance frequency	$\sqrt{\dfrac{1}{N_f}\sum_{f}^{N_f}(S(f)-FC)^2}$

Here is an input signal (i.e., optimized subband), N is the total number of samples, S(f) is the magnitude response of the fast Fourier transform of the input signal s, N_f is the total number of frequency bins, $\sigma = \sqrt{\frac{1}{N}\sum_{n=1}^{N}(s_n-\bar{s})^2}$, and $p_n = \dfrac{s_n^2}{\sum_{n=1}^{N} s_n^2}$.

The hyperplane, denoted as w, is determined as the maximized width of the margin and the minimized structural risk, given by:

$$(w,b) = \underset{w,b}{\arg\min}\frac{1}{2}w^T w + C\sum_{m=1}^{M} \xi_m, \tag{21}$$

subject to: $y_m(w^T\psi(x_m)+b) \geq 1 - \xi_m, \forall m = 1, 2, \ldots, M; -\xi_m \leq 0, \forall m = 1, 2, \ldots, M$

Here, b is bias, C is the trade-off parameter, $\xi = \{\xi_1, \xi_2, \ldots, \xi_N\}$ is the set of slack variables, and $\psi(.)$ is a feature vector in the expanded feature space. Equation (21) can be solved by applying the Lagrange duality solution [44] as shown below:

$$\underset{\alpha}{\arg\max}\, w(\alpha) = \sum_{m=1}^{M}\alpha_m - \frac{1}{2}\sum_{m=1}^{M}\sum_{k=1}^{M}\alpha_m \alpha_k y_m y_k \psi^T(x_m)\psi(x_k), \tag{22}$$

subject to: $\sum_{m=1}^{M}\alpha_m y_m = 0, 0 \leq b_m \leq C, \forall m = 1, 2, \ldots, M$ where, α_m and α_k are Lagrange multipliers, x_m and x_k are two input training vectors, and $K(x_m, x_k) = \psi^T(x_m)\psi(x_k)$ is a kernel function used to map the input data space into a higher-dimensional feature space. Several kernel functions, such as linear, polynomial, Gaussian, radial basis, and sigmoid functions, can be used in SVM classification methods. Countless classification applications have more than two classes in their datasets and thus require a solution beyond the binary SVM just described. Multiclass SVMs have been developed to classify datasets of N different classes (N > 2), and they use one of three structures: one-against-one, one-against-all, and hierarchical. Among those structures, OAOMCSVM requires more classifiers than the others, but it also has the most reliable classification accuracy [45]. Therefore, we use OAOMCSVM, illustrated in Figure 11, in the methodology proposed in this paper.

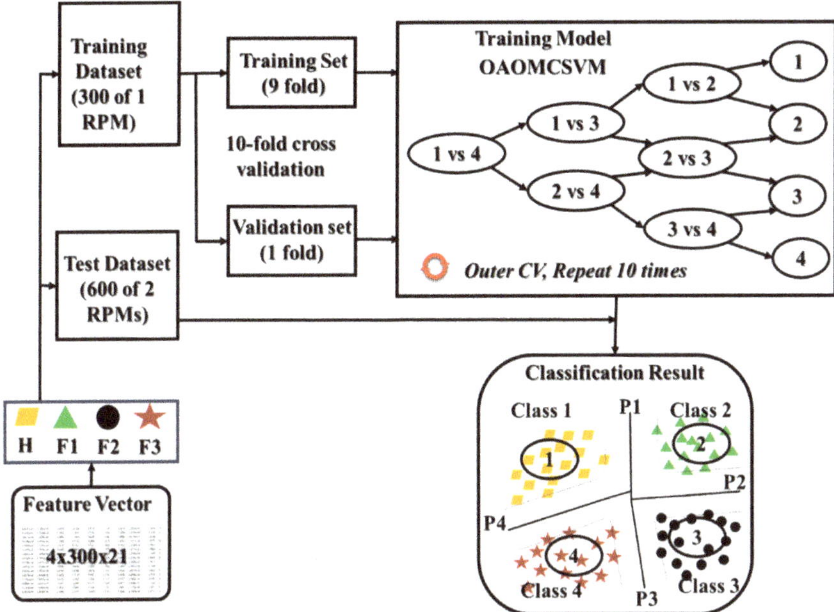

Figure 11. The classification methodology of OAOMCSVM.

4. Experimental Results

To verify the advantage of the ANR-GRS module in the proposed methodology, we implemented experiments in two technological zones, signal processing and features dataset classification, and compared our results with those from conventional methodologies.

4.1. Signal Processing Experimental Results

The 1-s vibration signals acquired using the experimental testbed described above contained the phase and amplitude modulation signals bearing information about the health states of a gearbox. To investigate the effectiveness of the proposed noise reduction technique, the experimental dataset was collected under various shaft rotation speeds that are equal to 300, 600, 900, and 1200 RPM, respectively. The vibration signals output from the accelerometer are analog, so they were digitized with a high holding sample frequency of 65536 Hz to gather as much information (and noise) as possible on the wideband PCI-based data acquisition board (Table 1). Each 1-s digital vibration signal was down-sampled three times, incorporating a lowpass filter for antialiasing to output a digital vibration signal realistically compatible with the working range frequency of the accelerometer (0–10 kHz, Table 1). Then, the vibration signal was input into the ANR-GRS module (Figure 5). The shaft rotation speed (RPM), measured by the displacement transducer, was observed by the ANR-GRS module according to appropriate vibration signal data to generate the Gaussian reference signal. The optimal subbands were the output of the ANR-GRS module.

To demonstrate the superiority of the ANR-GRS technique, we compared its optimized subbands with the outputs of other signal processing approaches for noise reduction: the Hilbert transform (HT), window bandpass filter (WBF), and wavelet transform with optimal subband-based maximum kurtosis (WTK). We tested those approaches by replacing the ANR-GRS module with them. Figure 12 illustrates the frequency spectra compared with the input vibration signal. Figure 12a shows the output of a lowpass filter that received a 1-s vibration sample with 900 RPM (15 Hz) of fault type 2 (meshing frequency, $f_M = P.RPM = 25.15 = 375$ Hz and sideband gear frequency, $f_G = P.RPM/G = 9.87$ Hz, shown as lpf(n) in Figure 5 and labeled as the OutLPF signal in Figure 12a). The output signals from the noise-reduction modules are shown in Figure 12b (OutHT signal), Figure 12c (OutWBF signal), Figure 12d (OutWTK signal), and Figure 13, the proposed ANR-GRS (OptANR signal). The three conventional methods (HT, WBF, WTK) changed the outLPF signal into different shapes and types (the outLPF signal is an amplitude and phase modulation signal) regardless of the fault information (meshing frequency and its harmonics and sideband gear frequencies). HT exalted the area of the low-frequency components, whereas WTK fortified the high-frequency components in the frequency spectrum (Figure 12b,d). WBF was better than the HT and WTK methods because it filtered the noise in some of the meshing frequency harmonics and sideband gear frequencies, but it also reduced or removed significantly informative frequency components (Figure 12c). The outANR signal (Figure 13), the output signal from the ANR-GRS module proposed here, fulfilled the needs of signal processing: reducing the noise components and preserving the original informative components. It made the vibration signal from the gearbox "cleaner" (lowered the noise) and approached the characteristics of the gearbox vibrations signal presented in Section 2.1. This comparison verifies that our accurate is a suitable technique for reducing the noise in gearbox vibration signals and returning an honest reflection of the health states of a gearbox along an electronic signal path.

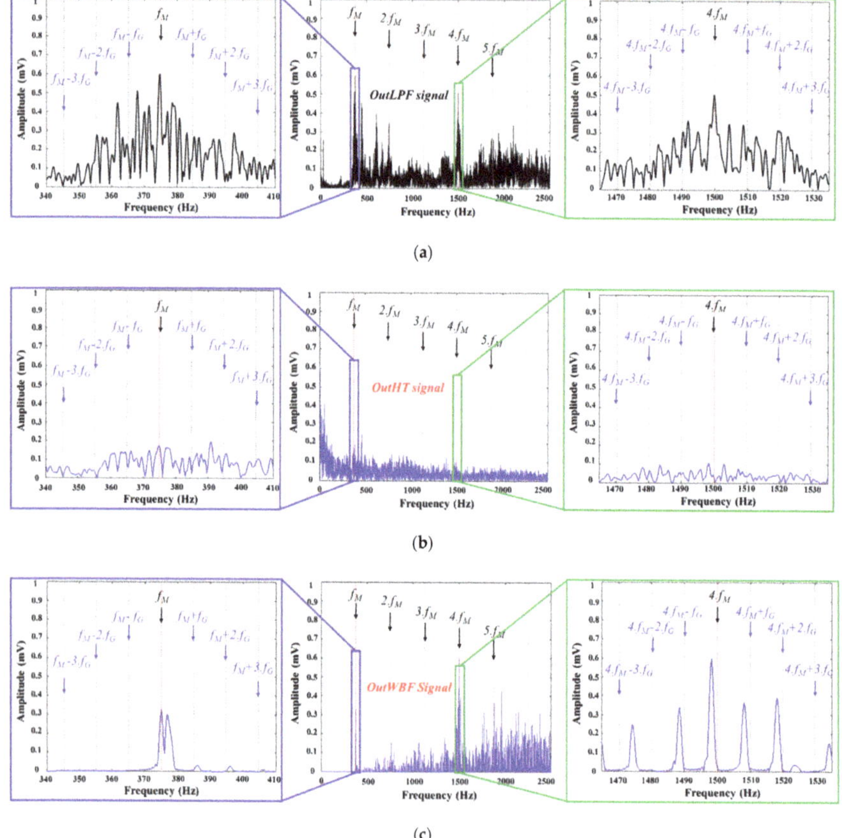

Figure 12. *Cont.*

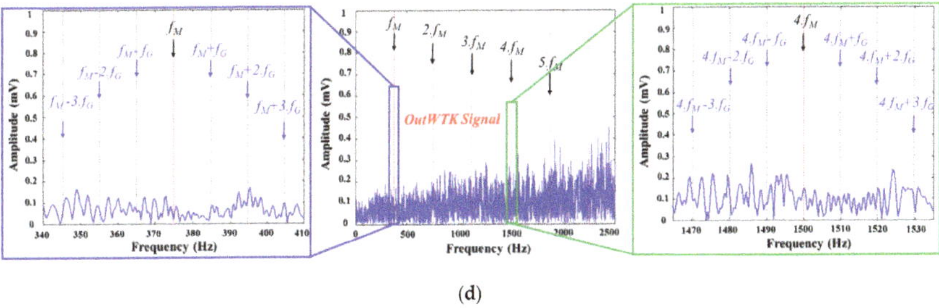

Figure 12. The frequency spectrum analysis for the state-of-the-art methodologies: (**a**) the input signal, (**b**) the output signal from the Hilbert transform module, (**c**) the output signal from the window bandpass filter module, and (**d**) the output signal from the wavelet transform WTK module.

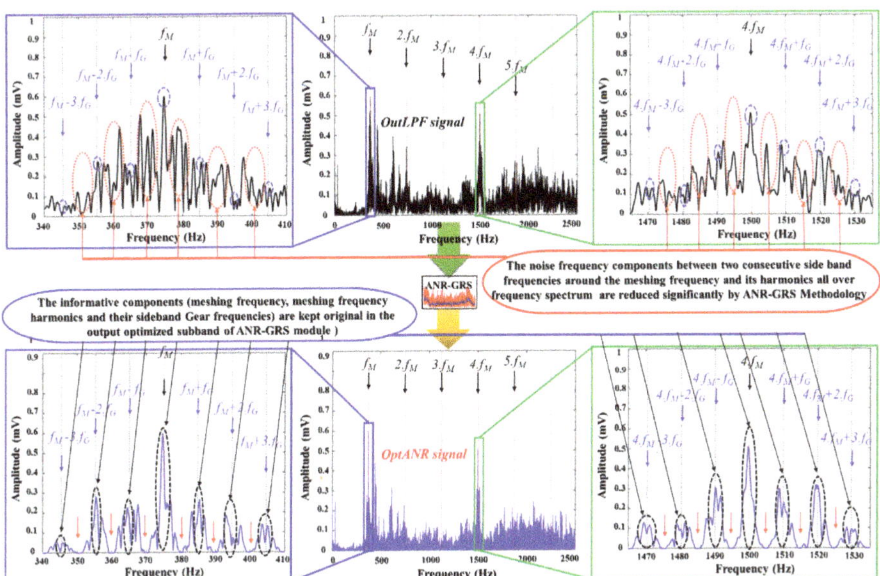

Figure 13. Frequency spectrum analysis of the input and output signal of the ANR-GRS module.

4.2. Classification Results

The classification performance of the proposed methodology is evaluated in two experiments. At the beginning, the feature sets of the dimensions $4 \times 300 \times 21$ (300 samples from each of four health states, as shown in Table 2) for each speed (in this experiment, 300 RPM, 600 RPM, 900 RPM, and 1200 RPM) were selected. Then, in the first experiment, the dataset is created by merging the data of four available health states under different rotating speeds resulting in the new feature set of the dimensionality 4800×21. This dataset was then randomly divided into a train and a test sets at ratio 8:2 for training and testing OAOMCSVM classifier and general evaluation of the proposed fault diagnosis methodology.

Furthermore, to prove the robustness of the proposed ANR-GRS technique, the second experiment is performed where the classifier is trained on data (i.e., feature sets) corresponding to single rotating speed and tested by data instances collected under other speeds. Specifically, feature set for a single rotating speed was used as a training set (for instance, feature set corresponding to 300 RPM with the dimensionality of $4 \times 300 \times 21$), and the feature sets corresponding to two other speeds were used for testing (for instance, feature set corresponding to two other speeds 600 RPM and 900 RPM with the dimensionality of $4 \times 600 \times 21$).

Those processes were run four times. To construct the training model for classification, k-fold cross-validation (k-cv) was used to estimate the accuracy of the generalized classification [52]. In k-cv, the set of samples in the feature vector is split randomly into k mutual folds (k=10 in this study), denoted as C1, C2, ... , Ck. The classification OAOMCSVM operates on k-times of the accuracy estimation.

Some folds {Cj} (a random subset from k folds) are used as a training set, and the rest are used as a validation set and alternative iteration k times. More specifically, for each speed, 300 feature vectors for each health state in the training set were partitioned into ten folds (each fold containing 30 randomly chosen feature vectors (30×21) for each health state); 9 of those folds were used for training, and the 1 remaining fold was used for validation. That process was repeated 10 times until all folds had been used as the validation set. The final measure of performance in the training model is the average value of the accuracies attained in each fold. These data are then used as the testing dataset (which was not used at all in the training process) to verify the OAOMCSVM method and provide the final classification result.

We also used the OAOMCSVM classification method to classify the feature pool configuration datasets extracted from the comparison signal processing methodologies: the raw vibration signal (lowpass filter output signal) extraction (methodology I), HT, (methodology II), WBF (methodology III), WTK (methodology IV), and the IMFs and residuals from the EMD (methodology V). The implementation of those methodologies for achieving the classification result for the four health states complied strictly with the conditions used with input from the ANR-GRS module just described. To estimate the classification result between methodologies (the proposed method and others), all twenty-one features of the vibration signal were used as input feature vectors for the OAOMCSVM module to ensure that the most informative features for and from each methodology were used fairly for the classification. The classification results of the state-of-the-art methodologies and the proposed ANR-GRS methodology obtained during two experiments are shown in Tables 4 and 5 and Figure 14 to visualize the results tabulated in Table 5. Those classification accuracies were computed as follows:

$$C_{accuracy} = \frac{\sum_L N_{TP}}{N_{samples}} \cdot 100\% \tag{23}$$

where L is the number of categories ($L = 4$ as four health states), N_{TP} is the number of true positives (the number of fault samples in category *i* that are correctly classified as class *i*), and $N_{samples}$ is the total number of samples used to estimate the performance of the proposed methodology.

Table 4. Classification results for state-of-the-art methodologies and the proposed ANR-GRS methodology by a combination dataset of different speeds.

Methodology	OAOMCSVM (4800 Samples)		Accuracy (%)				
	Training Set (80%)	**Test Set (20%)**	**Healthy**	**Fault Type 1**	**Fault Type 2**	**Fault Type 3**	**Overall (%)**
I	3840	960	59	73	69	75	69.0
II	3840	960	84	80	67	83	78.30
III	3840	960	92	89	76	83	84.6
IV	3840	960	85	87	58	74	73.10
V	3840	960	92	89	88	94	90.80
ANR-GRS	3840	960	100	99	99	100	99.70

Table 5. Classification results for state-of-the-art methodologies and the proposed ANR-GRS methodology by observation of separated speed dataset.

Methodology	OAOMCSVM (10-Fold Cross Validation)		Accuracy (%)				
	Training Set (300 Samples)	Test Set (600 Samples)	Healthy	Fault Type 1	Fault Type 2	Fault Type 3	Overall (%)
I	300 RPM	600 RPM, 900 RPM	53	78	69	52	63
	600 RPM	900 RPM, 1200 RPM	74	47	53	80	63.5
	900 RPM	600 RPM, 1200 RPM	54	46	64	81	61.25
	1200 RPM	300 RPM, 600 RPM	53	53	68	77	62.75
	Overall by health states		58.5	56	63.5	72.5	**62.63**
II	300 RPM	600 RPM, 900 RPM	51	99	63	85	74.5
	600 RPM	900 RPM, 1200 RPM	75	67	64	72	69.5
	900 RPM	600 RPM, 1200 RPM	75	48	70	83	69
	1200 RPM	300 RPM, 600 RPM	74	62	74	84	73.5
	Overall by health states		68.75	69	67.75	81	**71.63**
III	300 RPM	600 RPM, 900 RPM	75	58	69	93	73.75
	600 RPM	900 RPM, 1200 RPM	74	70	80	84	77
	900 RPM	600 RPM, 1200 RPM	70	49	72	63	63.5
	1200 RPM	300 RPM, 600 RPM	83	53	72	66	68.5
	Overall by health states		75.5	57.5	73.25	76.5	**70.69**
IV	300 RPM	600 RPM, 900 RPM	64	74	87	63	72
	600 RPM	900 RPM, 1200 RPM	82	49	72	64	66.75
	900 RPM	600 RPM, 1200 RPM	63	47	69	76	63.75
	1200 RPM	300 RPM, 600 RPM	63	49	70	67	62.25
	Overall by health states		68	54.75	74.5	67.5	**66.19**

Table 5. *Cont.*

Methodology	OAOMCSVM (10-Fold Cross Validation)		Accuracy (%)				
	Training Set (300 Samples)	Test Set (600 Samples)	Healthy	Fault Type 1	Fault Type 2	Fault Type 3	Overall (%)
V	300 RPM	600 RPM, 900 RPM	77	94	72	89	83
	600 RPM	900 RPM, 1200 RPM	90	82	91	82	86.25
	900 RPM	600 RPM, 1200 RPM	94	80	69	85	82
	1200 RPM	300 RPM, 600 RPM	98	65	69	83	78.75
	Overall by health states		89.75	80.25	75.25	84.75	**82.5**
ANR-GRS	300 RPM	600 RPM, 900 RPM	100	95	98	100	98.25
	600 RPM	900 RPM, 1200 RPM	98	99	99	100	99
	900 RPM	600 RPM, 1200 RPM	98	99	97	99	98.25
	1200 RPM	300 RPM, 600 RPM	99	98	95	99	97.75
	Overall by health states		98.75	97.75	97.25	99.5	**98.31**

Table 4 illustrates that the proposed technique significantly outperforms its counterparts when it is trained on the data instances corresponding to all available speeds and achieving the highest accuracy of 99.7%.

Table 5 demonstrates that the proposed approach using ANR-GRS also yielded the highest average classification accuracies (98.31%) in comparison with the other five state-of-the-art signal processing methodologies when it is trained and validated on datasets corresponding to separate rotating speeds.

The methodology I extracted the feature vectors of all four speeds for classification by the OAMCSVM directly from the raw vibration signal (OutLPF signal), in which non-linear and non-stationary signals drown out the informative signal. Accordingly, those results are distributed chaotically among the four classes, producing the lowest accuracy among the 6 methodologies (62.63%). For methodologies II, III, and IV, the vibration signals change with the different characteristics of the gearbox vibration signal (its amplitude and phase modulation signal), so their classification accuracy is also low, around 70%. Methodology V (the EMD technique) is outstanding in comparison with the first four approaches (82.5%) because it extracts IMFs, which contain fault-related information to better discriminate between classes. However, IMFs can be mistakenly extracted from noise components, which damaged the accuracy compared with the ANR-GRS technique by around 15%.

In addition, as a quantitative evaluation, we present the space distribution in a 3-dimensional visualization (Figure 15) of samples belonging to four classes based on some features extracted from the outLPF signal and the outANRsignal (signals before and after using the ANR-GRS technique, respectively). The features of the outANR signal show better separation and clustering for different health states of the gearbox fault diagnosis experimental scheme. Samples from the same class are more closely clustered, whereas samples from different classes are discriminated and easy to classify. On the contrary, before using the ANR-GRS, the features of different classes overlap, making it difficult to distinguish the fault classes.

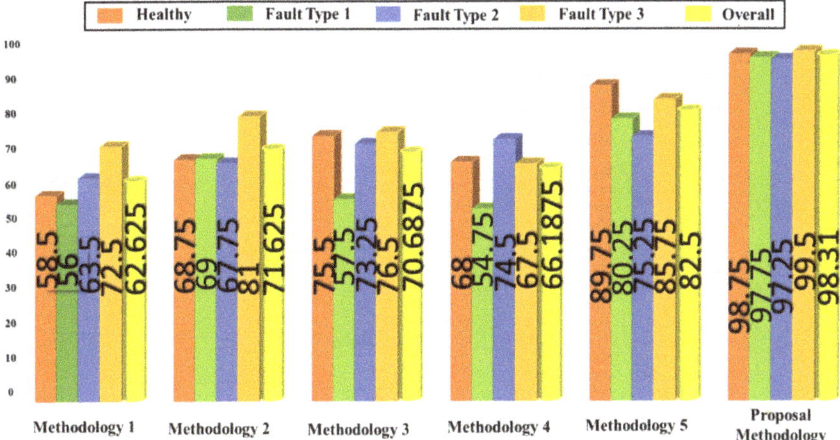

Figure 14. The accuracy of each class and the average accuracy of the state-of-the-art methodologies and the proposed ANR-GRS methodology.

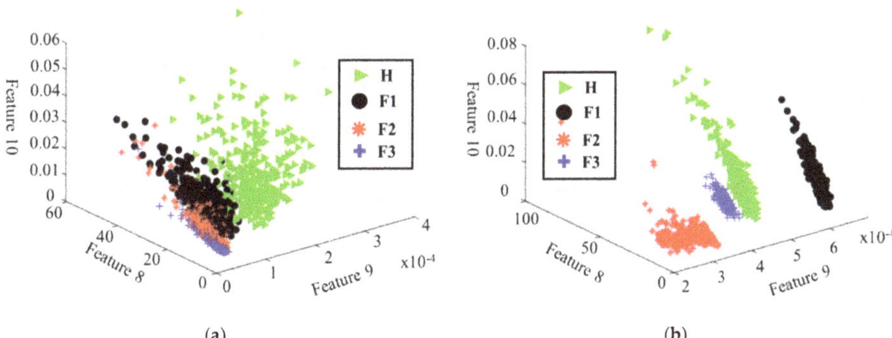

Figure 15. Three-dimensional visualization of features extracted from (**a**) the input signal of the ANR-GRS module and (**b**) the output signal of the ANR-GRS module.

Moreover, confusion matrixes are shown in Figure 16 to demonstrate the reliability of the varying-speed gearbox fault diagnosis methodology using the ANR-GRS module for effective noise reduction. Using real-time tracking of the rotation speed (RPM) of a gearbox system, the ANR-GRS generated speed-related function signals for real-time tracking of speed-dependent noise components, and the optimized output signal was unaffected by speed during classification.

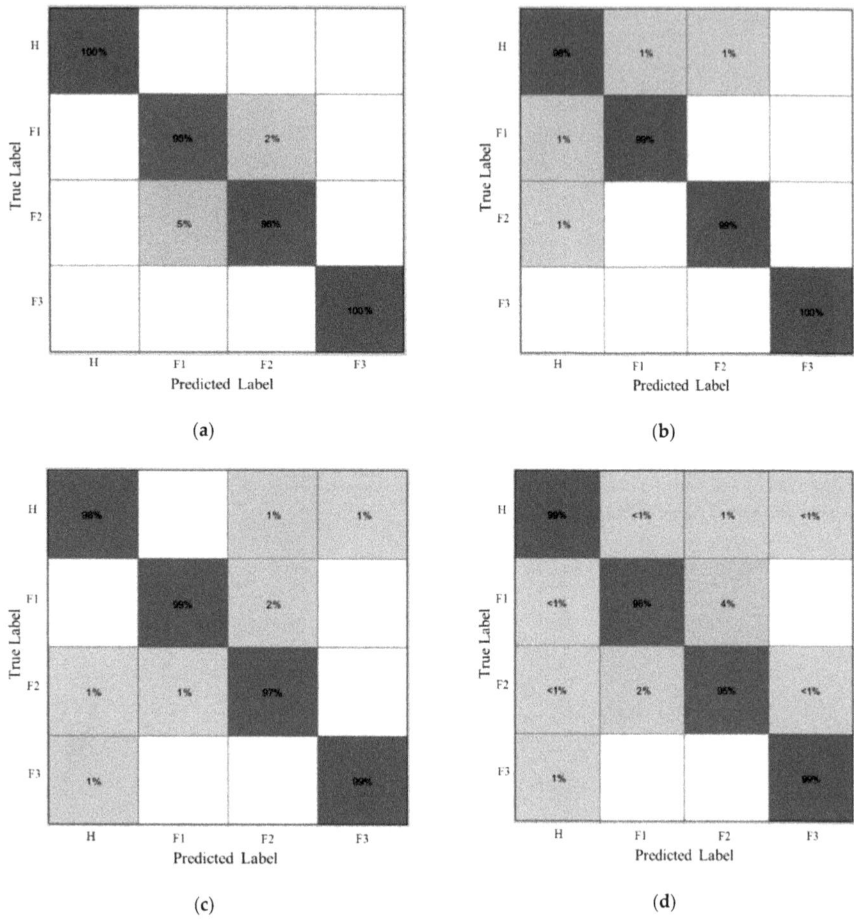

Figure 16. Confusion matrixes for four classification cases according to the speed dataset input for training: (**a**) 300 RPM, (**b**) 600 RPM, (**c**) 900 RPM, (**d**) 1200 RPM.

5. Conclusions

In this study we propose a reliable fault diagnosis methodology for gearbox systems under varying speed conditions. It integrates adaptive noise control to significantly reduce noise with machine learning classification to classify the fault states of the gearboxes. First, we created a set of Gaussian reference signals that are a function of the rotation speed and consist of many noise components such as white noise and band noise that are correlated to the parasitic noise in the vibration signals and independent of the intrinsic informative components. Then, we applied those GRSs to an adaptive noise control technique that produced an optimal sub-band as output for each 1-s vibration sample. The most optimal sub-bands were then used in the feature pool configuration to extract feature vectors, and an OAOMCSVM was used for classification. The experimental results indicated that the proposed gearbox fault diagnosis methodology achieved the highest classification accuracies in both experiments that are equal to 99.7% and 98.31% while significantly outperforming the counterpart state-of-the-art methodologies used for the comparison. In future research, we will continue improving the robustness of the proposed methodology and investigate it's applicability to the real-time fault diagnosis scenarios.

Author Contributions: All the authors contributed equally to the conception of the idea, the design of the experiments, the analysis and interpretation of results, and the writing and improvement of the manuscript. All authors have read and agreed to the published version of the manuscript.

Funding: This research was financially supported by the Ministry of Trade, Industry & Energy (MOTIE) of the Republic of Korea and Korea Institute for Advancement of Technology (KIAT) through the Encouragement Program for The Industries of Economic Cooperation Region. (P0006123)

Conflicts of Interest: The authors declare no conflict of interest.

References

1. MacFadden, P.D.; Smith, J.D. An Explanation for the Asymmetry of the Modulation Sidebands about the Tooth Meshing Frequency in Epicyclic Gear Vibration. *Proc. Inst. Mech. Eng. Part C J. Mech. Eng. Sci.* **1985**, *199*, 65–70. [CrossRef]
2. McNames, J. Fourier series analysis of epicyclic gearbox vibration. *J. Vib. Acoust. Trans. ASME* **2002**, *124*, 150–160. [CrossRef]
3. Fakhfakh, T.; Chaari, F.; Haddar, M. Numerical and experimental analysis of a gear system with teeth defects. *Int. J. Adv. Manuf. Technol.* **2005**, *25*, 542–550. [CrossRef]
4. Chaari, F.; Bartelmus, W.; Zimroz, R.; Fakhfakh, T.; Haddar, M. Gearbox vibration signal amplitude and frequency modulation. *Shock Vib.* **2012**, *19*, 635–652. [CrossRef]
5. Mills, R. An Introduction to Machinery Analysis and Monitoring. *Comput. Eng.* **1991**, *10*, 314–315. [CrossRef]
6. Ghodake, S.B.; Mishra, P.A.K.; Deokar, P.A.V. A Review on Fault Diagnosis of Gear-Box by Using Vibration Analysis Method. *IPASJ Int. J. Mech. Eng.* **2016**, *4*, 31–35.
7. Patil, C.R.; Kulkarni, P.P.; Sarode, N.N.; Shinde, K.U. Gearbox Noise & Vibration Prediction and Control. *Int. Res. J. Eng. Technol.* **2017**, *4*, 873–877.
8. Boyat, A.K.; Joshi, B.K. A Review Paper: Noise Models in Digital Image Processing. *Signal Image Process. Int. J.* **2015**, *6*, 63–75. [CrossRef]
9. Randall, R.B. *Frequency analysis*, 3rd ed.; Bruel & Kjaer: Nairobi, Denmark, 1987.
10. Matz, G.; Hlawatsch, F. Wigner distributions (nearly) everywhere: Time-frequency analysis of signals, systems, random processes, signal spaces, and frames. *Signal Process.* **2003**, *83*, 1355–1378. [CrossRef]
11. Zhang, W.; Wang, H.; Teng, R.; Xu, S. Application of rank-order morphological filter in vibration signal de-noising. In Proceedings of the 3rd International Congress on Image and Signal Processing, Yantai, China, 16–18 October 2010. [CrossRef]
12. Antoni, J. Cyclostationarity by examples. *Mech. Syst. Signal Process.* **2009**, *23*, 987–1036. [CrossRef]
13. Randall, R.B.; Antoni, J.; Chobsaard, S. The relationship between spectral correlation and envelope analysis in the diagnostics of bearing faults and other cyclostationary machine signals. *Mech. Syst. Signal Process.* **2001**, *15*, 945–962. [CrossRef]
14. Touti, W.; Salah, M.; Ben Salem, S.; Bacha, K.; Chaari, A. Spur gearbox mixed fault detection using vibration envelope and motor stator current signatures analysis. In Proceedings of the 17th International Conference on Sciences and Techniques of Automatic Control and Computer Engineering (STA), Sousse, Tunisia, 19–21 December 2016. [CrossRef]
15. Jiang, R.; Liu, S.; Tang, Y.; Liu, Y. A novel method of fault diagnosis for rolling element bearings based on the accumulated envelope spectrum of the wavelet packet. *JVC/Journal Vib. Control* **2015**, *21*, 1580–1593 [CrossRef]
16. Xie, H.; Lin, J.; Lei, Y.; Liao, Y. Fast-varying AM-FM components extraction based on an adaptive STFT. *Digit. Signal Process. A Rev. J.* **2012**, *22*, 664–670. [CrossRef]
17. Oppenheim, A.V.; Schafer, R.W. *Discrete-Time Signal Processing*; Prentice Hall: Englewood Cliffs, NJ, USA, 1989.
18. Vetterli, M.; Kovačević, J. *Wavelets and Subband Coding*; Prentice Hall PTR: Upper Saddle River, NJ, USA, 1995.
19. Olkkonen, J. *Discrete Wavelet Transforms—Theory and Applications*; Intech: Rijeka, Croatia, 2011; ISBN 9789533071855.
20. Liu, J. Shannon wavelet spectrum analysis on truncated vibration signals for machine incipient fault detection. *Meas. Sci. Technol.* **2012**, *23*, 055604. [CrossRef]
21. Aharamuthu, K.; Ayyasamy, E.P. Application of discrete wavelet transform and Zhao-Atlas-Marks transforms in non stationary gear fault diagnosis. *J. Mech. Sci. Technol.* **2013**, *27*, 641–647. [CrossRef]

22. Kang, M.; Kim, J.; Kim, J.M.; Tan, A.C.C.; Kim, E.Y.; Choi, B.K. Reliable fault diagnosis for low-speed bearings using individually trained support vector machines with kernel discriminative feature analysis. *IEEE Trans. Power Electron.* **2015**, *30*, 2786–2797. [CrossRef]

23. Huang, N.E.; Shen, Z.; Long, S.R.; Wu, M.C.; Snin, H.H.; Zheng, Q.; Yen, N.C.; Tung, C.C.; Liu, H.H. The empirical mode decomposition and the Hilbert spectrum for nonlinear and non-stationary time series analysis. *Proc. R. Soc. A Math. Phys. Eng. Sci.* **1998**, *454*, 903–995. [CrossRef]

24. Loutridis, S.J. Damage detection in gear systems using empirical mode decomposition. *Eng. Struct.* **2004**, *26*, 1833–1841. [CrossRef]

25. Zhang, C.; Peng, Z.; Chen, S.; Li, Z.; Wang, J. A gearbox fault diagnosis method based on frequency-modulated empirical mode decomposition and support vector machine. *Proc. Inst. Mech. Eng. Part C J. Mech. Eng. Sci.* **2018**, *232*, 369–380. [CrossRef]

26. Chen, K.; Zhou, X.C.; Fang, J.Q.; Zheng, P.F.; Wang, J. Fault Feature Extraction and Diagnosis of Gearbox Based on EEMD and Deep Briefs Network. *Int. J. Rotating Mach.* **2017**, *2017*. [CrossRef]

27. Buzzoni, M.; Mucchi, E.; D'Elia, G.; Dalpiaz, G. Diagnosis of Localized Faults in Multistage Gearboxes: A Vibrational Approach by Means of Automatic EMD-Based Algorithm. *Shock Vib.* **2017**, *2017*. [CrossRef]

28. Liu, B.; Riemenschneider, S.; Xu, Y. Gearbox fault diagnosis using empirical mode decomposition and Hilbert spectrum. *Mech. Syst. Signal Process.* **2006**, *20*, 718–734. [CrossRef]

29. Goharrizi, A.Y.; Sepehri, N. Internal leakage detection in hydraulic actuators using empirical mode decomposition and hilbert spectrum. *IEEE Trans. Instrum. Meas.* **2012**, *61*, 368–378. [CrossRef]

30. Han, G.D.; Wan, S.T.; Lv, Z.J.; Liu, R.H.; Wang, J.; Tang, G.J. The analysis of gearbox fault diagnosis research based on the EMD and hilbert envelope demodulation. *Adv. Mater. Res.* **2014**, *926–930*, 1800–1805. [CrossRef]

31. Peng, Z.K.; Tse, P.W.; Chu, F.L. A comparison study of improved Hilbert-Huang transform and wavelet transform: Application to fault diagnosis for rolling bearing. *Mech. Syst. Signal Process.* **2005**, *19*, 974–988. [CrossRef]

32. Lin, J.; Dou, C.; Wang, Q. Comparisons of MFDFA, EMD and WT by neural network, Mahalanobis distance and SVM in fault diagnosis of gearboxes. *Sound Vib.* **2018**, *52*, 11–15. [CrossRef]

33. Yang, Q.; An, D. EMD and wavelet transform based fault diagnosis for wind turbine gear box. *Adv. Mech. Eng.* **2013**, *2013*. [CrossRef]

34. Zamanian, A.H.; Ohadi, A. Gear fault diagnosis based on Gaussian correlation of vibrations signals and wavelet coefficients. *Appl. Soft Comput. J.* **2011**, *11*, 4807–4819. [CrossRef]

35. Van, M.; Kang, H.J.; Hin, K.S. Rolling element bearing fault diagnosis based on non-local means de-noising and empirical mode decomposition. *IET Sci. Meas. Technol.* **2014**, *8*, 571–578. [CrossRef]

36. Guo, Y.; Liu, Q.N.; Wu, X.; Na, J. Gear fault diagnosis based on narrowband demodulation with frequency shift and spectrum edit. *Int. J. Eng. Technol. Innov.* **2016**, *6*, 243–254.

37. Raj, A.S.; Murali, N. Morlet wavelet UDWT denoising and EMD based bearing fault diagnosis. *Electronics* **2013**, *17*, 1–8. [CrossRef]

38. Wang, X.; Zi, Y.; He, Z. Multiwavelet denoising with improved neighboring coefficients for application on rolling bearing fault diagnosis. *Mech. Syst. Signal Process.* **2011**, *25*, 285–304. [CrossRef]

39. Haykin, S.S. *Adaptive Filter Theory*, 4th ed.; Prentice Hall: Upper Saddle River, NJ, USA, 2002; ISBN 0130901261.

40. Lee, K.A.; Gan, W.S.; Kuo, S.M. *Subband Adaptive Filtering: Theory and Implementation*; John Wiley and Sons: Chichester, UK, 2009; ISBN 9780470516942.

41. Rauber, T.W.; De Assis Boldt, F.; Varejão, F.M. Heterogeneous feature models and feature selection applied to bearing fault diagnosis. *IEEE Trans. Ind. Electron.* **2015**, *62*, 637–646. [CrossRef]

42. Liu, R.; Yang, B.; Zio, E.; Chen, X. Artificial intelligence for fault diagnosis of rotating machinery: A review. *Mech. Syst. Signal Process.* **2018**, *108*, 33–47. [CrossRef]

43. Liu, J.; Zio, E. Feature vector regression with efficient hyperparameters tuning and geometric interpretation. *Neurocomputing* **2016**, *218*, 411–422. [CrossRef]

44. Widodo, A.; Kim, E.Y.; Son, J.D.; Yang, B.S.; Tan, A.C.C.; Gu, D.S.; Choi, B.K.; Mathew, J. Fault diagnosis of low speed bearing based on relevance vector machine and support vector machine. *Expert Syst. Appl.* **2009**, *36*, 7252–7261. [CrossRef]

45. Vapnik, V.N. *The Nature of Statistical Learning Theory*; Springer: New York, NY, USA, 1995.

46. Cristianini, N.; Shawe-Taylor, J. *An Introduction to Support Vector Machines and Other Kernel-Based Learning Methods*; Cambridge University Press: Cambridge, UK, 1999; ISBN 0521780195.

47. Hsu, C.W.; Lin, C.J. A comparison of methods for multiclass support vector machines. *IEEE Trans. Neural Networks* **2002**, *13*, 415–425. [CrossRef]
48. Fan, X.; Zuo, M.J. Gearbox fault detection using Hilbert and wavelet packet transform. *Mech. Syst. Signal Process.* **2006**, *20*, 966–982. [CrossRef]
49. Figliola, R.S.; Beasley, D.E. Sampling, digital devices, and data acquisition. In *Theory and Design for Mechanical Measurements*, 5th ed.; John Wiley & Sons: Hoboken, NJ, USA, 2011; pp. 260–303.
50. Widrow, B.; Stearns, S.D. *Adaptive Signal Processing*; Prentice Hall: Upper Saddle River, NJ, USA, 1985.
51. Caesarendra, W.; Tjahjowidodo, T. A review of feature extraction methods in vibration-based condition monitoring and its application for degradation trend estimation of low-speed slew bearing. *Machines* **2017**, *5*, 21. [CrossRef]
52. Rodríguez, J.D.; Pérez, A.; Lozano, J.A. Sensitivity Analysis of k-Fold Cross Validation in Prediction Error Estimation. *IEEE Trans. Pattern Anal. Mach. Intell.* **2010**, *32*, 569–575. [CrossRef]

Letter

Sensor Fault-Tolerant Control Design for Magnetic Brake System

Krzysztof Patan *, Maciej Patan and Kamil Klimkowicz

Institute of Control and Computation Engineering, University of Zielona Góra, 65-516 Zielona Góra, Poland;
M.Patan@issi.uz.zgora.pl (M.P.); k.klimkowicz@issi.uz.zgora.pl (K.K.)
* Correspondence: k.patan@issi.uz.zgora.pl

Received: 14 July 2020; Accepted: 14 August 2020; Published: 16 August 2020

Abstract: The purpose of the paper is to develop an efficient approach to fault-tolerant control for nonlinear systems of magnetic brakes. The challenging problems of accurate modeling, reliable fault detection and a control design able to compensate for potential sensor faults are addressed. The main idea here is to make use of the repetitive character of the control task and apply iterative learning control based on the observational data to accurately tune the system models for different states of the system. The proposed control scheme uses a learning controller built on a mixture of neural networks that estimate system responses for various operating points; it is then able to adapt to changing working conditions of the device. Then, using the tracking error norm as a sufficient statistic for detection of sensor fault, a simple thresholding technique is provided for verification of the hypothesis on abnormal sensor states. This also makes it possible to start the reconstruction of faulty sensor signals to properly compensate for the control of the system. The paper highlights the components of the complete iterative learning procedure including the system identification, fault detection and fault-tolerant control. Additionally, a series of experiments was conducted for the developed control strategy applied to a magnetic brake system to track the desired reference with the acceptable accuracy level, taking into account various fault scenarios.

Keywords: braking control; nonlinear systems; fault tolerant control; fault detection; iterative learning control; neural networks

---

1. Introduction

Modern industrial systems are usually complex and nonlinear. This leads directly to challenging problems related to control design as requirements imposed on the control quality and robustness continuously increase. Since it is not easy to deal with the nonlinear models, there is still a need for new systematic methods taking advantage of the specificity of a particular class of nonlinear systems.

A magnetic brake also called an eddy-current brake is a typical example of such nonlinear systems with tangible applications in various areas of the automotive industry, including high speed railways, big trucks and industrial elevators. Basically, in the regime of the high speed reached by current vehicles, the classical friction-based retarders become ineffective and insufficient. Therefore, magnetic or electro-magnetic brake control systems constitute an attractive alternative to ensure travelers and vehicle safety and reliability. These are characterized by the fast response time, a non-contact construction, and a relatively low level of failure rate in comparison to conventional brakes.

As for control design for magnetic brakes, the problem is considered to be difficult since the complex dynamics include not only temporal but also spatial effects. This excludes the direct extension of results from control theory for linear systems. Additionally, the system is characterized by a strict antisymmetry of control, as one can control only the rate of deceleration. Nevertheless, a few more or less successful attempts at adapting this theory have been reported by various authors in the control engineering literature.

In [1], the authors used an input–output feedback linearization scheme for speed control. However, they used a simplified control-affine model to derive the control law. In [2], the author used the classical proportional-integral (PI) controller with the current reference provided by means of the Kriging method. However, the optimization problem was solved using a genetic algorithm, which is a time-consuming method. In [3], the authors applied an adaptive control of vehicle speed. However, they assumed a low order lumped-parameter model of the plant. In [4], the authors applied a sliding mode control to design a speed control that is robust for changing road conditions, but again, they used a simplified approximated model as proposed in [1]. In turn, a torque tracking control by means of a sliding model was proposed in [5]. The author applied a simple static polynomial-based torque model, which leads directly to excessive sensitivity on the changes of the operating point of the control system during the implementation of this technique.

All reported papers used the simplified models of the magnetic brake and described closed-loop control schemes. The main problem that arises in the case of nonlinear control system design is derivation of the exact model of the plant considered. From a physical point of view, the magnetic brake is a distributed-parameter system, requiring sophisticated mathematical modeling that takes into account its spatiotemporal dynamics. This makes the accurate modeling numerically difficult and time-consuming, which constitutes one of crucial design difficulties.

In addition to this, there is a significant lack of contributions discussed in the context of fault-tolerant control. In fact, there are just a few papers covering this issue in the context of breaking control for electromagnetic systems. An adaptive fault-tolerant control based on a Takagi–Sugeno fuzzy model and observer design is reported in [6], but it consists of sequential recalibration of Kalman filters, reverting this approach to the linear approximation of the system. Another approach to achieve fault-tolerant brake control utilized the brake-by-wire strategy [7]. However, it was strongly dependent on preset recovery strategies. Additionally, it focused on the nonlinear effects of contact force control and did not take into account the specificity of the electromagnetic brakes, limiting its applicability in this area. In [8], the braking controller was constructed for vehicles with regenerative in-wheel electromagnetic brakes using sliding mode control and a fault-tolerant control architecture included as an adaptive scheme. However, in this case, the main stress was put on the vehicle dynamics and again the linear model of braking torque was applied. On the one hand, all of the approaches mentioned above allow to avoid significant perturbations in the control and provide satisfactory results in some particular situations; on the other hand, they do not take into account the effects of the complex dynamics of brakes, neglecting the nonlinearity that may introduce an additional level of uncertainty and significantly affect the reliability of fault detection. To fill this gap, this work focuses on an alternative control strategy driven by measurement data and developed on the basis of an iterative learning controller enhanced with statistical hypothesis testing for fault detection.

If an accurate mathematical model of the system is not available, the reasonable alternative is to design the model using input–output data recorded in the real plant during replications of the control task. In this context, iterative learning control (ILC) is a suitable technique to achieve ideal tracking of the given reference profile [9]. The learning controller calculates the current control signal based on the archive control data available from previous experiments. Most often, a learning controller has a fixed structure along the trial domain. However, in practice, a controller should be robust to disturbances and the faults within the infrastructure of the plant [10]. Therefore, a learning controller design should include the above circumstances into the analysis. In this paper, a neural network-based iterative learning control scheme is proposed, which is capable of adapting its own behavior to the changing working conditions of the plant (related to potential sensor faults) through a data-driven training process. In the background of

ILC, the additional advantage is that the squared tracking error norm stands for the convenient statistic for the fault detection task. It makes it possible to use classical parametric hypothesis testing on the abnormal state of the system, leading to a simple yet efficient thresholding rule for sensor fault detection.

Due to technological developments, modern industrial plants are more and more complex and are composed of an ever-growing number of interacting devices, including a huge number of measuring devices. That is the reason that industrial plants are becoming very vulnerable to any deviation from the so-called normal operating conditions caused by unpermitted deviations of plant characteristics or unexpected changes of system variables. Such a deviation is called a fault. Early detection of faults can provide a way to avoid a shut-down or even a failure of the system. More importantly, a proper fault accommodation is strongly related to avoiding the large financial losses and serious injuries or even death of personnel. On this basis, fault-tolerant control (FTC) has come into prominence and has received increased attention in the last decade [11–15]. The FTC approaches can be be split into passive and active ones. Passive FTC uses a priori knowledge about anticipated faults that can affect the system. In turn, active FTC methods use the information acquired from a fault diagnosis subsystem in order to accommodate a fault. An important class of fault tolerant systems is sensor FTC. The simple and intuitive way is to introduce hardware redundancy, which, however, complicates the construction of the system and increases maintenance costs. A more flexible solution is to use the idea of a virtual sensor [16–18]. The idea is to exclude the measurements given by the faulty sensor and replace them with data provided by the virtual sensor. The virtual sensor requires a fairly accurate plant model. In this paper, we propose to apply a mixture of neural network models, which renders it possible to effectively accommodate a sensor fault. This strategy is called a fault hiding approach [19], which renders it possible to apply the same controller to both faulty and fault-free cases.

The contributions of the paper can be stated as follows.

1. Proposing a sensor active FTC system for magnetic brakes based on iterative learning control.
2. Developing a model of a magnetic brake by means of the mixture of state-space neural network models and gain scheduling.
3. Performing fault accommodation analysis for various types of sensor faults.

The paper is organized as follows. In Section 2 the magnetic brake system is described. In Section 3 the control scheme using iterative learning control is provided and the control performance for normal operating conditions is also discussed. Section 4 is devoted to the modeling issues. State-space neural networks are employed to model the plant at different operating points and the overall model of the plant is achieved by means of gain scheduling in the form of a mixture of models. Section 5 introduces the idea of sensor active FTC, including fault detection and accommodation and FTC in case of different sensor faults. The last section summarizes the paper.

2. Magnetic Brake

A magnet brake, in simple terms, consists of a disk of conductive material and a magnet generating a magnetic field in which the disk is rotating. The simplest form of the device is depicted in Figure 1 (left panel).

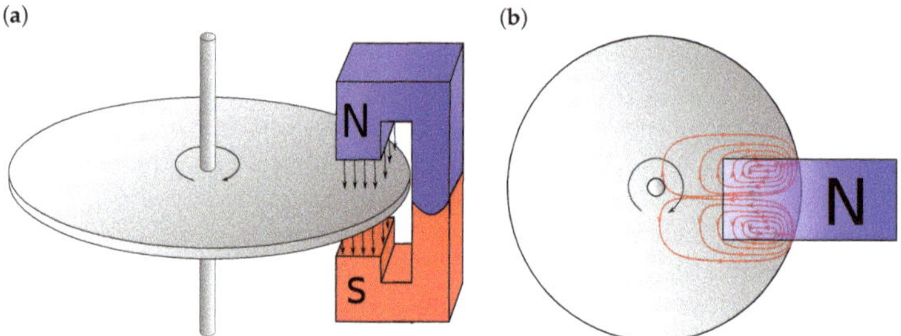

Figure 1. (**a**) Schematic view of a magnetic brake and (**b**) upper view with eddy currents field.

When a conductor moves in a magnetic field, it induces eddy currents, cf. Figure 1 (right panel), which interact with the magnetic flux to produce Lorentz forces resulting in a braking torque slowing down the disk. Let $\mathcal{D} \in \mathrm{R}^3$ denote a spatial domain representing the disk. Then, the evolution of the rotational velocity ω of the disk over the time interval $(0, t_f)$ can be described by the following initial value problem:

$$J\frac{\mathrm{d}\omega(t)}{\mathrm{d}t} = M(B_0(t), \omega(t)), \quad \omega(0) = \omega_0, \tag{1}$$

where J is a disk moment of inertia and $B_0(t)$ is the magnetic flux of the external magnet (or electromagnet). M is a total torque affecting the disk:

$$M = \int_{\mathcal{D}} T_q(B(x, t), B_0(t), \omega(t)) \, \mathrm{d}V, \tag{2}$$

where $B(x, t)$ is the spatiotemporal field of internal magnetic flux inside the disk, $\mathrm{d}V$ is a volume element of the spatial domain and T_q denotes a Lorentz force on the volume element related to the spatial location x.

It becomes clear that the eddy current brakes are nonlinear systems with responses depending on the spatiotemporal dynamics of changes in magnetic flux ([20], Chapter 9). It is strongly dependent on the shape of the domain $\mathcal{D}$ as well as the form of the external magnetic flux generated by the magnet. Hence, the modeling with typical methods for distributed-parameter systems, such as finite or boundary element methods, are not only challenging but also difficult to apply in practical control design, as any change to the spatial distribution of external magnetic fields (e.g., reallocating the magnets or changing the boundary conditions) requires a total rebuilding of the model.

In fact, the accurate estimation of the braking torque is one of the key challenging problems for both the design of control systems and fault detection and its compensation. Furthermore, the modeling uncertainty has to be taken into account, as it is an important factor in applications to real-world engineering systems. Fortunately, Equation (2) defining the torque integrated over the spatial domain of the disk can be approximated up to a satisfactory accuracy based on the measurement data. Here, we make use of the well-known ensemble averaging properties of neural networks. They not only provide an important alternative for accurate modeling of the nonlinear torque, but also deliver necessary robustness with respect to disturbances as they can generalize the system response. Additionally, in order to introduce these uncertainties into control synthesis, the iterative learning control is applied, being considered a well-known robust control design technique. This is especially valid in the efficient control design, which will be discussed in the following section.

For the purpose of the simulation study considered in this work, we used an aluminum disk with a radius of 10 cm and a thickness of $d = 1$ mm. The disk is moving in the external magnetic field generated by an electromagnet with a maximum flux of 0.1 T. This represents a typical configuration encountered in analog energy meters.

The control objective is to produce the input signal of magnetic flux $B_0(t)$ leading to the desired profile of output angular velocity $\omega(t)$ with the initial value of 200 rpm. As the control algorithm, the iterative learning control portrayed in detail in Section 3 was used. The exemplary spatial distribution of the magnetic flux inside the disk and the normalized reference profile are presented in Figure 2.

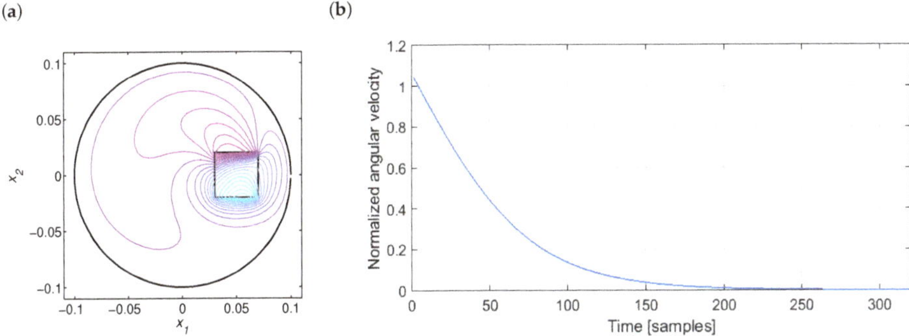

Figure 2. (**a**) Magnetic flux density B for $\omega = 200$ rpm and $B_0 = 0.1$ T, and (**b**) required reference profile.

3. Iterative Learning Control

Because of the sequential regime of the process of gathering the measurement data as a basic control scheme for the magnetic brake, a nonlinear iterative learning was employed. The choice of the ILC is justified not only by the repetitive character of the process, but is also dictated by its robustness to model inaccuracies and repetitive disturbances [9,21] via proper compensation with the use of the measurement data. This makes it especially useful, especially in combination with neural networks providing proven convergence for the considered class of systems [15].

Taking into account the fundamental property of the magnetic brake, namely the strictly energy dissipative character of the system (one cannot force the overshoot upwards), and in order to keep the complexity of the controller at a relatively low level, open-loop ILC was chosen so as to achieve the assumed control performance. Even though it is clear that the feedback control may improve the average performance, this may come at the cost of higher sensitivity to online disturbances.

Because of the specificity of the controller learning process and neural modeling, it can be useful to further discuss the discrete-time domain. Thus, the general idea of the applied nonlinear ILC is represented by the scheme depicted in Figure 3. The learning controller has the following form:

$$B_p(k) = f(B_{p-1}(k), e_{p-1}(k)). \tag{3}$$

where $f(\cdot, \cdot)$ is a nonlinear function representing the dynamics of the controller, p stands for the iteration (trial) number, k is a time instant, $e_p(k) = \omega_d(k) - \omega_p(k)$ is the tracking error and $\omega_d(k)$ is the reference signal. The system (3) constitutes the so-called P-type first-order learning controller [9]. Various choices exists for effective realization of $f(\cdot, \cdot)$. One of the possible solutions is to apply a feedforward neural network. The essential element of the learning algorithm is the necessity of estimating the output sensitivities of the magnetic brake with respect to the control signal. This is achieved using a model of the system as reported in previous works by the current authors [15,22,23]. The important feature of such a solution is the possibility to adapt its structure to the changing operation conditions of the control system. Due to the availability of representative measurement data, such an approach seems to be especially attractive here. As the details of the training procedure for the controller are beyond the scope of this paper, the interested reader can be referred to [15,22,23]. Although the physics of magnetic brakes can be discussed in terms of distributed parameter systems, such a class of systems can be accurately approximated with non-parametric models, such as a nonlinear state-space innovation form model (NSSIF):

$$
\begin{aligned}
x_p(k+1) &= g(x_p(k), B_p(k), \epsilon_p(k)), \\
\hat{\omega}_p(k) &= C x_p(k),
\end{aligned}
\tag{4}
$$

where $x_p(k)$, $B_p(k)$ and $\hat{\omega}_p(k)$ are the state-space, input, and predicted output vectors, respectively, $\epsilon_p(k) = \omega_p(k) - \hat{\omega}_p$ is the prediction error and C is the output (observation) matrix.

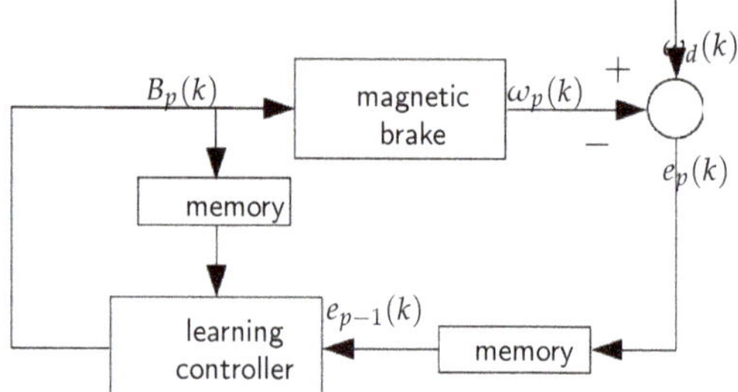

Figure 3. Iterative learning control scheme.

Figure 4 shows the control results relating to the tracking of the reference presented in Figure 2. Figure 4a illustrates the convergence of the tracking error norm for two cases: the untrained learning controller (red dashed line) and the preliminary trained controller (blue solid line). If we start with the randomly selected initial controller weights, the learning algorithm needs some time to adapt the parameters in order to follow the reference closely. Such an experiment is performed at the very beginning. After that we can use the preliminary trained neural controller, which is able to adapt to the current working conditions of the plant very quickly (blue-solid line). Figure 4b presents the quality of tracking of the reference, where the output of the the plant is marked with the red solid line while the reference is marked with the blue dashed line. Evidently, the applied ILC functions well and the plant output follows the reference closely.

The model (4) is in fact the state observer and works satisfactorily. However, in this paper we consider a fault-tolerant control design that relies on fault diagnosis. For the purposes of model-based fault diagnosis, a model of the system should be independent of the measured process variables. Unfortunately, the model (4) is dependent on the output of the system and thus cannot be used for fault detection and there a need for developing another model. This matter is taken up in the next section.

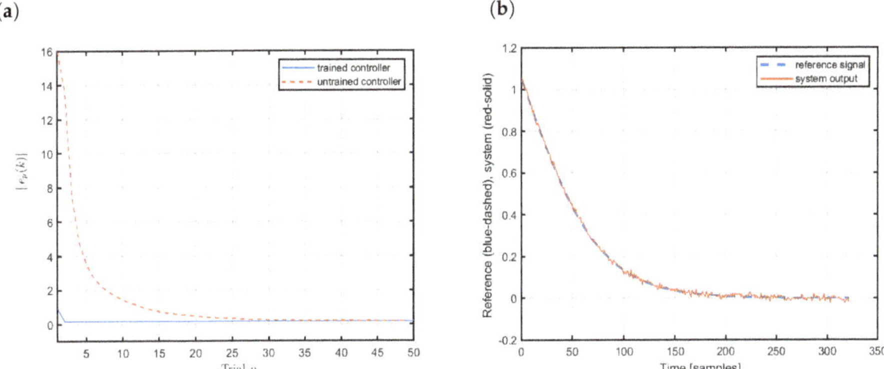

Figure 4. Control of the magnetic brake: (**a**) convergence of the tracking error norm and (**b**) tracking of the reference.

4. Model Design

As the NSSIF model cannot be used for fault diagnosis, our first attempt was aimed at the application of a state-space neural network represented by (4) assuming that the prediction error $\epsilon_p(k) = 0$. However, due to the complex characteristics of the magnetic brake portrayed in Section 2, just one state-space model is insufficient to accurately follow the dynamics of the system at different operating points.

Therefore, in order to overcome this impediment, a reasonable idea is to express the dynamics of the system as a mixture of multiple models, where each of them describes the system around an operating point. The idea is reminiscent of the so-called ensemble averaging from machine learning theory and is closely related to the gain scheduling approach to nonlinear control. Although the NSSIF is known to possess the universal approximation property [15], the proper generalization of the system response is often achieved at the cost of the bias. The mixture of models, if properly adopted, can reduce the distribution of the model output still keeping a bias on the low level. As a result, we are able to determine a good trade-off between uncertainty modeling and robustness with respect to disturbances. However, to achieve this balance we have to properly combine the components of the approach, which will be discussed in the following sections.

First, if not leading to ambiguity, the trial index p is omitted for the clarity of the presentation. Then the overall model is represented as follows:

$$\hat{\omega}(k) = \sum_{i-1}^{n} \hat{\omega}_i(k)\mu_i(\rho) \tag{5}$$

with

$$\sum_{i=1}^{n} \mu_i(\rho) = 1, \tag{6}$$

where μ_i are scaling functions, $\hat{\omega}_i(k)$ represents estimates of the system output at the i-th operating point, n is the number of operating points considered and ρ stands for the parameter embodying dependence of the combination of local models on the operating point that can be represented by input, state, etc. The formulation (5) is known as a nonlinear gain scheduling [24]. It can also be considered as an ensemble of models used for regression [25]. The functions μ_i can be perceived as a membership functions and then

(5) looks similar to the Takagi–Sugeno representation. The crucial parts of the modeling are: (i) a proper designing of local models and (ii) proper choice of the membership functions. The first problem can be readily solved using the state-space neural network of the form:

$$
\begin{aligned}
x_i(k+1) &= h_i(x_i(k), B(k)), \\
\hat{\omega}_i(k) &= C_i x_i(k),
\end{aligned}
\tag{7}
$$

where the function $h_i(\cdot, \cdot)$ describing the i-th neural model is represented by:

$$
x_i(k+1) = W_{i_2}\sigma_h(W_{i_1}^x x_i(k) + W_{i_1}^b B(k))
\tag{8}
$$

where $\sigma_h(\cdot)$ is the nonlinear activation function of the hidden layer, W_{i_1}, and $W_{i_2}^x$ and $W_{i_2}^b$ are the i-th neural model weight matrices subject to training. Each neural network (7) is trained using data recorded during control of the magnetic brake at the specific operating point.

The latter problem can be effectively solved by means of Gaussian membership functions properly distributed on the domain ρ. In this paper, ρ is selected to be the operating point depending on the initial control value B_0. This choice is motivated by the fact that the behavior of the magnetic brake strictly depends on the initial conditions. One of them is an initial control value. Then:

$$
\mu_i(B_0) = e^{-\frac{(|B_0|-B_i)^2}{2\sigma^2}}
\tag{9}
$$

where B_i is the i-th operating point and σ is the spread of the Gaussian function. In (9) the absolute value of u_0 is used because for the magnetic brake the total torque defined by (2) is invariant to the exchange of magnetic poles; thus we can restrict the considerations to nonnegative values of the magnetic flux only. Consequently, it was possible to reduce the number of operating points and then the number of local models required to develop the overall model (5). In the present work, it is assumed that each membership function has the same spread and they are uniformly distributed, cf. Figure 5. In this way, the condition (6) is always satisfied.

Throughout the empirical analysis of the magnetic brake system the effective range of the magnetic field varied from 0.003 to 0.095. After performing a series of preliminary experiments we decided to set the number of operating points to seven. The distribution of Gaussian membership functions is depicted in Figure 5.

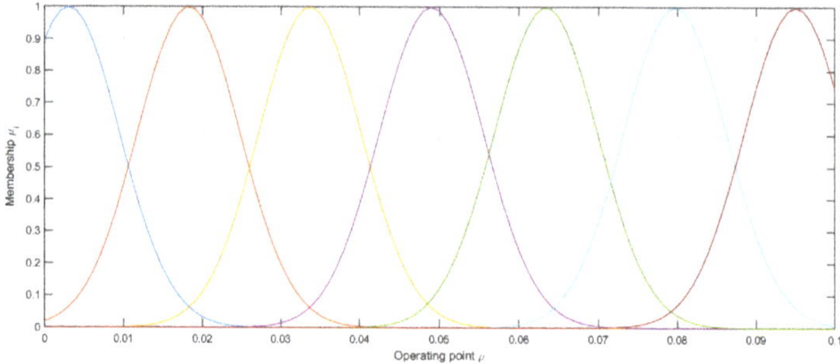

Figure 5. Distribution of the membership functions.

In order to satisfy condition (6), the spread of the functions was set to $\sigma = 0.006512$. The set of operating points was $\{0.003, 0.0183, 0.0337, 0.049, 0.0634, 0.0797, 0.095\}$. Data for model training were recorded during the normal work of the magnetic brake control system. For each operating point, one model represented by (7) was designed. The number of hidden neurons v as well as the order of the network n_x were selected experimentally through trial and error. The final configuration of the models is presented in Table 1. A very important observation is that each local model has a relatively simple structure including a lower number of hidden neurons (only five) and model order (the second or third).

Table 1. Models configuration.

Model no.	n_x	v	σ_h
1	2	5	hyperbolic tangent
2	2	5	hyperbolic tangent
3	3	5	hyperbolic tangent
4	3	5	hyperbolic tangent
5	3	5	hyperbolic tangent
6	3	5	hyperbolic tangent
7	3	5	hyperbolic tangent

For illustration, let us analyze the modeling results presented in Figure 6.

In the case considered, the initial control was $\rho = -0.0423$. Then, four membership functions were activated with the following membership values: $\mu_2 = 0.0011$, $\mu_3 = 0.4179$, $\mu_4 = 0.5892$ and $\mu_5 = 0.0053$. This means that the estimated system output was calculated using four models. The third and fourth models had the most significant impact. The first, sixth and seventh models were not used here. Figure 6a shows the control signal. In turn, Figure 6b presents the output of the system (solid blue line) and the overall estimated output (solid black line) along with the outputs of all local models used to develop the overall model. Clearly, the proposed modeling design works pretty well. The sum of squared errors between the output of the system and its estimation is 0.0984. The model was tested on 100 sequences with different initial control conditions and the mean value of the sum of squared errors was 0.1681.

(a)   **(b)**

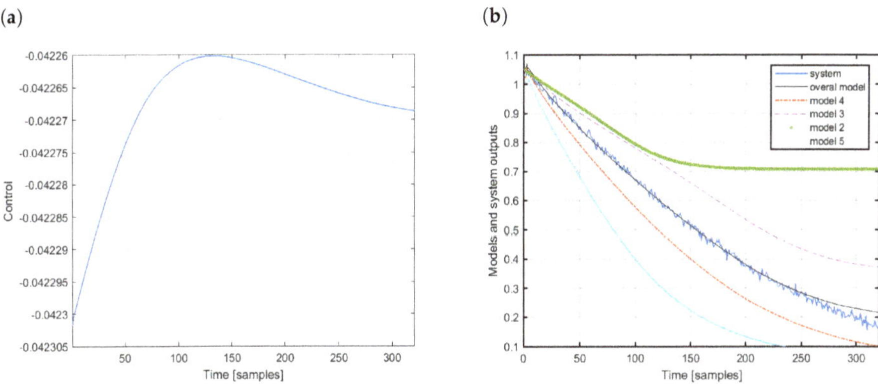

Figure 6. Modeling example: (**a**) Control signal and (**b**) output of models.

5. Sensor Fault-Tolerant Control

5.1. Fault Detection and Accommodation

Sensor fault detection is carried out on the ground of a residual signal. The residual is derived as a difference between the measurably available output of the system $w(k)$ and the output of the overall system model $\hat{w}(k)$:

$$r(k) = w(k) - \hat{w}(k). \tag{10}$$

Using (10) the diagnostic signal in the form of the squared norm is proposed:

$$d(r) = \|\tilde{r}(k)\|_n^2, \tag{11}$$

where n is the length of samples in r and

$$\tilde{r}(k) = (r(k) - \mu_r)/\sigma_r \tag{12}$$

with μ_r and σ_r denoting the expected value and the standard deviation of the residual signal, respectively.

From a statistical point of view, this constitutes a sufficient statistic preserving information about the state of the system. In fact, assuming that the measurement noise is a zero-mean Gaussian white process and according to the form of membership functions given by (9), the diagnostic signal (11) is a random variable distributed according to the χ^2 distribution with n degrees of freedom. In consequence, it is possible to use the diagnostic signal to construct a statistical parametric test to verify the null hypothesis about the excessive discrepancy of the norm from its nominal value. Thus, a decision about faults can be made. A very simple decision rule can be proposed, i.e., to compare the diagnostic signal $d(r)$ with the suitably selected threshold T:

$$s(r) = \begin{cases} 0 & \text{if } d(r) \leq T, \\ 1 & \text{otherwise.} \end{cases} \tag{13}$$

In the ideal conditions (absence of measurement and modeling uncertainty) the threshold is zero in the fault-free case and different from zero in the case of a fault. However, due to the modeling uncertainty, disturbances and the measurement noise, it is required to assign a threshold larger than zero to avoid excessive numbers of false alarms. In order to determine the threshold, let us introduce a significance level α that corresponds to the fixed a priori probability of a false alarm. Then the threshold T can be determined as some critical value being approximately equal to the $(1 - \alpha) \cdot 100\%$ percentile from the cumulative χ^2 distribution. When n is large, the random variable

$$\tilde{d}(r) = (d(r) - n)/\sqrt{2n} \tag{14}$$

weakly converges to a standard normal distribution, so in practice, the determination of the threshold is even easier, as it is a solution to the equation

$$\alpha = prob(\tilde{d}(r) > T), \tag{15}$$

which can be efficiently found from the tabulated values of the cumulative distribution function of $\mathcal{N}(0,1)$.

To make the threshold T applicable, it is required to determine the statistics μ_r and σ_r of the residual signal. These can be derived by analysis of the residual signal derived in the normal (fault-free or healthy) operating conditions of the magnetic brake, as portrayed in Section 3. The important index assessing the fault diagnosis quality is the time of fault detection. In this paper we deal with a dynamic system that is used in a repetitive manner. Then, we propose to introduce a similar quantity to the time of fault detection called the trial of fault detection p_d. Clearly, the trial of fault detection is the trial number at which the diagnostic signal (11) permanently exceeds the predefined threshold (15).

After detecting a sensor fault, it is time for reconstructing the value measured by the faulty sensor. To this end, the overall model of a magnetic brake is used. Since the fault is detected, the output of this model replaces the value measured by the faulty sensor. However, the perfectly fitted model of the system does not exist. Then, it is necessary to estimate the modeling uncertainty somehow. Here, we decided to apply a very simple strategy that is easy to implement, in which the modeling uncertainty becomes constant from trail to trail during the fault occurrence and is estimated as:

$$\Delta\omega(k) = \omega_{p_d}(k) - \hat{\omega}_{p_d}(k) \tag{16}$$

Finally, the reconstructed value of the system output can be derived as:

$$\omega_r(k) = \hat{\omega}(k) + \Delta\omega(k) \tag{17}$$

5.2. Fault-Tolerant Control

In this paper, both the abrupt and incipient sensor faults are investigated. The abrupt fault is simply a sudden change of variables describing a sensor. When a fault occurs the value of a parameter jumps to a new constant value. In turn, the incipient fault gradually develops to a larger and larger value. The strength f_s of the incipient fault can be modeled as follows:

$$f_s(t) = \frac{t - t_{from}}{t_{end} - t_{from}} val \tag{18}$$

where t_{from} is the fault start up time, f_{end} is the fault end time and val is the fault strength at t_{end}. For the system working repetitively, this means that an incipient fault can develop through a number of subsequent trials. The faults can also be split into multiplicative and additive ones. The multiplicative fault is simulated as:

$$\omega_m(t) = \omega(t)(1 + f_s(t)), \tag{19}$$

where $\omega(t)$ is the real value of the process variable, $\omega_m(t)$ is the measured value and f_s is the fault intensity. The additive fault is simulated adding some bias to the angular velocity sensor as follows:

$$\omega_m(t) = \omega(t) + f_s(t). \tag{20}$$

In the case study the following faulty scenarios were investigated:

- scenario f_1—abrupt fault, multiplicative type, fault intensity: $f_s(t) = 0.05$;
- scenario f_2—abrupt fault, additive type, fault intensity: $f_s(t) = +0.05$;
- scenario f_3—abrupt fault, additive type, fault intensity: $f_s(t) = -0.05$;
- scenario f_4—incipient fault (18), additive type (20), final fault intensity: $val = 0.05$;
- scenario f_5—incipient fault (18), multiplicative type (19), final fault intensity: $val = 0.05$.

Each abrupt fault was simulated at the 11th trial and lasted for the following 20 trials. Figure 7 shows the behavior of the control system without fault tolerant abilities. In Figure 7a one can see the control performance of the healthy system. ILC drives the system to follow the reference closely with the tracking error norm below 0.2. Figure 7b displays the results in the case of scenario f_1. Clearly, the abrupt multiplicative fault with the intensity $f_s(t) = 0.05$ significantly deteriorated the control performance. During the fault occurrence the tracking error norm increased to the level of more than 0.3. Even worse results were observed for the additive faults either with the positive or negative bias (Figure 7c,d). Clearly, any bias in the angular velocity sensor significantly brought down the control quality. The incipient faults were simulated at the 30th time instant of the 11th trial and lasted for the next 2000 time instants. This means that the incipient fault develops during the next seven trials. Figure 7e and Figure 7f show the control performance in case of the incipient scenarios f_4 and f_5, respectively. In general, additive faults had the stronger impact on the control performance relative to multiplicative ones. All presented scenarios clearly show that there is a need for applying a control system that renders it possible to accommodate faults and maintain the acceptable control performance in case of faults.

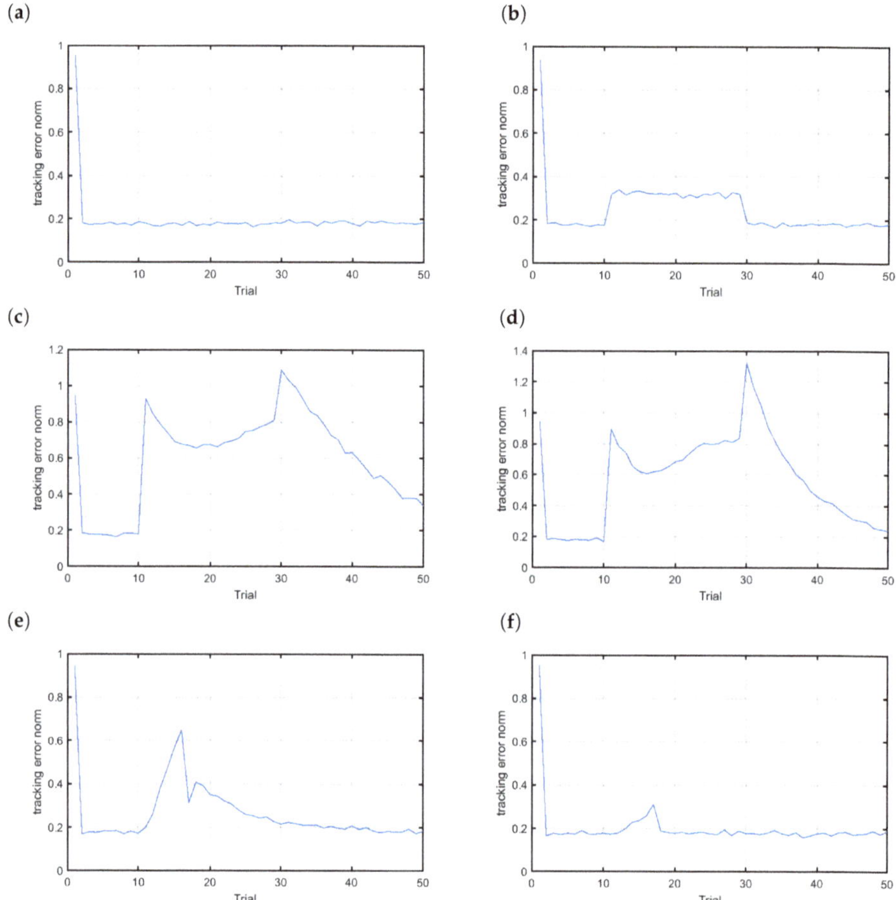

Figure 7. Control of the magnetic brake: (**a**) normal functioning, (**b**) scenario f_1, (**c**) scenario f_2, (**d**) scenario f_3, (**e**) scenario f_4 and (**f**) scenario f_5.

Abrupt faults. As mentioned earlier, each abrupt scenario was simulated at the 11th trial and lasted for 20 trials. Fault detection was carried out by means of the constant threshold (15). Then, the performance of the control system should first be evaluated during normal operation conditions. The control was carried out for the 100 trials, for each trial calculating the value of the diagnostic signal. After that the mean value as well as standard deviation were calculated to be:

$$\mu_r = -0.0226, \quad \sigma_r = 0.0344.$$

For the significance level $\alpha = 0.05$ the value of the threshold was $T = 1.644$.

In Figure 8a one can see the diagnostic signal (solid blue line) with the threshold marked with the dashed red line for the fault scenario f_1. Clearly, the fault is reliably detected with the trial detection index p_d equal to 1. Due to the fact that ILC works offline, i.e., the control signal is derived based on data recorded during the previous trial, the fault accommodation can be immediately launched. The fault-tolerant control results are presented in Figure 8b. The fault was completely accommodated and the tracking error norm remained on the acceptable level. Additionally, the neural model had the property of data filtering. During training, the model weights are derived in such a way as to make it possible for the model to approximate any nonlinear mapping. That is the reason for the flat region observed in the tracking error norm course between the 11th and 30th trial.

(a)   **(b)**

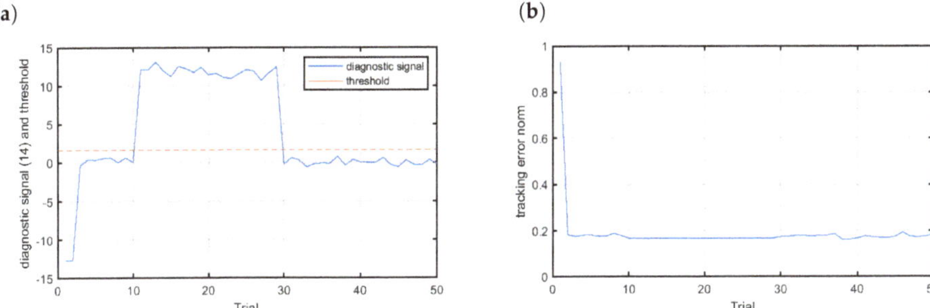

Figure 8. Fault-tolerant control—scenario f_1: (**a**) diagnostic signal (solid blue line) and the threshold (dashed red line), (**b**) control results.

Figure 9 shows the results for the faulty scenario f_2. The changes observed in the diagnostic signal (Figure 9a) were not as large as in the case of the multiplicative fault; however, this fault was also quickly detected with $p_d = 1$, which renders possible the fast reconstruction of the signal measured by the faulty sensor while keeping high quality control of the magnetic brake.

The last abrupt scenario was the additive fault with negative fault intensity (Figure 10). Contrary to the previously analyzed scenarios, this time the fault effect was distinctly visible in the diagnostic signal course (see Figure 10a). As with the previous abrupt scenarios the fault accommodation was immediately launched. The fault was detected within one trial. The fault-tolerant control was satisfactory, as presented in Figure 10b.

(**a**)   (**b**)

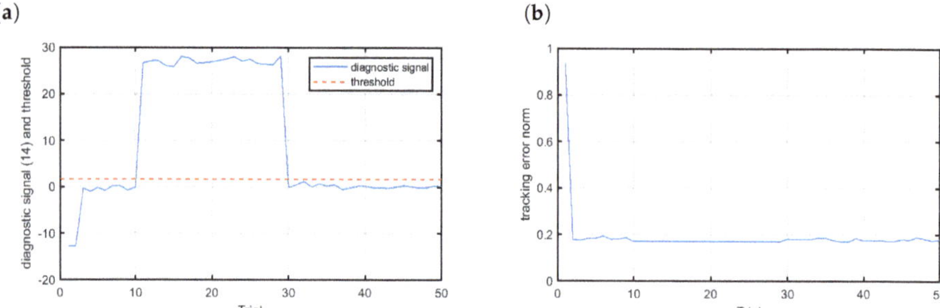

Figure 9. Fault-tolerant control—scenario f_2: (**a**) diagnostic signal (solid blue line) and the threshold (dashed red line), (**b**) control results.

(**a**)   (**b**)

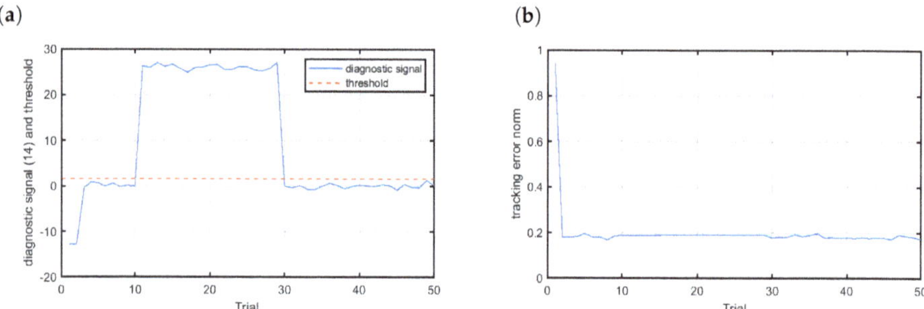

Figure 10. Fault-tolerant control—scenario f_3: (**a**) diagnostic signal (solid blue line) and the threshold (dashed red line), (**b**) control results.

Incipient faults. The incipient faults were introduced at the 11th trial. The fault startup time was set to the 30th time instant and the fault end time was the 2030th time instant. The full strength of incipient faults was achieved at the 17th trial. The threshold level used was the same as in case of abrupt faults.

Figure 11 presents the results achieved for the additive scenario f_4. As the fault developed slowly its detection took some time. In Figure 11a one can see that the diagnostic signal started to increase when the fault affected the velocity sensor. The diagnostic signal crossed the threshold at the 12th trial. This time fault accommodation launched one trial later than in the case of the abrupt faults. Nonetheless, the fault-tolerant control worked with a quality similar to the normal operating conditions. As a matter of fact, a slight increase of the tracking error norm was observable, but this change was not significant, contrary to the control scheme without FTC (compare with Figure 7e).

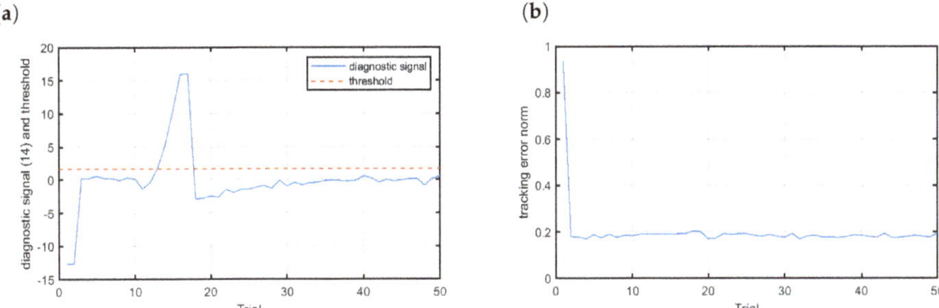

Figure 11. Fault-tolerant control—scenario f_4: (**a**) diagnostic signal (solid blue line), (**b**) control results.

The last investigated scenario was the incipient multiplicative sensor fault. The obtained results are shown in Figure 12. It is clear that the evolution of the diagnostic signal was quite similar to that observed for the incipient additive fault f_4. The fault was accommodated at the 12th trial (Figure 12a). In means that the trail of fault detection was equal to 2. Despite the one trail delay of the reconstruction of the signal measured by the velocity sensor, the fault did not affect the quality of the control at all (see Figure 12b). This phenomenon can be easily explained by analyzing the evolution of the incipient multiplicative fault. From (18) we know that after one trial the fault intensity growed up to the value of 0.0073. For the multiplicative fault this means that the value measured by the velocity sensor increased by $\approx 0.007\%$. Such a fault intensity is not meaningful to the work of the control system.

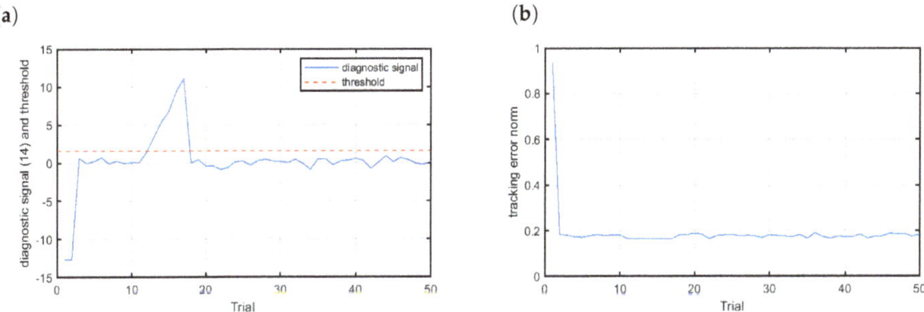

Figure 12. Fault-tolerant control—scenario f_5: (**a**) residual, (**b**) control results.

The main objective of the FTC system is to maintain the current performance of the system as close as possible to the desirable one in the presence of faults. Then, the main performance index pointing out the quality of the proposed FTC system is the value of the tracking error. All reported experimental results clearly show that sensor faults were detected fast enough to assure high quality reference tracking. As shown in Figures 8b, 9b, 10b, 11b and 12b, the tracking error norm in case of a fault was kept on the level observed for the nominal operation conditions (see Figure 7a).

Fault size influence. To verify the performance as well as the sensitivity of the proposed FTC scheme the influence of the fault size on the control system performance was investigated. Key results are shown in Table 2. The abrupt/additive fault with a size smaller than 0.01 stayed undetected. For the abrupt/multiplicative fault slight problems were observed for the fault size $val = 0.01$. In this case it was observed that for several trials the value of the diagnostic signal was below the threshold. However, the fault was marked as detected because during a predominant number of trials the diagnostic signal crossed the threshold. However, for a fault size equal to 0.008 serious problems were already observed as the diagnostic signal oscillated around the threshold and the fault was not detected anymore. For a fault size smaller than 0.006, the diagnostic signal remained below the threshold, indicating that the fault was undetected. In the case of an incipient fault, the smaller was the fault size, the longer was the fault detection time, which is expected behavior. For a fault size smaller than 0.007, the fault was not detectable. Generally speaking, the portrayed results clearly show that faults with an intensity of less than 0.01 were not detected by the proposed fault diagnosis procedure. Thus, the crucial question arises of what is the quality of the control when a fault occurred in a sensor and it was not detected. The simple verification is to compare the tracking error norm in the case of control with the presence of undetected fault with control at the nominal operating conditions. For example, for an incipient fault with a size equal to 0.007, the tracking error norm took the values from the interval $[0.1666, 0.1783]$. Similar values were observed for the control at nominal operating conditions (see Figure 7a). Clearly, for the fault-free case, the tracking error norm converged to the level of the norm of disturbances acting on the magnetic brake. Then, a fault with an intensity of less than 0.01 influenced the system similarly to the external disturbance and had no adverse impact on the control system performance.

Table 2. Influence of the fault size on the fault diagnosis performance.

Scenario	Type	Size *Val*	P_d	Remarks
f_1	abrupt/multiplicative	0.01	1	At some trials the diagnostic signal was below the threshold
f_1	abrupt/multiplicative	0.008	undetected	Oscillations around threshold
f_1	abrupt/multiplicative	0.006	undetected	The diagnostic signal was permanently below the threshold
f_3	abrupt/additive	0.01	undetected	The diagnostic signal was permanently below the threshold
f_5	incipient/additive	0.02	5	—
f_5	incipient/additive	0.01	6	—
f_5	incipient/additive	0.007	undetected	The diagnostic signal was permanently below the threshold

6. Concluding Remarks

The proposed method of sensor fault-tolerant control was found to be very efficient in the case of the class of nonlinear systems represented by the example of a magnetic brake. The advantage of the iterative control scheme is that the approach offers great flexibility to reduce the modeling and process uncertainty of nonlinear control design using the observational data available from previous trials. Despite its somewhat complex mathematical foundations, the resulting fault detection scheme is very easy to implement in practical conditions.

There is still space for further improvements, for example, to incorporate a feedback controller so as to increase the control performance with respect to online disturbances. More precisely, the offline ILC is not suitable for accurate compensation of random disturbances. In addition, the identification within a feedback loop is more difficult. As for a potential solution, the approach reported in [23] developed in the context of a similar class of nonlinear systems and feedback control can be extended to the online procedure.

Another potential future research direction that is very important in the context of neural modeling is the proper choice of representative observational data. A successful application of experimental design theory to solve this problem was developed in [22,26,27] but not in the context of FTC. The efficient adaptation of this approach establishes an open problem, which requires further investigation.

Yet another problem is an extension towards robust solutions with additional constraints imposed on the control design, e.g., related to the actuator faults or uncertainties in the material parameters. This can be solved within the framework of the proposed ILC scheme via exploration of dedicated structures of neural controllers.

Author Contributions: Concept: K.P. and M.P.; methodology: K.P. and M.P.; software: K.P. and M.P.; validation: K.P., M.P.; formal analysis: K.P., M.P. and K.K.; writing—original draft preparation: K.P., M.P. and K.K.; visualization: K.P. and K.K. All authors have read and agreed to the published version of the manuscript.

Funding: This research received no external funding.

Acknowledgments: The work was supported by the National Science Center in Poland, grant No. 2017/27/B/ST7/01874.

Conflicts of Interest: The authors declare no conflict of interest.

References

1. Simeu, E.; Georges, D. Modeling and control of an eddy current brake. *Control. Eng. Pract.* **1996**, *4*, 19–26. [CrossRef]
2. Zamani, A. Design of a controller for rail eddy current brake system. *IET Electr. Syst. Transp.* **2014**, *4*, 38–44. [CrossRef]
3. Yang, J.; Yi, F.; Wang, J. Model-based adaptive control of eddy current retarder. In Proceedings of the 30th Chinese Control and Decision Conference, Shenyang, China, 9–11 June 2018; pp. 1889–1891.
4. Lee, K.; Park, K. Optimal robust control of a contactless brake system using an eddy current. *Mechatronics* **1999**, *9*, 615–631. [CrossRef]
5. Anwar, S. A torque based sliding model control of an eddy current braking system for automotive applications. In Proceedings of the ASME International Mechanical Engineering Congress and Exposition, Orlando, FL, USA, 5–11 November 2005; pp. 297–302.
6. Xu, Y.N.; Deng, W.W. Research of Multiple Sensors Adaptive Fault-Tolerant Control Based on T-S Fuzzy Model for EMB System. *Int. J. Eng. Technol.* **2015**, *7*, 65–67. [CrossRef]
7. Huang, S.; Zhou, C.; Yang, L.; Qinab, Y.; Huang, X.; Huab, B. Fault-tolerant braking control with integerated EMBs and regenerative in-wheel motors. *Reliab. Eng. Syst. Saf.* **2016**, *149*, 148–163. [CrossRef]
8. Kim, S.; Huh, K. Fault-tolerant braking control with integerated EMBs and regenerative in-wheel motors. *Int. J. Automot. Technol.* **2016**, *17*, 923–936. [CrossRef]
9. Bristow, D.A.; Tharayil, M.; Alleyne, A.G. A survey of Iterative Learning Control: A learning-based method for high-performance tracking control. *IEEE Control Syst. Mag.* **2006**, *26*, 96–114.
10. Chien, C.J. A discrete iterative learning control for a class of nonlinear time-varying systems. *IEEE Trans. Autom. Control* **1998**, *43*, 748–752. [CrossRef]
11. Blanke, M.; Kinnaert, M.; Lunze, J.; Staroswiecki, M. *Diagnosis and Fault-Tolerant Control*; Springer: New York, NY, USA, 2016.
12. Noura, H.; Theilliol, D.; Ponsart, J.; Chamseddine, A. *Fault-Tolerant Control Systems: Design and Practical Applications*; Springer: Berlin, Germany, 2003.
13. Ducard, G.J.J. *Fault-Tolerant Flight Control and Guidance Systems: Practical Methods for Small Unmanned Aerial Vehicles*; Advances in Industrial Control; Springer: London, UK, 2009.
14. Mahmoud, M.; Jiang, J.; Zhang, Y. *Active Fault Tolerant Control Systems: Stochastic Analysis and Synthesis*; Springer: Berlin, Germany, 2003.
15. Patan, K.; Patan, M. Neural-network-based high-order iterative learning control. In Proceedings of the 2019 American Control Conference (ACC), Philadelphia, PA, USA, 9–11 July 2019; pp. 2873–2878. [CrossRef]
16. Ponsart, J.C.; Theilliol, D.; Aubrun, C. Virtual sensors design for active fault tolerant control system applied to a winding machine. *Control Eng. Pract.* **2010**, *18*, 1037–1044. [CrossRef]

17. Rotondo, D.; Nejjari, F.; Puig, V. A virtual actuator and sensor approach for fault tolerant control of LPV systems. *J. Process. Control.* **2014**, *24*, 203–222. [CrossRef]

18. Tabbache, B.; Benbouzid, M.E.H.; Kheloui, A.; Bourgeot, J.M. Virtual-sensor-based maximum-likelihood voting approach for fault-tolerant control of electric vehicle powertrains. *IEEE Trans. Veh. Technol.* **2012**, *62*, 1075–1083. [CrossRef]

19. Sami, M.; Patton, R.J. Wind turbine sensor fault tolerant control via a multiple-model approach. In Proceedings of the 2012 UKACC International Conference on Control, IEEE, Cardiff, UK, 3–5 September 2012; pp. 114–119.

20. Patan, M. *Optimal Sensor Networks Scheduling in Identification of Distributed Parameter Systems*; Springer Science & Business Media: Berlin/Heidelberg, Germany, 2012; Volume 425.

21. Ahn, H.S.; Moore, K.L.; Chen, Y. *Iterative Learning Control. Robustness and Monotonic Convergence for Interval Systems*; Communications and Control Engineering; Springer: London, UK, 2007.

22. Patan, K.; Patan, M. Design and convergence of iterative learning control based on neural networks. In Proceedings of the European Control Conference, ECC 2018, Limassol, Cyprus, 12–15 June 2018; pp. 3161–3166. [CrossRef]

23. Patan, K.; Patan, M. Neural-network-based iterative learning control of nonlinear systems. *ISA Trans.* **2020**, *98*, 445–453. [CrossRef] [PubMed]

24. Leith, D.J.; Leithead, W.E. Survey of gain-scheduling analysis and design. *Int. J. Control* **2000**, *73*, 1001–1025. [CrossRef]

25. Sollich, P.; Krogh, A. Learning with ensembles: How over-fitting can be useful. *Adv. Neural Inf. Process. Syst.* **1996**, *9*, 190–196.

26. Patan, K.; Patan, M.; Kowalów, D. Optimal sensor selection for model identification in iterative learning control of spatiotemporal systems. In Proceedings of the 55th IEEE Conference on Decision and Control (CDC), Las Vegas, NV, USA, 12–14 December 2016.

27. Patan, K.; Patan, M.; Kowalów, D. Neural networks in design of iterative learning control for nonlinear systems. *IFAC PapersOnLine* **2017**, *50*, 13402–13407. [CrossRef]

Article

Towards Simultaneous Actuator and Sensor Faults Estimation for a Class of Takagi-Sugeno Fuzzy Systems: A Twin-Rotor System Application

Marcin Pazera, Marcin Witczak, Norbert Kukurowski and Mariusz Buciakowski *

Institute of Control and Computation Engineering, University of Zielona Góra, ul. Szafrana 2,
65-516 Zielona Góra, Poland; m.pazera@issi.uz.zgora.pl (M.P.); m.witczak@issi.uz.zgora.pl (M.W.);
n.kukurowski@issi.uz.zgora.pl (N.K.)
* Correspondence: m.buciakowski@issi.uz.zgora.pl

Received: 6 May 2020; Accepted: 17 June 2020; Published: 19 June 2020

Abstract: The paper is devoted to the problem of estimating simultaneously states, as well as actuator and sensor faults for Takagi–Sugeno systems. The proposed scheme is intended to cope with multiple sensor and actuator faults. To achieve such a goal, the original Takagi–Sugeno system is transformed into a descriptor one containing all state and fault variables within an extended state vector. Moreover, to facilitate the overall design procedure an auxiliary fault vector is introduced. In comparison to the approaches proposed in the literature, a usual restrictive assumption concerning fixed fault rate of change is removed. Finally, the robust convergence of the whole observer is guaranteed by the so-called quadratic boundedness approach which assumes that process and measurement uncertainties are unknown but bounded within an ellipsoid. The last part of the paper portrays an exemplary application concerning a nonlinear twin-rotor system.

Keywords: fault detection and diagnosis; faults estimation; actuator and sensor fault; observer design; Takagi-Sugeno fuzzy systems

---

1. Introduction

Nowadays, owing to the interest in highly efficient systems, industrial companies permanently extend the number of sensors and actuators being used. Indeed, with the advent of IoT the number of components is systematically proliferating. This is mainly due to the fact of their relative low cost and wide accessibility. Irrespective of such an appealing effect, such a growth can increase the chance that actuator and sensor faults appear simultaneously. Moreover, the probability of multiple actuator and sensor faults is also increased. Thus, fault estimation is an important actor in modern Fault Diagnosis (FD) [1–4]. Indeed, it may give a knowledge about the presence, location and size of the fault. Such information is also a necessary ingredient of the active Fault-Tolerant Control (FTC) [5–8], which can accommodate the fault using appropriate control-oriented recovery actions.

There is no doubt that the problem of fault estimation was approached from many different angles. However, most of them focus on estimating either actuator or sensor faults. For a good representative example of recent developments in the area of actuator fault estimation the reader is referred to [9,10]. The case of sensor fault estimation can be realized similarly (cf. Pazera et al. [6] and the references therein).

The number of observers for simultaneous actuator and sensor faults is proliferating as well. However, most of them suffer due to the assumption of a limited fault rate, i.e., the fault derivative (or difference in consecutive discrete samples) is assumed to be close to zero.

Among the existing fault estimation strategies, the crucial attention is focused on the ones based on the sliding-mode observers [11,12], as well as the Kalman filter [13,14] and the related applications.

Taking into account the above discussion, it is also natural that the problem of simultaneous actuator and sensor fault estimation receives growing research attention [8,15–18]. However, the researches are developed for linear systems while the number of strategies capable of handling some classes of nonlinear systems is rather limited. Indeed, a recent state-of-the-art overview clearly indicates the trends for handling multiple simultaneous actuator and sensor fault estimation for nonlinear systems. Indeed, Gu et al. [19] converted an original Lipschitz system into a Linear Parameter Varying (LPV) one by a suitable reformulation of the Lipschitz property. The resulting design approach provides optimal fault estimates according to $\mathcal{H}_\infty$ criterion within the frequency domain. Another appealing strategy for Lipschitz class of systems was introduced in Abdollahi [20]. It transforms the system into two subsystems while each of them is affected by either a sensor or an actuator fault, respectively. Finally, two separate Sliding Mode Observers (SMOs) are designed. This scheme suffers from the fact that an asymptotically convergence is guaranteed without considering any robustness to modelling uncertainities and/or external disturbances. A similar SMO was also developed in Hmidi et al. [21] and the authors remove the above drawback by ensuring robustness to bounded disturbances.

An important group of strategies convert the original nonlinear system into an equivalent Takagi-Sugeno one. A representative example of such strategies is presented in Bounemeur et al. [22]. As a result, an adaptive fuzzy estimator is developed capable of handling numerous fault scenarios. Unfortunately, it is devoted to deterministic systems neglecting uncertainty factors. Another strategy is developed in Fu et al. [23] and tackles the systems with switching non-linearities. Unfortunately, it also does not provide appropriate means for settling the robustness issue. Finally, in Shaker [24] a novel multiple integral unknown input observer is developed that is capable of decoupling disturbances. The final group of approaches are the ones for polynomial systems [25,26]. In this case, the design methodology reduces to converting the system by augmenting the state vector with fault variables. This strategy can also cope with process disturbances.

The unappealing feature of the existing approaches to simultaneous actuator and sensor fault estimation is that they do not provide sufficient information about the estimation quality. To handle this issue, two kind of approaches can be utilized: the first one uses post-fault measurements [27] and the second predicts the fault based on its historical values and past measurements [28]. It should be noted that the latter is the only scheme which can be applied in the active FTC. Unfortunately, such schemes struggle with the problem of minimising the effect of fault prediction [9,28].

The scheme proposed in this paper can handle the above mentioned issues, which constitutes the novelty of the proposed approach:

- the problem of one-step fault prediction and the related fault rate of change is removed by transforming the system into the descriptor Takagi-Sugeno one whose state vector contains both original states and the faults.
- the original system with the actuator fault with a time-varying distribution matrix is transformed into an equivalent one with a constant distribution matrix and the so-called auxiliary fault vector. This strategy naturally reduces the design conservatives.
- the effect of external disturbances is tackled with the so-called quadratic boundedness. As a result, the estimation quality can be assessed by the so-called uncertainty intervals of both states and faults. Thus, they can be perceived as possible worst cases of the unknown faults and states.

The paper is organised as follows. In Section 2, a simultaneous estimation of actuator and sensor faults problem is described along with suitable feasibility assumptions. Furthermore, a suitable convergence analysis is performed. Section 3 exhibits an illustrative example pertaining the twin-rotor system. Finally, Section 4 concludes the paper.

2. Observer Design

Let us start from the formulation of a Takagi-Sugeno (T-S) fuzzy system:

$$x_{k+1} = A\left(v_k\right)x_k + B\left(v_k\right)u_k + B_f\left(v_k\right)\bar{f}_{a,k} + W_1 w_{1,k}$$
$$= \sum_{i=1}^{M} h_i\left(v_k\right)\left[A^i x_k + B^i u_k + B_f^i \bar{f}_{a,k}\right] + W_1 w_{1,k}, \tag{1}$$

$$y_k = C x_k + C_f f_{s,k} + W_2 w_{2,k}, \tag{2}$$

with

$$h_i(v_k) \geq 0, \qquad \forall i = 1, \ldots, M_m, \qquad \sum_{i=1}^{M_m} h_i(v_k) = 1, \tag{3}$$

where k stands for the discrete time and $x_k \in \mathbb{X} \subset \mathbb{R}^n$, $u_k \in \mathbb{R}^r$, $y_k \in \mathbb{R}^m$ are state, input and output vectors, respectively. Furthermore, $\bar{f}_{a,k} \in \mathbb{F}_a \subset \mathbb{R}^{n_a}$ and $f_{s,k} \in \mathbb{F}_s \subset \mathbb{R}^{n_s}$ signify the actuator and sensor fault vectors, respectively, where n_a and n_s stand for the number of actuator and sensor faults, respectively. Matrices A, B and C are the known state, input and output matrices, respectively, while M_m stands for the number of submodels. Thus, C_f denotes the sensor fault distribution matrix, where $\text{rank}(C_f) = n_s$ and $\text{rank}(B_f(v_k)) = n_a$ satisfy the inequality $n_a + n_s \leq m$. This means that it is impossible to estimate more faults than there are measured outputs. Finally, W_1 and W_2 are the distribution matrices of $w_{1,k}$ and $w_{2,k}$, which are exogenous disturbance vectors for the process and measurement uncertainties, respectively. The activation functions $h_i(\cdot)$ depend on the vector of premise variables $v_k = \left[v_k^1, v_k^2, \ldots, v_k^p\right]^T$, which is assumed to depend on measurable variables, e.g., system outputs and known inputs [29]. However, an extension towards unmeasurable premise variables is possible via direct applications of the solution proposed, e.g., in Ichalal et al. [30].

Finally, in the remaining part of the paper, the following notation is used $X\left(v_k\right) = \sum_{i=1}^{M} h_i\left(v_k\right)X^i$. First, let us start with transforming the state Equation (1) into an equivalent form

$$x_{k+1} = \sum_{i=1}^{M} h_i\left(v_k\right)\left[A^i x_k + B^i u_k\right] + B_f f_{a,k} + W_1 w_{1,k}, \tag{4}$$

with an auxiliary matrix B_f satisfying $\text{rank}(B_f) = n_a$ and an auxiliary actuator fault vector $f_{a,k}$. Comparing Equation (1) and Equation (4), it can be observed that

$$B_f(v_k)\bar{f}_{a,k} = B_f f_{a,k}. \tag{5}$$

Thus, having f_k, the original fault vector can be determined with

$$\bar{f}_{a,k} = (B_f(v_k))^\dagger B_f f_{a,k}. \tag{6}$$

where † stands for the pseudo inverse operator. Thus, the objective of further deliberations is to propose a novel observer capable of providing x_k, $f_{a,k}$ and $f_{s,k}$ estimates simultaneously. It should be also noted that the selection of B_f is not critical, i.e., it can be set as $B_f = \frac{1}{M}\sum_{i=1}^{M} B_f^i$, which clearly preserves $\text{rank}(B_f) = n_a$.

The proposed strategy starts with transforming Equation (4) and Equation (2) into an equivalent descriptor form with the following state variable

$$\bar{x}_k = \left[x_k^T \quad f_{a,k-1}^T \quad f_{s,k}^T\right]^T. \tag{7}$$

Using the above state vector Equation (7), the system Equation (4) and Equation (2) can be rewritten as:

$$E\bar{x}_{k+1} = \bar{A}\left(v_k\right)\bar{x}_k + \bar{B}\left(v_k\right)u_k + \bar{W}_1\bar{w}_k, \tag{8}$$

$$y_k = \bar{C}x_k + \bar{W}_2\bar{w}_k, \tag{9}$$

with:

$$E = \begin{bmatrix} I_n & -B_f & 0 \\ 0 & 0 & 0 \\ 0 & 0 & 0 \end{bmatrix}, \quad \bar{A}\left(v_k\right) = \begin{bmatrix} A\left(v_k\right) & 0 & 0 \\ 0 & 0 & 0 \\ 0 & 0 & 0 \end{bmatrix}, \quad \bar{B}\left(v_k\right) = \begin{bmatrix} B\left(v_k\right) \\ 0 \\ 0 \end{bmatrix},$$

$$\bar{W}_1 = \begin{bmatrix} W_1 & 0 \\ 0 & 0 \\ 0 & 0 \end{bmatrix}, \quad \bar{W}_2 = \begin{bmatrix} 0 & W_2 \end{bmatrix}, \quad \bar{w}_k = \begin{bmatrix} w_{1,k}^T & w_{2,k}^T \end{bmatrix}^T, \quad \bar{C} = \begin{bmatrix} C & 0 & C_f \end{bmatrix}.$$

This means that both faults as well as the state of the system are incorporated into a super state vector $\bar{x}_k$. Thus, for the purpose of the state estimation obeying Equation (8) and Equation (9), the following observer is proposed:

$$z_{k+1} = N\left(v_k\right)z_k + M\left(v_k\right)u_k + L\left(v_k\right)y_k, \tag{10}$$

$$\hat{\bar{x}}_k = z_k + T_2 y_k, \tag{11}$$

where $z_k \in \mathbb{R}^{n+n_a+n_s}$ is the internal state of the estimator while $\hat{\bar{x}}_k \in \mathbb{R}^{n+n_a+n_s}$ stands for the estimate of Equation (7).

It should be noted that the proposed approach eliminates the usual assumption concerning a bounded rate of change of actuator and sensor fault, which increase the conservatives level of the approaches proposed in the literature (see, e.g., Pazera et al. [28] and the references therein.)

Let us start with assuming that there exist matrices T_1 and T_2 such that

$$T_1 E + T_2 \bar{C} = I, \tag{12}$$

or

$$\begin{bmatrix} T_1 & T_2 \end{bmatrix} \begin{bmatrix} E \\ \bar{C} \end{bmatrix} = I, \tag{13}$$

which yields Equation (10) and Equation (11) design condition

$$\mathrm{rank}\left(\begin{bmatrix} E \\ \bar{C} \end{bmatrix}\right) = n + n_a + n_s. \tag{14}$$

Under the above assumption, it is possible to derive the error of state estimation with Equation (11), i.e.,

$$e_k = \bar{x}_k - \hat{\bar{x}}_k = \bar{x}_k - z_k - T_2 y_k = \left(I - T_2\bar{C}\right)\bar{x}_k - z_k - T_2\bar{W}_2\bar{w}_k, \tag{15}$$

which, by using Equation (12), boils down to

$$e_k = T_1 E\bar{x}_k - z_k - T_2\bar{W}_2\bar{w}_k. \tag{16}$$

Thus, by substituting Equations (8)–(10), the dynamics of state estimation error obeys

$$
\begin{aligned}
e_{k+1} &= T_1 E \bar{x}_{k+1} - z_{k+1} - T_2 \bar{W}_2 \bar{w}_{k+1} = T_1 \bar{A}\left(v_k\right) \bar{x}_k + T_1 \bar{B}\left(v_k\right) u_k \\
&\quad + T_1 \bar{W}_1 \bar{w}_k - N\left(v_k\right) z_k - M\left(v_k\right) u_k - L\left(v_k\right) \bar{C} \bar{x}_k - L\left(v_k\right) \bar{W}_2 \bar{w}_k - T_2 \bar{W}_2 \bar{w}_{k+1} \\
&= \left(T_1 \bar{A}\left(v_k\right) - L\left(v_k\right) \bar{C} - \left(v_k\right) T_1 E\right) \bar{x}_k + \left(T_1 \bar{B}\left(v_k\right) - M\left(v_k\right)\right) u_k \\
&\quad + N\left(v_k\right) e_k + \left(T_1 \bar{W}_1 - L\left(v_k\right) \bar{W}_2 + N\left(v_k\right) T_2 \bar{W}_2\right) \bar{w}_k - T_2 \bar{W}_2 \bar{w}_{k+1}.
\end{aligned}
\tag{17}
$$

From Equation (17), it is evident that the following relations should be satisfied:

$$
T_1 \bar{A}\left(v_k\right) - N\left(v_k\right) T_1 E - L\left(v_k\right) \bar{C} = 0,
\tag{18}
$$

$$
T_1 \bar{B}\left(v_k\right) - M\left(v_k\right) = 0.
\tag{19}
$$

Indeed, by satisfying Equation (18), the term related to $\bar{x}_k$ vanished from Equation (17). Similarly, satisfying Equationv Equation (19) means that Equation (18) is no longer dependent on the system input u_k. Applying Equation (12) to Equation (18) gives

$$
T_1 \bar{A}\left(v_k\right) - N\left(v_k\right)\left(I - T_2 \bar{C}\right) - L\left(v_k\right) \bar{C} = 0,
\tag{20}
$$

or

$$
N\left(v_k\right) = T_1 \bar{A}\left(v_k\right) - \left(L\left(v_k\right) - N\left(v_k\right) T_2\right) \bar{C}.
\tag{21}
$$

Finally, by defining

$$
K\left(v_k\right) = L\left(v_k\right) - N\left(v_k\right) T_2,
\tag{22}
$$

equality Equation (21) boils down to

$$
N\left(v_k\right) = T_1 \bar{A}\left(v_k\right) - K\left(v_k\right) \bar{C},
\tag{23}
$$

which makes it possible to transform Equation (17) into

$$
\begin{aligned}
e_{k+1} &= N\left(v_k\right) e_k + \left(T_1 \bar{W}_1 - L\left(v_k\right) \bar{W}_2 + N\left(v_k\right) T_2 \bar{W}_2\right) \bar{w}_k - T_2 \bar{W}_2 \bar{w}_{k+1} \\
&= \left(T_1 \bar{A}\left(v_k\right) - K\left(v_k\right) \bar{C}\right) e_k + T_1 \bar{W}_1 \bar{w}_k - K\left(v_k\right) \bar{W}_2 \bar{w}_k - N\left(v_k\right) T_2 \bar{W}_2 \bar{w}_k \\
&\quad + N\left(v_k\right) T_2 \bar{W}_2 \bar{w}_k - T_2 \bar{W}_2 \bar{w}_{k+1} = \left(T_1 \bar{A}\left(v_k\right) - K\left(v_k\right) \bar{C}\right) e_k \\
&\quad + T_1 \bar{W}_1 \bar{w}_k - K\left(v_k\right) \bar{W}_2 \bar{w}_k - T_2 \bar{W}_2 \bar{w}_{k+1}.
\end{aligned}
\tag{24}
$$

Subsequently, by defining the following super-vector

$$
\tilde{w}_k = \begin{bmatrix} \bar{w}_k \\ \bar{w}_{k+1} \end{bmatrix},
\tag{25}
$$

equality Equation (24) can be rewritten into a simpler form

$$
e_{k+1} = X\left(v_k\right) e_k + Z\left(v_k\right) \tilde{w}_k,
\tag{26}
$$

with:

$$
X\left(v_k\right) = \tilde{A}\left(v_k\right) - K\left(v_k\right) \bar{C}, \qquad Z\left(v_k\right) = \tilde{W}_1 - K\left(v_k\right) \tilde{W}_2,
$$

where:

$$
\tilde{W}_1\left(v_k\right) = \begin{bmatrix} T_1 \bar{W}_1 & -T_2 \bar{W}_2 \end{bmatrix}, \quad \tilde{W}_2 = \begin{bmatrix} \bar{W}_2 & 0 \end{bmatrix}, \quad \tilde{A}\left(v_k\right) = T_1 \bar{A}\left(v_k\right).
$$

For the purpose of further convergence analysis, let us start with reminding the Finsler's Lemma [31]:

Lemma 1. *The following expressions are equivalent:*

1. $\tilde{x}_k^T Q \tilde{x}_k < 0, \qquad \forall \tilde{x} \in \{\tilde{x} \in \mathbb{R}^{n_x} | \tilde{x} \neq 0, R\tilde{x} = 0\},$
2. $\exists \bar{M} \in \mathbb{R}^{n+ns \times m}$ *such that* $Q + \bar{M}R + R^T \bar{M}^T < 0.$

Let us also define the Lyapunov function

$$V_k = e_k^T P e_k, \tag{27}$$

with $P \succ 0$. Furthermore, the estimation error Equation (26) can be rewritten in an alternative form

$$X(v_k) e_k + Z(v_k) \tilde{w}_k - e_{k+1} = 0, \tag{28}$$

which implies that the following extended-vector can be defined

$$\tilde{x}_k = \begin{bmatrix} e_k^T & \tilde{w}_k^T & e_{k+1}^T \end{bmatrix}^T, \tag{29}$$

and as a consequence the following statement can be formulated

$$R(v_k) \begin{bmatrix} e_k \\ \tilde{w}_k \\ e_{k+1} \end{bmatrix} = 0. \tag{30}$$

Thus, based on Equation (28), Equation (29) and Equation (30), it can be shown that

$$R(v_k) = [X(v_k) \; Z(v_k) \; -I], \tag{31}$$

where

$$R(v_k) \tilde{x}_k = 0, \tag{32}$$

in order to satisfy Equation (28). The convergence of the proposed observer is to be determined with the so-called Quadratic Boundedness (QB) approach [32]. This technique can be perceived as an extension of the usual Lyapunov approach towards the systems with external bounded disturbances. The usefulness of QB approach was proven in many papers while in Pazera et al. [28] it was proven that the standard $\mathcal{H}_\infty$ framework can be perceived as a special case of QB. To use the QB approach, it is necessary to assume that $\tilde{w}_k$ is bounded by the following ellipsoid:

$$\mathbb{E}_w = \{\tilde{w}_k : \; \tilde{w}_k^T Q_w \tilde{w}_k \leq 1\}, \tag{33}$$

with $Q_w \succ 0$. Under such assumptions, the following definition can be recalled:

Definition 1. *The system Equation (26) is strictly quadratically bounded for all $\tilde{w}_k \in \mathbb{E}_w$, $k \geq 0$, if $V_k > 1 \implies V_{k+1} - V_k < 0$ for any $\tilde{w}_k \in \mathbb{E}_w$.*

As shown in Alessandri et al. [32] and Pazera et al. [28], the stability condition associated with

$$V_{k+1} - (1 - \alpha) V_k - \alpha \tilde{w}_k^T Q_w \tilde{w}_k < 0, \tag{34}$$

with $0 < \alpha < 1$. Based on the above considerations, the following theorem is established:

Theorem 1. *The observer-based system Equation (26) is strictly quadratically bounded for all $\tilde{w}_k \in \mathbb{E}_w$ if there exist matrices $P \succ 0$, U, $\bar{N}$ as well as $\alpha \in (0,1)$ such that the following holds:*

$$\begin{bmatrix} \alpha P - P & 0 & \tilde{A}^T(v_k)U^T - \bar{C}^T\bar{N}^T(v_k) \\ 0 & -\alpha Q_w & \tilde{W}_1^T U^T - \tilde{W}_2^T \bar{N}^T(v_k) \\ U\tilde{A}(v_k) - \bar{N}(v_k)\bar{C} & U\tilde{W}_1 - \bar{N}(v_k)\tilde{W}_2 & P - U - U^T \end{bmatrix} \prec 0, \tag{35}$$

Proof. Using Equation (34) and setting

$$Q = \begin{bmatrix} \alpha P - P & 0 & 0 \\ 0 & -\alpha Q_w & 0 \\ 0 & 0 & P \end{bmatrix}, \tag{36}$$

$$\bar{M} = \begin{bmatrix} 0^T & 0^T & U^T \end{bmatrix}^T. \tag{37}$$

along with Lemma 1 leads to

$$\begin{bmatrix} \alpha P - P & 0 & X^T(v_k)U^T \\ 0 & -\alpha Q_w & Z^T(v_k)U^T \\ UX(v_k) & UZ(v_k) & P - U - U^T \end{bmatrix} \prec 0. \tag{38}$$

Setting

$$UX(v_k) = U\left(\tilde{A}(v_k) - K(v_k)\bar{C}\right) = U\tilde{A}(v_k) - \bar{N}(v_k)\bar{C}, \tag{39}$$

$$UZ(v_k) = U\left(\tilde{W}_1 - K(v_k)\tilde{W}_2\right) = U\tilde{W}_1 - \bar{N}(v_k)\tilde{W}_2, \tag{40}$$

into Equation (38) completes the proof. $\square$

Irrespective of the incontestable appeal of the proposed solution Equation (35), its numerical tractability is feasible only if it is transformed to the set of linear matrix inequalities under fixed α:

$$\begin{bmatrix} \alpha P - P & 0 & \tilde{A}^{i^T}U^T - \bar{C}^T\bar{N}^{i^T} \\ 0 & -\alpha Q_w & \tilde{W}_1^T U^T - \tilde{W}_2^T \bar{N}^{i^T} \\ U\tilde{A}^i - \bar{N}^i\bar{C} & U\tilde{W}_1 - \bar{N}^i\tilde{W}_2 & P - U - U^T \end{bmatrix} \prec 0, \quad i = 1, \ldots, M. \tag{41}$$

Finally, the design procedure boils down to:

Step 0: Determine T_1 and T_2 by solving Equation (12).
Step 1: Set α, $0 < \alpha < 1$ and determine $\bar{N}^i$ and U by solving Equation (41).
Step 2: Calculate:

$$K^i = U^{-1}\bar{N}^i, \tag{42}$$

$$N^i = T_1\tilde{A}^i - K^i\bar{C}, \tag{43}$$

$$M^i = T_1\bar{B}^i, \tag{44}$$

$$L^i = K^i + N^iT_2. \tag{45}$$

While the application procedure leads to:

Step 0: Substitute $k = 0$ and set the initial conditions z_0.
Step 1: Calculate z_{k+1} and $\hat{x}_k$ with Equations (10) and (11).
Step 2: Set $k = k + 1$ and go to *Step 1*.

As demonstrated in Pazera et al. [28], the value of α has a direct influence of the convergence rate of the underlying observer. On the other hand, matrix P shapes the ellipsoidal bound of the estimation error:

$$e_k^T P e_k \le \zeta_k, \quad \zeta_k = 1 + (1-\alpha)^k (1 - e_0^T P e_0), \tag{46}$$

which can be directly used to determine the uncertainty intervals of the state variables [28]:

$$\hat{\bar{x}}_{k,i} - \sqrt{\zeta_k c_i^T P^{-1} c_i} \le \bar{x}_{k,i} \le \hat{\bar{x}}_{k,i} + \sqrt{\zeta_k c_i^T P^{-1} c_i} \tag{47}$$

where c^i signifies ith column of an identity matrix of an appropriate dimension.

3. Case Study: Twin-Rotor System

The aim of this section is to validate the performance and correctness of the novel observer. Accordingly, the proposed actuator and sensor fault estimation method has been implemented to the Twin-Rotor System (TRS) [28], illustrated in Figure 1. Note that all equations and a detailed nonlinear model of the TRS can be found in [28]. This model was further transformed into the Takagi-Sugeno form using the dedicated methodology proposed in Rotondo et al. [33]. The matrices shaping the model Equation (1) and Equation (2) are presented in the Appendix A.

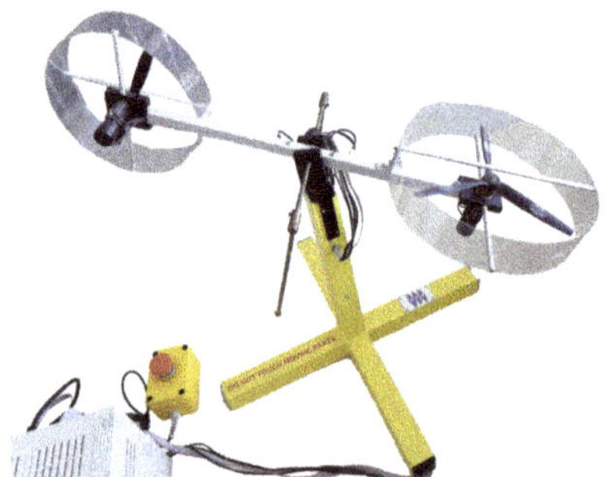

Figure 1. A nonlinear twin-rotor MIMO system.

The state vector of the system is given by

$$x = \begin{bmatrix} \omega_v & \Omega_v & \theta_v & \omega_h & \Omega_h & \theta_h \end{bmatrix}^T, \tag{48}$$

hence the input vector is described by

$$u = \begin{bmatrix} u_v & u_h \end{bmatrix}^T, \tag{49}$$

where ω_v and Ω_v stand for the rotational and angular velocities of the main rotor while θ_v signifies the pitch angle of the beam. Furthermore, ω_h and Ω_h are the tail rotor rotational as well as angular velocities whilst θ_h is the yaw angle of the beam. Finally, u_v and u_h stand for actuation due to the main and tail DC motors, respectively. It should be noted that both inputs u_v and u_h operate in range $(-1, 1)$. Moreover, all details and parameters can be found in manufacturer's user manual [34] and in Pazera et al. [28].

The following fault scenarios have been considered:

$$f_{a,1,k} = \begin{cases} -0.1 \cdot u_{1,k} & 3000 \leq k \leq 6500 \\ 0 & \text{otherwise,} \end{cases} \tag{50}$$

$$f_{a,2,k} = \begin{cases} 0.1 \cdot u_{2,k} & 4500 \leq k \leq 8000 \\ 0 & \text{otherwise,} \end{cases} \tag{51}$$

$$f_{s,1,k} = \begin{cases} y_k - 2.5 & 5000 \leq k \leq 7000 \\ 0 & \text{otherwise,} \end{cases} \tag{52}$$

$$f_{s,2,k} = \begin{cases} y_k + 1.6 & 6000 \leq k \leq 9500 \\ 0 & \text{otherwise.} \end{cases} \tag{53}$$

According to the approach proposed along with Equation (6):

$$B_f = \frac{1}{M_m} \sum_{i=1}^{M_m} B_f^i = \begin{bmatrix} 49.4084 \cdot 10^{-6} & -66.2480 \cdot 10^{-6} \\ 14.3375 \cdot 10^{-3} & -13.2367 \cdot 10^{-3} \\ 303.3781 & 0 \\ 6.9349 \cdot 10^{-6} & 3.6497 \cdot 10^{-6} \\ 1.3860 \cdot 10^{-3} & 1.0854 \cdot 10^{-3} \\ 0 & 58.0016 \end{bmatrix}, \tag{54}$$

which means that all actuators are considered as possibly faulty and their faults have to be estimated. Whilst the sensor fault distribution matrix is obtained from matrix C (see Appendix A, Equation (A25)) by extracting the rows corresponding to the first and second sensor, respectively and it is given as follows

$$C_f = \begin{bmatrix} 0 & 0 & 1 & 0 & 0 \\ 0 & 0 & 0 & 1 & 0 \end{bmatrix}^T. \tag{55}$$

It can be clearly viewed that within this scenario, both actuator and sensor faults are partially at the same time. Moreover, the considered faults are constant biases, which are either positive or negative. From Equations (50)–(53), it can be clearly viewed that actuator and sensor faults appear partially at the same time. The considered actuator fault Equation (50) pertains a 10% performance decrease of the first actuator, while Equation (51) concerns a 10% performance increase of the second actuator. Moreover, the considered sensor faults Equation (52) and Equation (53) are represented by either positive or negative constant biases equal to 1.6 or −2.5, respectively. Thus, they denote a significant sensor readings inaccuracies. However, the distribution matrix of sensor fault denotes that they influence the measurements of the angular velocity of the tail rotor Ω_h as well as the angular velocity of the main rotor Ω_v. Firstly, let us present a comparison between nonlinear estimation approach described in Pazera et al. [28] and the proposed Takagi-Sugeno scheme, which is shown in Figure 2a,b. As it can be observed, the Takagi-Sugeno model-based approach follows the response of the original nonlinear model-based one, and hence, it does not impair the quality of state and fault estimates significantly.

Figure 3a,b illustrate the real actuator faults of the main $f_{a,1}$ and tail $f_{a,2}$ rotor with blue dash-dotted lines along with their estimates given with red dashed lines. Additionally, the real and estimated values are overbounded by uncertainty intervals which are indicated with black dashed lines. Moreover, the real sensor faults $f_{s,1}$ and $f_{s,2}$ are presented in Figure 4a,b with blue dash-dotted lines as well as their estimates depicted by red dashed lines along with the uncertainty intervals indicated with black dashed lines. From these figures it is easily seen that the actuator and sensor faults were estimated with a very good accuracy under the measurement uncertainties. Consequently, the system states have been correctly reconstructed.

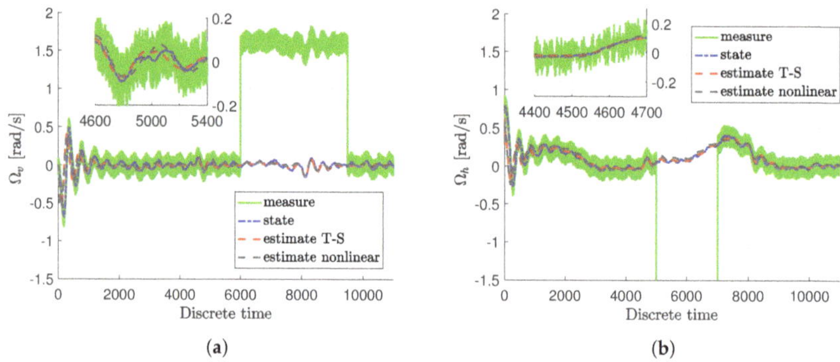

Figure 2. Comparison between nonlinear and Takagi-Sugeno response of the system: angular velocity of the main rotor Ω_h (**a**) and angular velocity of the tail rotor Ω_v (**b**).

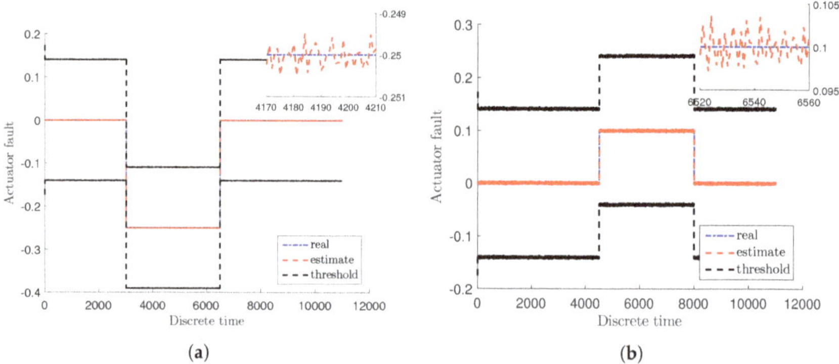

Figure 3. Actuator faults $f_{a,1}$ (**a**) and $f_{a,2}$ (**b**).

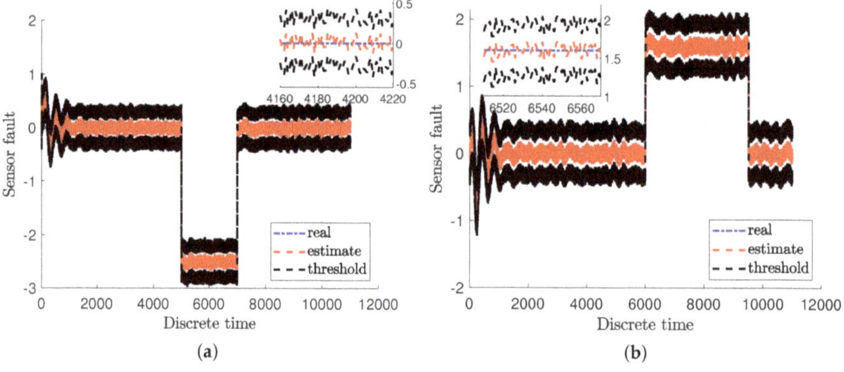

Figure 4. Sensor faults $f_{s,1}$ (**a**) and $f_{s,2}$ (**b**).

Figure 5a,b show the real rotational velocities of the main ω_v and tail ω_h rotor with blue dash-dotted lines. Their measured outputs are presented by light green solid lines along with their estimates given with red dashed lines as well as the uncertainty intervals indicated with black dashed lines. Moreover, Figure 6a,b present the angular velocities of the main Ω_v and of the tail Ω_h rotor with blue dash-dotted lines whilst their estimates are given with red dashed lines as well as the measured outputs depicted by light green solid lines. Additionally, the uncertainty intervals are indicated with black dashed lines. As can be observed in Figure 6a,b, irrespective of the intermittent fault (marked in light green) the state estimates are very close to the original states. The rotational velocities of the main and tail rotor and the pitch angle of the beam have been accurately estimated even if the actuator and sensor faults occurred simultaneously. This fact is illustrated in Figure 7a,b and Figure 8, which show the evolution of the state estimation error for the above variables. The figures clearly show that the states are properly reconstructed under the actuator glitch along with positive and negative incorrect readings of the sensor.

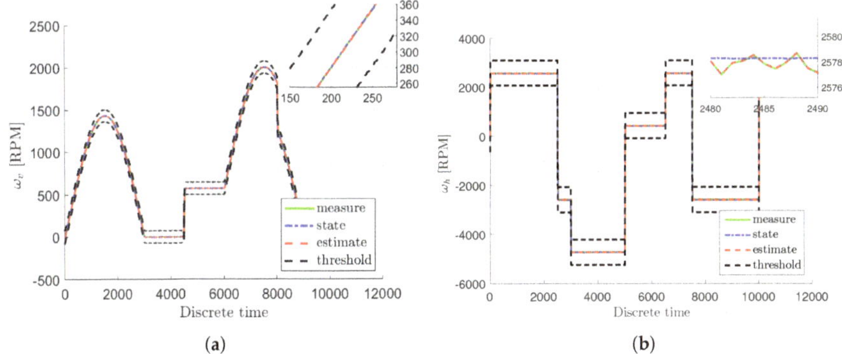

Figure 5. State variables ω_v (**a**) and ω_h (**b**).

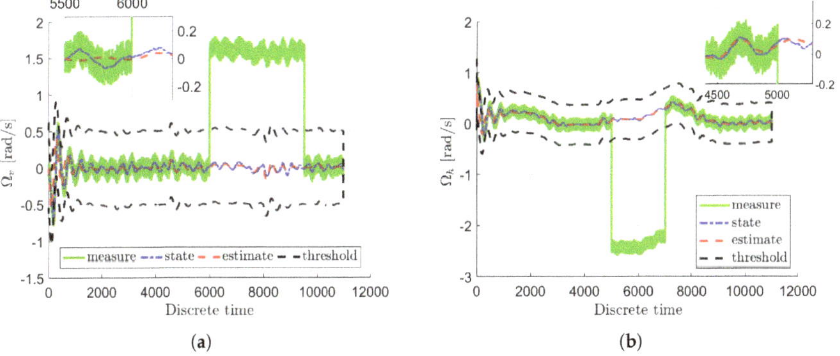

Figure 6. State variables Ω_v (**a**) and Ω_h (**b**).

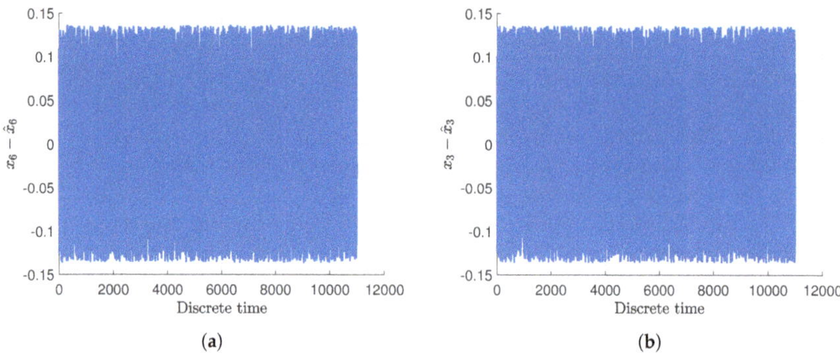

(a)   (b)

Figure 7. Evolution of the state estimation error for the rotational velocities of the main rotor (**a**) and of the tail rotor (**b**).

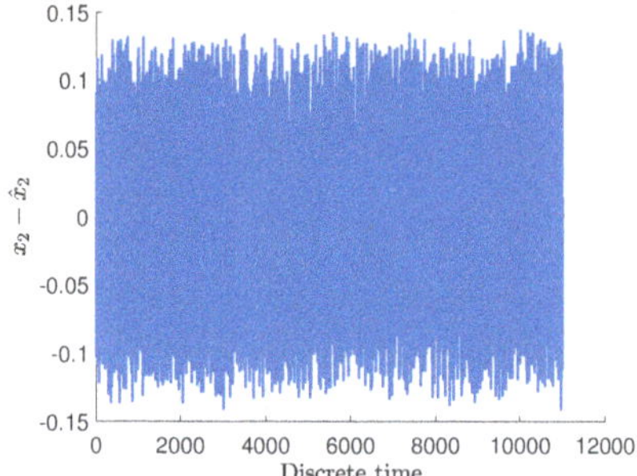

Figure 8. Evolution of the state estimation error for the pitch angle of the beam.

4. Concluding Remarks

An important aspect of the paper was to develop an observer for the Takagi-Sugeno systems which would be capable of estimating both state with both actuator and sensor. The first part of the proposed strategy was developed in such a way that no adaptive observer is required to obtain the sensor fault estimates. In other words, it is obtained directly by simply transforming an output equation. The second part of the observer concerns the adaptive actuator fault estimation and the state as well. In order to achieve robustness against process and measurement uncertainties and to ensure the stability performance a Quadratic Boundedness strategy is employed. For the purpose of obtaining the observer gain matrices, a set of LMIs is needed to be solved. The other part of the paper presents a performance validation. It has been achieved by implementing the approach to a nonlinear Twin-Rotor MIMO System. The obtained results plainly confirm that the desired properties of the observer have been accomplished. The state has been estimated properly even in the case of both the actuator and sensor fault case. While analysing an actuator fault, it can be concluded that its estimation error is at a very low level in both cases: fault and fault-free as well. Furthermore, the obtained sensor fault estimates quality strongly depends on the quality of the state estimation. The future work will be devoted to applying such an approach to a fault-tolerant control scheme for a T–S fuzzy systems.

Author Contributions: M.P. derived the design procedure for fault estimation. M.W. contributed to the development of the observer structure and determination of the uncertainty interval calculation rules. N.K. contributed to the experiments and described the experimental results. M.B. contributed to the development of the convergence conditions and review of the literature. All authors have read and agreed to the published version of the manuscript.

Funding: The work was supported by the National Science Centre, Poland under Grant: UMO-2017/27/B/ST7/00620.

Conflicts of Interest: The authors declare no conflict of interest.

Appendix A

$$
A^1 = \begin{bmatrix}
1 & 9.9709 \cdot 10^{-3} & 4.5486 \cdot 10^{-7} & -5.1930 \cdot 10^{-7} & -1.7300 \cdot 10^{-9} & -1.5629 \cdot 10^{-14} \\
0 & 9.9418 \cdot 10^{-1} & 8.5060 \cdot 10^{-5} & -1.0376 \cdot 10^{-4} & -5.1856 \cdot 10^{-7} & -6.2171 \cdot 10^{-12} \\
0 & 0 & 6.6291 \cdot 10^{-1} & 0 & 0 & 0 \\
0 & 0 & 0 & 9.9988 \cdot 10^{-1} & 9.9785 \cdot 10^{-3} & 1.7793 \cdot 10^{-7} \\
0 & 0 & 0 & -2.4758 \cdot 10^{-2} & 9.9566 \cdot 10^{-1} & 3.4967 \cdot 10^{-5} \\
0 & 0 & 0 & 0 & 0 & 9.0333 \cdot 10^{-1}
\end{bmatrix}, \quad (A1)
$$

$$
A^2 = \begin{bmatrix}
1 & 9.9709 \cdot 10^{-3} & 4.5486 \cdot 10^{-7} & -4.6737 \cdot 10^{-7} & -1.5570 \cdot 10^{-9} & -1.4066 \cdot 10^{-14} \\
0 & 9.9418 \cdot 10^{-1} & 8.5060 \cdot 10^{-5} & -9.3380 \cdot 10^{-5} & -4.6671 \cdot 10^{-7} & -5.5954 \cdot 10^{-12} \\
0 & 0 & 6.6291 \cdot 10^{-1} & 0 & 0 & 0 \\
0 & 0 & 0 & 9.9988 \cdot 10^{-1} & 9.9785 \cdot 10^{-3} & 1.7793 \cdot 10^{-7} \\
0 & 0 & 0 & -2.4758 \cdot 10^{-2} & 9.9566 \cdot 10^{-1} & 3.4967 \cdot 10^{-5} \\
0 & 0 & 0 & 0 & 0 & 9.0333 \cdot 10^{-1}
\end{bmatrix}, \quad (A2)
$$

$$
A^3 = \begin{bmatrix}
1 & 9.9707 \cdot 10^{-3} & 4.5690 \cdot 10^{-7} & -5.2162 \cdot 10^{-7} & -1.7378 \cdot 10^{-9} & -1.5699 \cdot 10^{-14} \\
0 & 9.9415 \cdot 10^{-1} & 8.5440 \cdot 10^{-5} & -1.0422 \cdot 10^{-4} & -5.2088 \cdot 10^{-7} & -6.2449 \cdot 10^{-12} \\
0 & 0 & 6.6291 \cdot 10^{-1} & 0 & 0 & 0 \\
0 & 0 & 0 & 9.9988 \cdot 10^{-1} & 9.9785 \cdot 10^{-3} & 1.7793 \cdot 10^{-7} \\
0 & 0 & 0 & -2.4758 \cdot 10^{-2} & 9.9566 \cdot 10^{-1} & 3.4967 \cdot 10^{-5} \\
0 & 0 & 0 & 0 & 0 & 9.0333 \cdot 10^{-1}
\end{bmatrix}, \quad (A3)
$$

$$
A^4 = \begin{bmatrix}
1 & 9.9707 \cdot 10^{-3} & 4.5690 \cdot 10^{-7} & -4.6946 \cdot 10^{-7} & -1.5640 \cdot 10^{-9} & -1.4129 \cdot 10^{-14} \\
0 & 9.9415 \cdot 10^{-1} & 8.5440 \cdot 10^{-5} & -9.3798 \cdot 10^{-5} & -4.6879 \cdot 10^{-7} & -5.6204 \cdot 10^{-12} \\
0 & 0 & 6.6291 \cdot 10^{-1} & 0 & 0 & 0 \\
0 & 0 & 0 & 9.9988 \cdot 10^{-1} & 9.9785 \cdot 10^{-3} & 1.7793 \cdot 10^{-7} \\
0 & 0 & 0 & -2.4758 \cdot 10^{-2} & 9.9566 \cdot 10^{-1} & 3.4967 \cdot 10^{-5} \\
0 & 0 & 0 & 0 & 0 & 9.0333 \cdot 10^{-1}
\end{bmatrix}, \quad (A4)
$$

$$
A^5 = \begin{bmatrix}
1 & 9.9709 \cdot 10^{-3} & 4.5486 \cdot 10^{-7} & -5.1929 \cdot 10^{-7} & -1.7300 \cdot 10^{-9} & -1.5629 \cdot 10^{-14} \\
0 & 9.9418 \cdot 10^{-1} & 8.5060 \cdot 10^{-5} & -1.0375 \cdot 10^{-4} & -5.1856 \cdot 10^{-7} & -6.2171 \cdot 10^{-12} \\
0 & 0 & 6.6291 \cdot 10^{-1} & 0 & 0 & 0 \\
0 & 0 & 0 & 9.9981 \cdot 10^{-1} & 9.9783 \cdot 10^{-3} & 1.7793 \cdot 10^{-7} \\
0 & 0 & 0 & -3.7387 \cdot 10^{-2} & 9.9559 \cdot 10^{-1} & 3.4966 \cdot 10^{-5} \\
0 & 0 & 0 & 0 & 0 & 9.0333 \cdot 10^{-1}
\end{bmatrix}, \quad (A5)
$$

$$
A^6 = \begin{bmatrix}
1 & 9.9709 \cdot 10^{-3} & 4.5486 \cdot 10^{-7} & -4.6736 \cdot 10^{-7} & -1.5570 \cdot 10^{-9} & -1.4066 \cdot 10^{-14} \\
0 & 9.9418 \cdot 10^{-1} & 8.5060 \cdot 10^{-5} & -9.3378 \cdot 10^{-5} & -4.6670 \cdot 10^{-7} & -5.5954 \cdot 10^{-12} \\
0 & 0 & 6.6291 \cdot 10^{-1} & 0 & 0 & 0 \\
0 & 0 & 0 & 9.9981 \cdot 10^{-1} & 9.9783 \cdot 10^{-3} & 1.7793 \cdot 10^{-7} \\
0 & 0 & 0 & -3.7387 \cdot 10^{-2} & 9.9559 \cdot 10^{-1} & 3.4966 \cdot 10^{-5} \\
0 & 0 & 0 & 0 & 0 & 9.0333 \cdot 10^{-1}
\end{bmatrix}, \quad (A6)
$$

$$A^7 = \begin{bmatrix} 1 & 9.9707 \cdot 10^{-3} & 4.5690 \cdot 10^{-7} & -5.2161 \cdot 10^{-7} & -1.7377 \cdot 10^{-9} & -1.5699 \cdot 10^{-14} \\ 0 & 9.9415 \cdot 10^{-1} & 8.5440 \cdot 10^{-5} & -1.0422 \cdot 10^{-4} & -5.2088 \cdot 10^{-7} & -6.2449 \cdot 10^{-12} \\ 0 & 0 & 6.6291 \cdot 10^{-1} & 0 & 0 & 0 \\ 0 & 0 & 0 & 9.9981 \cdot 10^{-1} & 9.9783 \cdot 10^{-3} & 1.7793 \cdot 10^{-7} \\ 0 & 0 & 0 & -3.7387 \cdot 10^{-2} & 9.9559 \cdot 10^{-1} & 3.4966 \cdot 10^{-5} \\ 0 & 0 & 0 & 0 & 0 & 9.0333 \cdot 10^{-1} \end{bmatrix}, \quad (A7)$$

$$A^8 = \begin{bmatrix} 1 & 9.9707 \cdot 10^{-3} & 4.5690 \cdot 10^{-7} & -4.6945 \cdot 10^{-7} & -1.5640 \cdot 10^{-9} & -1.4129 \cdot 10^{-14} \\ 0 & 9.9415 \cdot 10^{-1} & 8.5440 \cdot 10^{-5} & -9.3796 \cdot 10^{-5} & -4.6879 \cdot 10^{-7} & -5.6204 \cdot 10^{-12} \\ 0 & 0 & 6.6291 \cdot 10^{-1} & 0 & 0 & 0 \\ 0 & 0 & 0 & 9.9981 \cdot 10^{-1} & 9.9783 \cdot 10^{-3} & 1.7793 \cdot 10^{-7} \\ 0 & 0 & 0 & -3.7387 \cdot 10^{-2} & 9.9559 \cdot 10^{-1} & 3.4966 \cdot 10^{-5} \\ 0 & 0 & 0 & 0 & 0 & 9.0333 \cdot 10^{-1} \end{bmatrix}, \quad (A8)$$

$$A^9 = \begin{bmatrix} 1 & 9.9709 \cdot 10^{-3} & 3.1840 \cdot 10^{-7} & -5.1930 \cdot 10^{-7} & -1.7300 \cdot 10^{-9} & -1.5629 \cdot 10^{-14} \\ 0 & 9.9418 \cdot 10^{-1} & 5.9542 \cdot 10^{-5} & -1.0376 \cdot 10^{-4} & -5.1857 \cdot 10^{-7} & -6.2172 \cdot 10^{-12} \\ 0 & 0 & 6.6291 \cdot 10^{-1} & 0 & 0 & 0 \\ 0 & 0 & 0 & 9.9995 \cdot 10^{-1} & 9.9787 \cdot 10^{-3} & 1.7793 \cdot 10^{-7} \\ 0 & 0 & 0 & -9.2220 \cdot 10^{-3} & 9.9573 \cdot 10^{-1} & 3.4968 \cdot 10^{-5} \\ 0 & 0 & 0 & 0 & 0 & 9.0333 \cdot 10^{-1} \end{bmatrix}, \quad (A9)$$

$$A^{10} = \begin{bmatrix} 1 & 9.9709 \cdot 10^{-3} & 3.1840 \cdot 10^{-7} & -4.6737 \cdot 10^{-7} & -1.5570 \cdot 10^{-9} & -1.4066 \cdot 10^{-14} \\ 0 & 9.9418 \cdot 10^{-1} & 5.9542 \cdot 10^{-5} & -9.3383 \cdot 10^{-5} & -4.6671 \cdot 10^{-7} & -5.5955 \cdot 10^{-12} \\ 0 & 0 & 6.6291 \cdot 10^{-1} & 0 & 0 & 0 \\ 0 & 0 & 0 & 9.9995 \cdot 10^{-1} & 9.9787 \cdot 10^{-3} & 1.7793 \cdot 10^{-7} \\ 0 & 0 & 0 & -9.2220 \cdot 10^{-3} & 9.9573 \cdot 10^{-1} & 3.4968 \cdot 10^{-5} \\ 0 & 0 & 0 & 0 & 0 & 9.0333 \cdot 10^{-1} \end{bmatrix}, \quad (A10)$$

$$A^{11} = \begin{bmatrix} 1 & 9.9707 \cdot 10^{-3} & 3.1983 \cdot 10^{-7} & -5.2162 \cdot 10^{-7} & -1.7378 \cdot 10^{-9} & -1.5699 \cdot 10^{-14} \\ 0 & 9.9415 \cdot 10^{-1} & 5.9808 \cdot 10^{-5} & -1.0422 \cdot 10^{-4} & -5.2089 \cdot 10^{-7} & -6.2450 \cdot 10^{-12} \\ 0 & 0 & 6.6291 \cdot 10^{-1} & 0 & 0 & 0 \\ 0 & 0 & 0 & 9.9995 \cdot 10^{-1} & 9.9787 \cdot 10^{-3} & 1.7793 \cdot 10^{-7} \\ 0 & 0 & 0 & -9.2220 \cdot 10^{-3} & 9.9573 \cdot 10^{-1} & 3.4968 \cdot 10^{-5} \\ 0 & 0 & 0 & 0 & 0 & 9.0333 \cdot 10^{-1} \end{bmatrix}, \quad (A11)$$

$$A^{12} = \begin{bmatrix} 1 & 9.9707 \cdot 10^{-3} & 3.1983 \cdot 10^{-7} & -4.6946 \cdot 10^{-7} & -1.5640 \cdot 10^{-9} & -1.4129 \cdot 10^{-14} \\ 0 & 9.9415 \cdot 10^{-1} & 5.9808 \cdot 10^{-5} & -9.3800 \cdot 10^{-5} & -4.6880 \cdot 10^{-7} & -5.6205 \cdot 10^{-12} \\ 0 & 0 & 6.6291 \cdot 10^{-1} & 0 & 0 & 0 \\ 0 & 0 & 0 & 9.9995 \cdot 10^{-1} & 9.9787 \cdot 10^{-3} & 1.7793 \cdot 10^{-7} \\ 0 & 0 & 0 & -9.2220 \cdot 10^{-3} & 9.9573 \cdot 10^{-1} & 3.4968 \cdot 10^{-5} \\ 0 & 0 & 0 & 0 & 0 & 9.0333 \cdot 10^{-1} \end{bmatrix}, \quad (A12)$$

$$A^{13} = \begin{bmatrix} 1 & 9.9709 \cdot 10^{-3} & 3.1840 \cdot 10^{-7} & -5.1930 \cdot 10^{-7} & -1.7300 \cdot 10^{-9} & -1.5629 \cdot 10^{-14} \\ 0 & 9.9418 \cdot 10^{-1} & 5.9542 \cdot 10^{-5} & -1.0376 \cdot 10^{-4} & -5.1857 \cdot 10^{-7} & -6.2171 \cdot 10^{-12} \\ 0 & 0 & 6.6291 \cdot 10^{-1} & 0 & 0 & 0 \\ 0 & 0 & 0 & 9.9989 \cdot 10^{-1} & 9.9785 \cdot 10^{-3} & 1.7793 \cdot 10^{-7} \\ 0 & 0 & 0 & -2.1851 \cdot 10^{-2} & 9.9567 \cdot 10^{-1} & 3.4967 \cdot 10^{-5} \\ 0 & 0 & 0 & 0 & 0 & 9.0333 \cdot 10^{-1} \end{bmatrix}, \quad (A13)$$

$$
A^{14} = \begin{bmatrix}
1 & 9.9709 \cdot 10^{-3} & 3.1840 \cdot 10^{-7} & -4.6737 \cdot 10^{-7} & -1.5570 \cdot 10^{-9} & -1.4066 \cdot 10^{-14} \\
0 & 9.9418 \cdot 10^{-1} & 5.9542 \cdot 10^{-5} & -9.3381 \cdot 10^{-5} & -4.6671 \cdot 10^{-7} & -5.5954 \cdot 10^{-12} \\
0 & 0 & 6.6291 \cdot 10^{-1} & 0 & 0 & 0 \\
0 & 0 & 0 & 9.9989 \cdot 10^{-1} & 9.9785 \cdot 10^{-3} & 1.7793 \cdot 10^{-7} \\
0 & 0 & 0 & -2.1851 \cdot 10^{-2} & 9.9567 \cdot 10^{-1} & 3.4967 \cdot 10^{-5} \\
0 & 0 & 0 & 0 & 0 & 9.0333 \cdot 10^{-1}
\end{bmatrix}, \quad (A14)
$$

$$
A^{15} = \begin{bmatrix}
1 & 9.9707 \cdot 10^{-3} & 3.1983 \cdot 10^{-7} & -5.2162 \cdot 10^{-7} & -1.7378 \cdot 10^{-9} & -1.5699 \cdot 10^{-14} \\
0 & 9.9415 \cdot 10^{-1} & 5.9808 \cdot 10^{-5} & -1.0422 \cdot 10^{-4} & -5.2088 \cdot 10^{-7} & -6.2449 \cdot 10^{-12} \\
0 & 0 & 6.6291 \cdot 10^{-1} & 0 & 0 & 0 \\
0 & 0 & 0 & 9.9989 \cdot 10^{-1} & 9.9785 \cdot 10^{-3} & 1.7793 \cdot 10^{-7} \\
0 & 0 & 0 & -2.1851 \cdot 10^{-2} & 9.9567 \cdot 10^{-1} & 3.4967 \cdot 10^{-5} \\
0 & 0 & 0 & 0 & 0 & 9.0333 \cdot 10^{-1}
\end{bmatrix}, \quad (A15)
$$

$$
A^{16} = \begin{bmatrix}
1 & 9.9707 \cdot 10^{-3} & 3.1983 \cdot 10^{-7} & -4.6946 \cdot 10^{-7} & -1.5640 \cdot 10^{-9} & -1.4129 \cdot 10^{-14} \\
0 & 9.9415 \cdot 10^{-1} & 5.9808 \cdot 10^{-5} & -9.3798 \cdot 10^{-5} & -4.6880 \cdot 10^{-7} & -5.6205 \cdot 10^{-12} \\
0 & 0 & 6.6291 \cdot 10^{-1} & 0 & 0 & 0 \\
0 & 0 & 0 & 9.9989 \cdot 10^{-1} & 9.9785 \cdot 10^{-3} & 1.7793 \cdot 10^{-7} \\
0 & 0 & 0 & -2.1851 \cdot 10^{-2} & 9.9567 \cdot 10^{-1} & 3.4967 \cdot 10^{-5} \\
0 & 0 & 0 & 0 & 0 & 9.0333 \cdot 10^{-1}
\end{bmatrix}, \quad (A16)
$$

$$
B^{1} = \begin{bmatrix}
5.7998 \cdot 10^{-5} & -5.4643 \cdot 10^{-5} \\
1.6830 \cdot 10^{-2} & -1.0918 \cdot 10^{-2} \\
3.0338 \cdot 10^{2} & 0 \\
6.9349 \cdot 10^{-6} & 3.6497 \cdot 10^{-6} \\
1.3860 \cdot 10^{-3} & 1.0854 \cdot 10^{-3} \\
0 & 58.002
\end{bmatrix}, \quad
B^{2} = \begin{bmatrix}
5.7998 \cdot 10^{-5} & -5.4643 \cdot 10^{-5} \\
1.6830 \cdot 10^{-2} & -1.0918 \cdot 10^{-2} \\
3.0338 \cdot 10^{2} & 0 \\
6.9349 \cdot 10^{-6} & 3.6497 \cdot 10^{-6} \\
1.3860 \cdot 10^{-3} & 1.0854 \cdot 10^{-3} \\
0 & 58.002
\end{bmatrix}, \quad (A17)
$$

$$
B^{3} = \begin{bmatrix}
5.8257 \cdot 10^{-5} & -5.4887 \cdot 10^{-5} \\
1.6905 \cdot 10^{-2} & -1.0967 \cdot 10^{-2} \\
3.0338 \cdot 10^{2} & 0 \\
6.9349 \cdot 10^{-6} & 3.6497 \cdot 10^{-6} \\
1.3860 \cdot 10^{-3} & 1.0854 \cdot 10^{-3} \\
0 & 58.002
\end{bmatrix}, \quad
B^{4} = \begin{bmatrix}
5.8257 \cdot 10^{-5} & -5.4887 \cdot 10^{-5} \\
1.6905 \cdot 10^{-2} & -1.0967 \cdot 10^{-2} \\
3.0338 \cdot 10^{2} & 0 \\
6.9349 \cdot 10^{-6} & 3.6497 \cdot 10^{-6} \\
1.3860 \cdot 10^{-3} & 1.0854 \cdot 10^{-3} \\
0 & 58.002
\end{bmatrix}, \quad (A18)
$$

$$
B^{5} = \begin{bmatrix}
5.7998 \cdot 10^{-5} & -1.2162 \cdot 10^{-4} \\
1.6830 \cdot 10^{-2} & -2.4301 \cdot 10^{-2} \\
3.0338 \cdot 10^{2} & 0 \\
6.9348 \cdot 10^{-6} & 3.6497 \cdot 10^{-6} \\
1.3859 \cdot 10^{-3} & 1.0854 \cdot 10^{-3} \\
0 & 58.002
\end{bmatrix}, \quad
B^{6} = \begin{bmatrix}
5.7998 \cdot 10^{-5} & -1.2162 \cdot 10^{-4} \\
1.6830 \cdot 10^{-2} & -2.4301 \cdot 10^{-2} \\
3.0338 \cdot 10^{2} & 0 \\
6.9348 \cdot 10^{-6} & 3.6497 \cdot 10^{-6} \\
1.3859 \cdot 10^{-3} & 1.0854 \cdot 10^{-3} \\
0 & 58.002
\end{bmatrix}, \quad (A19)
$$

$$
B^{7} = \begin{bmatrix}
5.8257 \cdot 10^{-5} & -1.2217 \cdot 10^{-4} \\
1.6905 \cdot 10^{-2} & -2.4410 \cdot 10^{-2} \\
3.0338 \cdot 10^{2} & 0 \\
6.9348 \cdot 10^{-6} & 3.6497 \cdot 10^{-6} \\
1.3859 \cdot 10^{-3} & 1.0854 \cdot 10^{-3} \\
0 & 58.002
\end{bmatrix}, \quad
B^{8} = \begin{bmatrix}
5.8257 \cdot 10^{-5} & -1.2217 \cdot 10^{-4} \\
1.6905 \cdot 10^{-2} & -2.4410 \cdot 10^{-2} \\
3.0338 \cdot 10^{2} & 0 \\
6.9348 \cdot 10^{-6} & 3.6497 \cdot 10^{-6} \\
1.3859 \cdot 10^{-3} & 1.0854 \cdot 10^{-3} \\
0 & 58.002
\end{bmatrix}, \quad (A20)
$$

$$
B^{9} = \begin{bmatrix}
4.0598 \cdot 10^{-5} & -1.0576 \cdot 10^{-5} \\
1.1781 \cdot 10^{-2} & -2.1132 \cdot 10^{-3} \\
3.0338 \cdot 10^{2} & 0 \\
6.9350 \cdot 10^{-6} & 3.6497 \cdot 10^{-6} \\
1.3860 \cdot 10^{-3} & 1.0854 \cdot 10^{-3} \\
0 & 58.002
\end{bmatrix}, \quad
B^{10} = \begin{bmatrix}
4.0598 \cdot 10^{-5} & -1.0576 \cdot 10^{-5} \\
1.1781 \cdot 10^{-2} & -2.1132 \cdot 10^{-3} \\
3.0338 \cdot 10^{2} & 0 \\
6.9350 \cdot 10^{-6} & 3.6497 \cdot 10^{-6} \\
1.3860 \cdot 10^{-3} & 1.0854 \cdot 10^{-3} \\
0 & 58.002
\end{bmatrix}, \quad (A21)
$$

$$
B^{11} = \begin{bmatrix} 4.0780 \cdot 10^{-5} & -1.0623 \cdot 10^{-5} \\ 1.1834 \cdot 10^{-2} & -2.1226 \cdot 10^{-3} \\ 3.0338 \cdot 10^{2} & 0 \\ 6.9350 \cdot 10^{-6} & 3.6497 \cdot 10^{-6} \\ 1.3860 \cdot 10^{-3} & 1.0854 \cdot 10^{-3} \\ 0 & 58.002 \end{bmatrix}, \quad B^{12} = \begin{bmatrix} 4.0780 \cdot 10^{-5} & -1.0623 \cdot 10^{-5} \\ 1.1834 \cdot 10^{-2} & -2.1226 \cdot 10^{-3} \\ 3.0338 \cdot 10^{2} & 0 \\ 6.9350 \cdot 10^{-6} & 3.6497 \cdot 10^{-6} \\ 1.3860 \cdot 10^{-3} & 1.0854 \cdot 10^{-3} \\ 0 & 58.002 \end{bmatrix}, \tag{A22}
$$

$$
B^{13} = \begin{bmatrix} 4.0598 \cdot 10^{-5} & -7.7558 \cdot 10^{-5} \\ 1.1781 \cdot 10^{-2} & -1.5496 \cdot 10^{-2} \\ 3.0338 \cdot 10^{2} & 0 \\ 6.9349 \cdot 10^{-6} & 3.6497 \cdot 10^{-6} \\ 1.3860 \cdot 10^{-3} & 1.0854 \cdot 10^{-3} \\ 0 & 58.002 \end{bmatrix}, \quad B^{14} = \begin{bmatrix} 4.0598 \cdot 10^{-5} & -7.7558 \cdot 10^{-5} \\ 1.1781 \cdot 10^{-2} & -1.5496 \cdot 10^{-2} \\ 3.0338 \cdot 10^{2} & 0 \\ 6.9349 \cdot 10^{-6} & 3.6497 \cdot 10^{-6} \\ 1.3860 \cdot 10^{-3} & 1.0854 \cdot 10^{-3} \\ 0 & 58.002 \end{bmatrix}, \tag{A23}
$$

$$
B^{15} = \begin{bmatrix} 4.0780 \cdot 10^{-5} & -7.7904 \cdot 10^{-5} \\ 1.1834 \cdot 10^{-2} & -1.5566 \cdot 10^{-2} \\ 3.0338 \cdot 10^{2} & 0 \\ 6.9349 \cdot 10^{-6} & 3.6497 \cdot 10^{-6} \\ 1.3860 \cdot 10^{-3} & 1.0854 \cdot 10^{-3} \\ 0 & 58.002 \end{bmatrix}, \quad B^{16} = \begin{bmatrix} 4.0780 \cdot 10^{-5} & -7.7904 \cdot 10^{-5} \\ 1.1834 \cdot 10^{-2} & -1.5566 \cdot 10^{-2} \\ 3.0338 \cdot 10^{2} & 0 \\ 6.9349 \cdot 10^{-6} & 3.6497 \cdot 10^{-6} \\ 1.3860 \cdot 10^{-3} & 1.0854 \cdot 10^{-3} \\ 0 & 58.002 \end{bmatrix}, \tag{A24}
$$

$$
W_1 = 0.01I, \qquad W_2 = 0.05I, \qquad C = \begin{bmatrix} 1 & 0 & 0 & 0 & 0 & 0 \\ 0 & 0 & 1 & 0 & 0 & 0 \\ 0 & 0 & 0 & 1 & 0 & 0 \\ 0 & 0 & 0 & 0 & 1 & 0 \\ 0 & 0 & 0 & 0 & 0 & 1 \end{bmatrix}. \tag{A25}
$$

References

1. Rodrigues, M.; Hamdi, H.; Theilliol, D.; Mechmeche, C.; BenHadj Braiek, N. Actuator fault estimation based adaptive polytopic observer for a class of LPV descriptor systems. *Int. J. Robust Nonlinear Control* **2015**, *25*, 673–688. [CrossRef]
2. Witczak, M.; Mrugalski, M.; Korbicz, J. Towards robust neural-network-based sensor and actuator fault diagnosis: Application to a tunnel furnace. *Neural Process. Lett.* **2015**, *42*, 71–87. [CrossRef]
3. Shao, H.; Jiang, H.; Zhang, X.; Niu, M. Rolling bearing fault diagnosis using an optimization deep belief network. *Meas. Sci. Technol.* **2015**, *26*, 115002. [CrossRef]
4. Zhirabok, A.; Shumsky, A. Fault diagnosis in nonlinear hybrid systems. *Int. J. Appl. Math. Comput. Sci.* **2018**, *28*, 635–648. [CrossRef]
5. Yao, L.; Feng, L. Fault diagnosis and fault tolerant tracking control for the non-Gaussian singular time-delayed stochastic distribution system with PDF approximation error. *Neurocomputing* **2016**, *175 Pt A*, 538–543. [CrossRef]
6. Pazera, M.; Buciakowski, M.; Witczak, M. Robust Multiple Sensor Fault–Tolerant Control For Dynamic Non–Linear Systems: Application To The Aerodynamical Twin–Rotor System. *Int. J. Appl. Math. Comput. Sci.* **2018**, *28*, 297–308. [CrossRef]
7. Rotondo, D.; Ponsart, J.C.; Theilliol, D.; Nejjari, F.; Puig, V. A virtual actuator approach for the fault tolerant control of unstable linear systems subject to actuator saturation and fault isolation delay. *Annu. Rev. Control* **2015**, *39*, 68–80. [CrossRef]
8. Liu, X.; Gao, Z.; Zhang, A. Robust Fault Tolerant Control for Discrete-Time Dynamic Systems With Applications to Aero Engineering Systems. *IEEE Access* **2018**, *6*, 18832–18847. [CrossRef]
9. Lan, J.; Patton, R.J. A decoupling approach to integrated fault-tolerant control for linear systems with unmatched non-differentiable faults. *Automatica* **2018**, *89*, 290–299. [CrossRef]

10. Rotondo, D.; Witczak, M.; Puig, V.; Nejjari, F.; Pazera, M. Robust unknown input observer for state and fault estimation in discrete-time Takagi–Sugeno systems. *Int. J. Syst. Sci.* **2016**, *47*, 3409–3424. [CrossRef]

11. Xia, J.; Guo, Y.; Dai, B.; Zhang, X. Sensor fault diagnosis and system reconfiguration approach for an electric traction PWM rectifier based on sliding mode observer. *IEEE Trans. Ind. Appl.* **2017**, *53*, 4768–4778. [CrossRef]

12. Li, J.; Pan, K.; Su, Q. Sensor fault detection and estimation for switched power electronics systems based on sliding mode observer. *Appl. Math. Comput.* **2019**, *353*, 282–294. [CrossRef]

13. Zhang, Q. Adaptive Kalman filter for actuator fault diagnosis. *Automatica* **2018**, *93*, 333–342. [CrossRef]

14. Gou, L.; Zhou, Z.; Liang, A.; Wang, L.; Liu, Z. Dynamic Threshold Design Based on Kalman Filter in Multiple Fault Diagnosis. In Proceedings of the 2018 37th Chinese Control Conference (CCC), Wuhan, China, 25–27 July 2018; pp. 6105–6109.

15. Youssef, T.; Chadli, M.; Karimi, H.; Wang, R. Actuator and sensor faults estimation based on proportional integral observer for TS fuzzy model. *J. Frankl. Inst.* **2017**, *354*, 2524–2542. [CrossRef]

16. Lan, J.; Patton, R. A new strategy for integration of fault estimation within fault-tolerant control. *Automatica* **2016**, *69*, 48–59. [CrossRef]

17. Chaves, E.; André, F.; Maitelli, A. Robust Observer-Based Actuator and Sensor Fault Estimation for Discrete-Time Systems. *J. Control. Autom. Electr. Syst.* **2019**, *30*, 160–169. [CrossRef]

18. Liu, M.; Cao, X.; Shi, P. Fuzzy-model-based fault-tolerant design for nonlinear stochastic systems against simultaneous sensor and actuator faults. *IEEE Trans. Fuzzy Syst.* **2013**, *21*, 789–799. [CrossRef]

19. Gu, Y.; Yang, G. Simultaneous actuator and sensor fault estimation for discrete-time Lipschitz nonlinear systems in finite-frequency domain. *Optimal Control Appl. Methods* **2018**, *39*, 410–423. [CrossRef]

20. Abdollahi, M. Simultaneous sensor and actuator fault detection, isolation and estimation of nonlinear Euler-Lagrange systems using sliding mode observers. In Proceedings of the 2018 IEEE Conference on Control Technology and Applications (CCTA), Copenhagen, Denmark, 21–24 August 2018; pp. 392–397.

21. Hmidi, R.; Brahim, A.; Hmida, F.; Sellami, A. Robust fault tolerant control for Lipschitz nonlinear systems with simultaneous actuator and sensor faults. In Proceedings of the 2018 International Conference on Advanced Systems and Electric Technologies (IC_ASET), Hammamet, Tunisia, 22–25 March 2018; pp. 277–283.

22. Bounemeur, A.; Chemachema, M.; Essounbouli, N. Indirect adaptive fuzzy fault-tolerant tracking control for MIMO nonlinear systems with actuator and sensor failures. *ISA Trans.* **2018**, *79*, 45–61. [CrossRef]

23. Fu, S.; Qiu, J.; Chen, L.; Mou, S. Adaptive fuzzy observer design for a class of switched nonlinear systems with actuator and sensor faults. *IEEE Trans. Fuzzy Syst.* **2018**, *26*, 3730–3742. [CrossRef]

24. Shaker, M. Hybrid approach to design Takagi-Sugeno observer-based FTC for non-linear systems affected by simultaneous time varying actuator and sensor faults. *IET Control Theory Appl.* **2019**, *13*, 632–641. [CrossRef]

25. Brizuela Mendoza, J.; Sorcia Vázquez, F.; Guzmán Valdivia, C.; Osorio Sánchez, R.; Martínez García, M. Observer design for sensor and actuator fault estimation applied to polynomial LPV systems: A riderless bicycle study case. *Int. J. Syst. Sci.* **2018**, *49*, 2996–3006. [CrossRef]

26. Li, X.; Ahn, C.; Lu, D.; Guo, S. Robust simultaneous fault estimation and nonfragile output feedback fault-tolerant control for Markovian jump systems. *IEEE Trans. Syst. Man, Cybern. Syst.* **2018**, *49*, 1769–1776. [CrossRef]

27. Gillijns, S.; De Moor, B. Unbiased minimum-variance input and state estimation for linear discrete-time systems. *Automatica* **2007**, *43*, 111–116. [CrossRef]

28. Pazera, M.; Witczak, M. Towards robust simultaneous actuator and sensor fault estimation for a class of nonlinear systems: Design and comparison. *IEEE Access* **2019**, *7*, 97143–97158. [CrossRef]

29. Takagi, T.; Sugeno, M. Fuzzy identification of systems and its application to modeling and control. *IEEE Trans. Syst. Man Cybern.* **1985**, *15*, 116–132. [CrossRef]

30. Ichalal, D.; Marx, B.; Mammar, S.; Maquin, D.; Ragot, J. How to cope with unmeasurable premise variables in Takagi–Sugeno observer design: Dynamic extension approach. *Eng. Appl. Artif. Intell.* **2018**, *67*, 430–435. [CrossRef]

31. Skelton, R.; Iwasaki, T.; Grigoriadis, D. *A Unified Algebraic Approach to Control Design*; CRC Press: Boca Raton, FL, USA, 1997.

32. Alessandri, A.; Baglietto, M.; Battistelli, G. Design of state estimators for uncertain linear systems using quadratic boundedness. *Automatica* **2006**, *42*, 497–502. [CrossRef]

33. Rotondo, D.; Puig, V.; Nejjari, F.; Witczak, M. Automated generation and comparison of Takagi-Sugeno and polytopic quasi-LPV models. *Fuzzy Sets Syst.* **2015**, *277*, 44–64. [CrossRef]
34. INTECO. *Two Rotor Aerodynamical System, User's Manual*; INTECO: Kraków, Poland, 2007.

Article

Learning Attention Representation with a Multi-Scale CNN for Gear Fault Diagnosis under Different Working Conditions

Yong Yao [1,*], Sen Zhang [2,3], Suixian Yang [1,*] and Gui Gui [4]

[1] School of Mechanical Engineering, Sichuan University, Chengdu 610065, China
[2] University of Chinese Academy of Sciences, Beijing 100049, China; 20120061@git.edu.cn
[3] Chengdu Institute of Computer Application, Chinese Academy of Sciences, Chengdu 610041, China
[4] National institute of Measurement and Testing Technology, Chengdu 610021, China; dg881005@163.com
* Correspondence: yao_yong92@163.com (Y.Y.); yangsuixian@scu.edu.cn (S.Y.)

Received: 30 January 2020; Accepted: 21 February 2020; Published: 24 February 2020

Abstract: The gear fault signal under different working conditions is non-linear and non-stationary, which makes it difficult to distinguish faulty signals from normal signals. Currently, gear fault diagnosis under different working conditions is mainly based on vibration signals. However, vibration signal acquisition is limited by its requirement for contact measurement, while vibration signal analysis methods relies heavily on diagnostic expertise and prior knowledge of signal processing technology. To solve this problem, a novel acoustic-based diagnosis (ABD) method for gear fault diagnosis under different working conditions based on a multi-scale convolutional learning structure and attention mechanism is proposed in this paper. The multi-scale convolutional learning structure was designed to automatically mine multiple scale features using different filter banks from raw acoustic signals. Subsequently, the novel attention mechanism, which was based on a multi-scale convolutional learning structure, was established to adaptively allow the multi-scale network to focus on relevant fault pattern information under different working conditions. Finally, a stacked convolutional neural network (CNN) model was proposed to detect the fault mode of gears. The experimental results show that our method achieved much better performance in acoustic based gear fault diagnosis under different working conditions compared with a standard CNN model (without an attention mechanism), an end-to-end CNN model based on time and frequency domain signals, and other traditional fault diagnosis methods involving feature engineering.

Keywords: acoustic-based diagnosis; gear fault diagnosis; attention mechanism; convolutional neural network

---

1. Introduction

As a one of the most important components in transmission systems, gears are widely used in many types of machinery, such as wind turbines, construction machinery, automobiles, and other fields [1], thanks to their unique merits, such as large transmission ratio, high efficiency, and heavy load capacity [2,3]. The working performance of gears directly influences the operational reliability of the whole machinery [4]. However, due to poor environmental conditions and the intensive impact load operational condition of transmission systems, gears are vulnerable to display some faults and cause the machine to break down. This may lead to significant economic losses [5]. Therefore, research on fault diagnosis for gears can effectively avoid catastrophic failure and reduce economic loss.

Recently, the fault diagnosis of gears has been extensively studied by researchers. However, most current studies focus on mainly stable working conditions. In the real world, gears usually work under variable and fluctuant operation conditions [6]. As such, the nonlinear and non-stationary characteristics of signals under variable conditions exhibit many unique characteristics, such as strong nonstationary, frequency mixing, and modulated phenomena [7]. Traditional fault diagnosis methods and technologies, which are only applicable to gears under stationary conditions [1], are incapable of detecting and identifying gear fault patterns under variable conditions.

To solve this issue, gear fault diagnosis under variable conditions has become the subject of extensive research and has aroused researchers' great concern in the past few years. Liu et al. [8] proposed a method for gear fault diagnosis under slight variations in working conditions via empirical mode decomposition (EMD) and multi-fractal detrended cross-correlation analysis (MFDCCA). By using EMD and MFDCCA methods, the multi-fractal fault features can effectively extract and distinguish fault modes. In order to avoid mode mixing, Chen et al. [9] proposed to use complementary ensemble empirical mode decomposition (CEEMD) technology to decompose the raw vibration signals and select the intrinsic mode functions (IMFs) using a correlation analysis algorithm (CorAA) for a probabilistic neural network to classify the gear fault patterns under different working conditions. Xing et al. [10] adopted the intrinsic time-scale decomposition (ITD) and singular value decomposition methods to improve the robustness of gear fault feature extraction under variable conditions. Zhang et al. [11] proposed a method for gear fault diagnosis under different working conditions based on local characteristic-scale decomposition (LCD) denoising and the vector mutual information method. Chen et al. [12] performed gearbox fault diagnosis under variable speed conditions via analysis of the torsional vibration signals in the time-frequency domain. Though these studies, rich methods, and technologies for gear fault diagnosis have been accumulated and provide a pivotal function under variable conditions, most of the methods typically use vibration signals as the main measurement values to diagnose gear faults in variable working conditions for use in vibration analysis [13]. In many practical conditions, the installation of vibration sensors is constrained by some working conditions and the complex structure of the equipment themselves, which makes the signal acquisition inconvenient. Moreover, vibration signals are easily masked in some special environments, such as high humidity, high temperature, and high corrosion; therefore, the application of vibration signal analysis methods for gear fault diagnosis under variable conditions is limited due to the requirement of contacted measuring. Meanwhile, those studies that adopt vibration analysis methods, usually rely on signal processing technology to decompose raw vibration signals into several proper signal components to extract valuable features for distinguishing gear fault patterns under different working conditions. Although all these vibration signal analysis methods can work well in fault mode detection tasks, they rely heavily on diagnostic expertise and prior knowledge of signal processing technology [14], which may lead to tedious and inefficient procedures in practical diagnosis tasks. Considering the existing issues, the effective methods and technologies of gear fault diagnosis under variable conditions still needs to be further developed.

As a typical non-contact measurement, acoustic-based diagnosis (ABD) methods, which have the capability to overcome the limitation of vibration measurement, are widely used in the fault diagnosis field. Lu et al. [15–17] proposed an acoustic-based fault diagnosis method based on near-field acoustic holography for detecting gear fault patterns under stationary working conditions. Glowacz [18,19] design several acoustic-based diagnosis methods with novelty acoustic features to detect the fault of commutator motors, electric impact drills, and coffee grinders. By combing time-frequency data fusion technology and the Doppler feature matching search (DFMS) algorithm, Zhang et al. [20] proposed a train bearings fault diagnosis method, which is based on wayside acoustic signals. Inspired by their study, Zhang et al. [21] designed an improved singular value decomposition with a resonance-based wayside acoustic signal sparse decomposition technique as an adaptive form of train bearings fault feature extraction. However, like the vibration-based diagnosis method, all the existing acoustic-based methods are also heavily rely on prior knowledge of signal processing technology rather than utilizing intelligent fault diagnosis techniques. This is because the fault data distribution that we obtain in one working condition are not consistent in another different working condition in real applications [22], which means the distribution difference between training data and test data changes as the working condition varies, which can lead to a dramatic drop in performance.

To manage the obstacles, we considered the role of the attention mechanism. As a novel intelligent method, the attention mechanism, which has the capability to adaptively capture temporal correlations between different sequences [23] and allows for feature extraction networks to focus on the relevant characteristics without signal processing technology and feature engineering, are commonly explored in various structural prediction tasks, such as document classification [24], speech recognition [25–27], and environmental classification [28,29]. Therefore, in this paper, we propose a novel ABD method for gear fault diagnosis under different working conditions based on a multi-scale convolutional learning structure and attention mechanism. In our methods, a multi-scale convolutional learning structure was designed to automatically mine multi-scale features using different filter banks from raw acoustic signals. Then, a novel attention mechanism, which was based on a multi-scale convolutional learning structure, was established to adaptively allow the multi-scale network to focus on relevant fault pattern information under different working conditions. Finally, a stacked convolutional neural network (CNN) model was proposed to detect the fault mode of the gears.

The main contributions of this paper are as follows:

1. We are the first to propose an acoustic-based diagnosis method to detect the fault patterns under different working conditions, where this method obtains information directly from raw acoustical signals without manual signal processing and feature engineering.
2. We are the first to introduce the attention mechanism theory into the acoustic-based diagnosis field to address gear fault pattern recognition under different working conditions by designing a novel attention-based mechanism that is based on a multi-scale convolutional learning structure to adaptively extract relevant fault patterns information and reduce data distribution variation under different working conditions.
3. We designed a novel attention-based, multi-scale CNN model based on the two innovations above. It outperformed a single-scale network and multi-scale network without attention mechanism, and achieved favorable results relative to other methods using manual feature engineering based on the function of multi-scale structure and an attention mechanism.

2. Model Building

In this section, we briefly introduce the mathematical model of our acoustic-based gear fault diagnosis method, which can be roughly divided into three parts. In the first part, a multi-scale convolutional layer operates directly on raw acoustic signals and automatically mines fault features using different filter sizes and strides to construct feature vectors. In the second part, an attention mechanism is adopted to obtain reasonable attention weight vectors from the convolutional layer, which are multiplied with each feature vector of the pooling layer. In the last part, the multi-dimensional attention output matrix, which is concatenated with the multi-scale attention structure, is constructed as a stacked CNN input to train the network. The block diagram of the proposed method is shown in Figure 1.

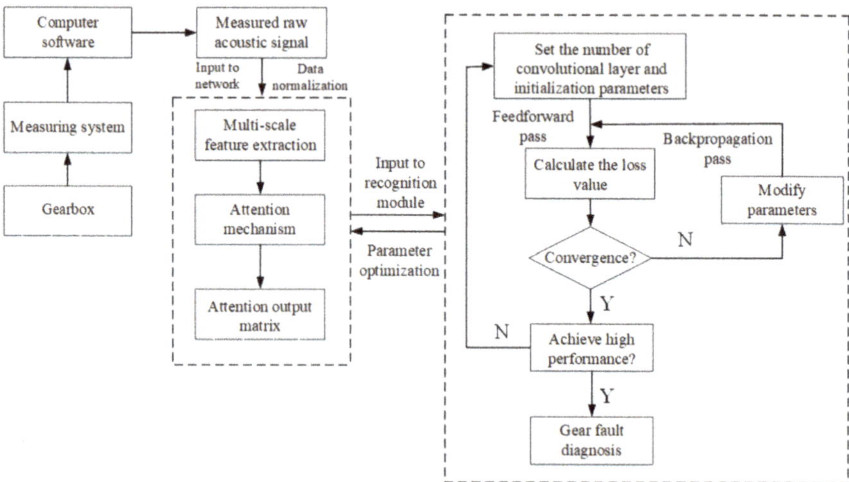

Figure 1. Block diagram of the proposed method.

2.1. Multi-Scale Convolution Operation

A commonly used approach for an end-to-end neural network is to pass the raw acoustic signal through a 1D convolutional layer, which has a fixed filter size and stride length to create invariance to phase shifts and further down-sample the signals. However, those methods are still constrained in various prediction tasks for two reasons: (1) There is always a trade-off when choosing the filter size. A high-scale filter size may have a good frequency resolution but does not have a sufficient filter for location in the high frequency area. A low-scale filter size, on the contrary, focuses on more frequency bands but has a low resolution [30,31]. (2) Features extracted using a fixed filter size cannot make full use of the raw signal information to build a discriminative representation for different patterns. Considering this, a multi-scale convolutional function, which has the capability to learn discrepant features, has been applied to address the obstacles. By extracting features with multiple different scale filter banks and splitting responsibilities based on what filter banks can efficiently represent, multi-scale convolutions have already been successfully used in various recognition fields, such as image classification [32], environmental sound classification [30], and speech recognition [33].

Inspired by their work, we designed a multi-scale convolutional learning structure to extract multi-scale fault features from raw acoustic signals. The structure is shown in Figure 2.

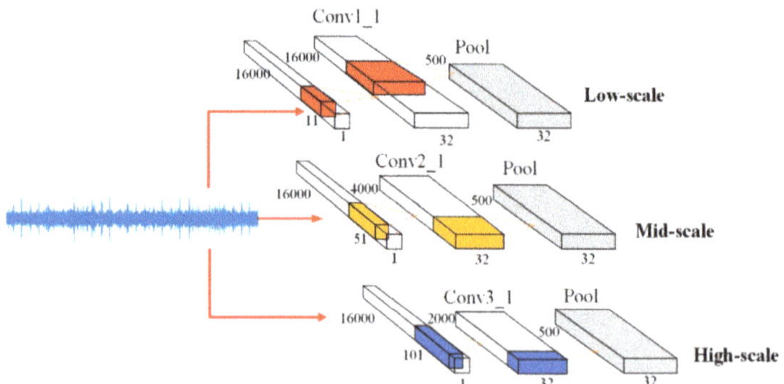

Figure 2. Multi-scale feature extraction mechanism.

Three scale convolution kernels were used to operate the input signal vector to extract different features and increase the bias to achieve a result. The function is defined as followed:

$$x_j^l = f\left(x_k \cdot \omega_j^l + b_j^l\right),\tag{1}$$

where x_k represents the input raw signal vector; j represents the three different convolutional scales ($j = 1, 2, 3$), corresponding to low, mid, and high filter sizes; ω_j^l represents the convolutional operation between the input vector and output feature map l at different scales; b_j^l is a bias that corresponds to the output vector $x_{j'}^l$ which represent the convolution operation result of the feature map l in scale j; and f is an activation function.

Then, each of the three different output vectors were subsampled by the max pooling layer in turn such that vectors of different sizes were rescaled to the same size.

2.2. Temporal Attention Mechanism

The acoustic signals obtained in one working condition may not follow the same temporal structure in another working condition and those signals are often masked by noises that are generated from the gearbox parts and transmitted via an elastic medium, i.e., through the air. We designed a novel temporal attention mechanism that puts more attention on the relevant information frames and suppresses noise ones to provide acoustic-based fault diagnosis under different working conditions to overcome those limitations.

In order to reduce the impact of channel information, we first used a 1×1 kernel size with one channel to aggregate the feature maps along the channel dimension to produce a multi-scale convolutional learning structure. Then, we adopted different convolutional operations for multi-scale vectors to transform the features map into the same scale and generate a temporal attention map through the softmax activation function. Finally, we multiplied the attention map with the feature vector of the pooling layer to obtain the attention output matrix. The detailed structure of the temporal attention mechanism is shown in Figure 3 and detailed information about the operation process are given below.

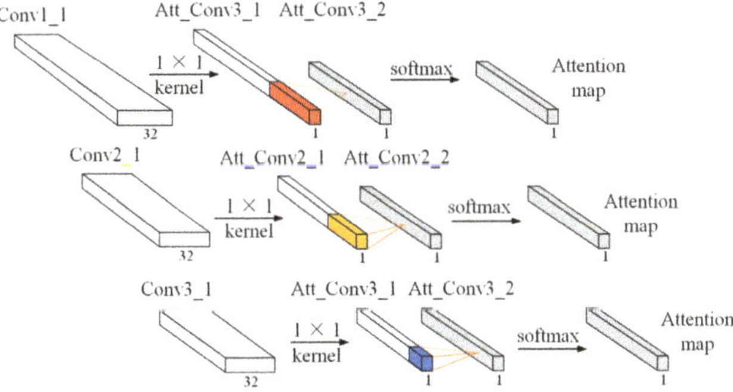

Figure 3. Temporal attention mechanism.

Let $Convi\_1 (i = 1, 2, 3)$ denote the feature vector of the multi-scale convolution learning structure. We first operate a 1×1 kernel size over $Convi\_1 \in R^{T \times 1 \times C}$ to generate one channel feature map $Att\_Convi\_1 \in R^{T \times 1 \times 1}$. Then, multi-scale 2-D convolution with a different kernel size was adopted to learn the hidden representation and compress the features map into the same scale using different strides. The softmax activation function was applied to normalize the attention weight of $Convi\_1$ and produce the temporal attention map. The mathematical equations are expressed as:

$$Att\_Convi\_1 = Conv^{1 \times 1}(Convi\_1) \; i = 1, 2, 3,\tag{2}$$

$$Att\_Convi\_2 = Att\_Convi\_1 \cdot \omega_i + b_i \; i = 1, 2, 3, \tag{3}$$

$$A_{tm} = softmax(Att\_Convi\_2) \; i = 1, 2, 3. \tag{4}$$

Finally, by multiplying the attention map A_{tm} with the feature vector of the pooling layer, we obtained the attention output matrix A_{to}. The equation is defined as followed:

$$A_{ot}^i = \left(x_j^l\right)_{pool} \times A_{tm}^i, \tag{5}$$

where $\left(x_j^l\right)_{pool}$ represents the *l*th feature map of the *j*th pooling layer $(j = 1, 2, 3)$. The j value refers to what is described in Section 2.1.

2.3. Fault Pattern Recognition Based on a CNN

We adopted a stacked convolutional neural network as a base structure for the recognition of gear fault patterns under different working conditions. The network consisted of four functional layers: the convolutional layer, the batch normalization layer (BN), the pooling layer, and the fully connected layer.

The attention output matrix from three different scales were concatenated along the channel dimension as a multi-dimensional matrix input into the convolutional layer. The stacked convolutional layer can be viewed as a fault pattern recognition module in an attention-based multi-scale CNN model. Through repetitive convolution operations, the network has the ability to learn high-level representation from the inputted multi-dimensional matrix. The process can be expressed as follows:

$$y_j^l = \sum_{i=t}^{T} y_i^{l-1} \cdot \omega_{ij}^l + b_j^l, \tag{6}$$

where y_i^{l-1} represents the *i*th output feature map of the former convolutional layer, and ω_{ij}^l represents the convolutional kernel, which is used to operate between the *i*th feature map of the former layer and the *j*th feature map of layer *l*. *T* represents the feature atlas of the former layer and b_j^l represents the bias of layer *l* corresponding to the output matrix y_i^l, which represents the convolution operation result of the *j*th feature map in layer *l*.

The output matrix of the convolutional layer is normalized by the BN layer such that the mean and variance of the feature become 0 and 1, respectively. Then, we used functions to transform and reconstruct a certain level of features to maintain the data distribution. Those equations can be expressed as:

$$y_2 = \frac{y_1 - \mu}{\sqrt{\sigma^2 + \varepsilon}}, \tag{7}$$

$$y_3 = f(\gamma y_2 + \beta), \tag{8}$$

where μ and σ^2 represent the mean and variance of the mini-batch in Equation (7), and γ and β are the two parameters in Equation (8), which can be learned by the training network. f represents the activation function, which is used to analyze nonlinear information for the output features of the BN layer.

Then, the max pooling layer was applied to the subsample feature information of the BN layer to prevent overfitting.

Finally, through a combination of these high-level representations in a nonlinear way, a fully connected layer that recognizes gear fault patterns under different working condition was produced. The mathematical equations of the fully connected layer can be expressed as:

$$h\left(y^l\right) = f\left(wy^{l-1} + b\right), \tag{9}$$

where y^{l-1} represents high-level information of the former layer, $h(y^l)$ represents the output nonlinear information from the fully connected layer l, and ω and b represent the weight and bias, respectively.

2.4. Architecture and Parameters of the Attention-Based Multi-Scale CNN Model

The detailed information of the architecture and parameters of the attention-based multi-scale CNN model are shown in Figure 4 and Table 1, respectively. The architecture of the attention-based multi-scale CNN model consists of two parts: (a) the attention-based multi-scale feature extraction module and (b) the fault pattern recognition module.

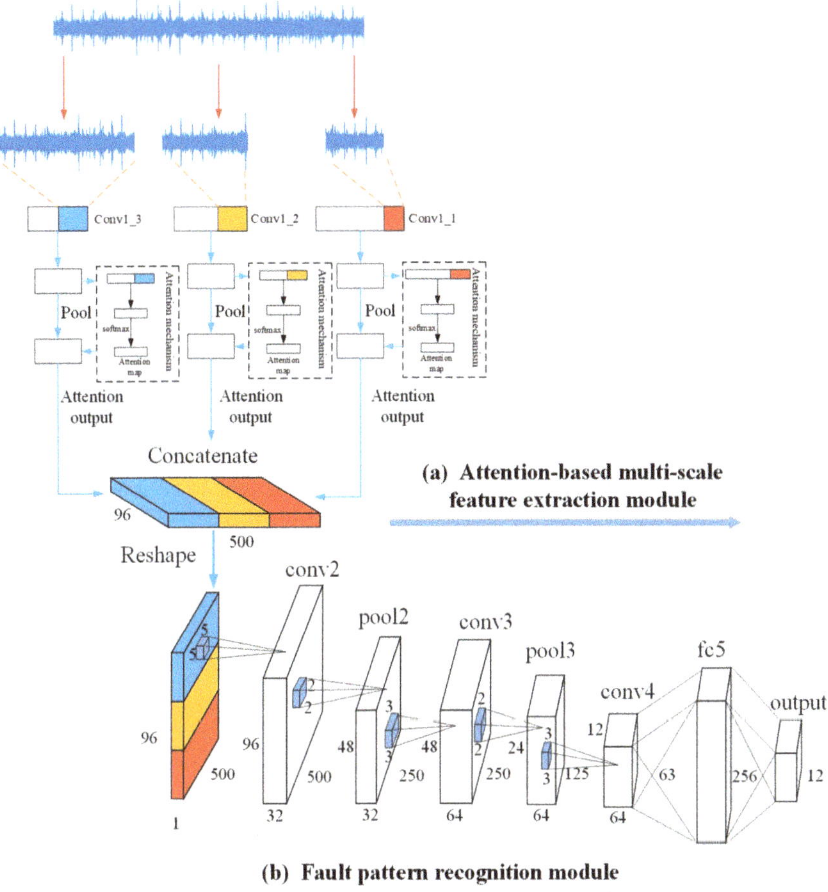

Figure 4. Architecture of the attention-based multi-scale convolutional neural network (CNN) model. It contains (**a**) an attention-based multi-scale feature extraction module and (**b**) a fault pattern recognition module.

These two modules can be divided from the concatenate layer. The attention-based multi-scale feature extraction module contains a multi-scale convolutional learning structure with an attention mechanism. The network operates on an input acoustic signal, which consists of 16,000 sampling points using three different scale convolutional neural networks of different scales. The three chosen scales are low-scale (size 11, stride1), mid-scale (size 51, stride 4), and high-scale (size 101, stride 8). Each scale has 32 filters in their convolutional layer. As an independent part in the network, the attention mechanism contains two convolutional layers to adaptively pay more attention to the relevant information frames from the output of the three convolutional layers. The kernel size of the two convolutional layers were set as (1, 1) and (64, 1) in the low scale, (1, 1) and (32, 1) in the middle scale, and (1, 1) and (16, 1) in the high scale. Then, the attention output matrix is concatenated along the channel dimension as a multi-dimensional matrix $500 \times 1 \times 96$ (two-dimensional feature matrix $\times$ channels). Finally, in order to convolve the feature matrix from time and frequency in module (b), the attention output matrix is reshaped from $500 \times 1 \times 96$ to $500 \times 96 \times 1$ before being used as an input to "conv2" for further processing.

The fault pattern recognition module contains three convolutional layers, two pooling layers, and one fully connected layer. The first convolutional layer uses 32 filters with a (5,5) kernel size and stride of (1,1). The second and third convolutional layers repeatedly use 64 filters with a (3,3) kernel size and the strides were set to (1,1) and (2,2), respectively. The pooling layer is applied to subsample feature information after the first and second convolutional layers. All the kernel sizes and strides for the pooling layer were set to (2,2). In addition, the BN layer is used as a function layer, which can alleviate the problem of exploding and vanishing gradients following each convolutional layer (before the pooling layer). Finally, the output feature of the BN layer is flattened and passed to the fully connected layer with 256 nodes and the softmax activated output layer for final recognition.

Table 1. Parameters of the attention-based multi-scale CNN model.

Layer	Input Shape	Filter	Kernel Size	Stride	Output Shape
Conv1_1	[batch,16000,1,4]	32	(11,1)	(1,1)	[batch,16000,1,32]
Conv2_1	[batch,16000,1,4]	32	(51,1)	(4,1)	[batch,4000,1,32]
Conv3_1	[batch,16000,1,4]	32	(101,1)	(8,1)	[batch,2000,1,32]
Conv2	[batch,500,96,1]	32	(5,5)	(1,1)	[batch,500,96,32]
Pool2	[batch,500,96,32]	-	(2,2)	(2,2)	[batch,250,48,32]
Conv3	[batch,250,48,32]	64	(3,3)	(1,1)	[batch,250,48,64]
Pool3	[batch,250,48,64]	-	(2,2)	(2,2)	[batch,125,24,64]
Conv4	[batch,125,24,64]	64	(3,3)	(2,2)	[batch,63,12,64]
Fc5	[batch,48384]	-	-	-	[batch,256]
Output	[batch,256]	-	-	-	[batch,12]

3. Experimental Setup and Datasets

3.1. Experimental System

Our study was a preliminary attempt at going from the theory and simulation experiment to the practical engineering application. In order to obtain pure acoustic signals that were not disturbed by environmental noise, the gear fault diagnosis experiments were conducted in a semi-anechoic chamber. The experimental system that we designed can be divided into three parts: the experiment table, the measuring system, and the data recoding software. The experiment table, as shown in the top-left corner of Figure 5 [34], consisted of the following equipment: a variable frequency motor, a two-stage gearbox, a tension controller, a frequency converter, and a magnetic brake. By adjusting the frequency converter and the tension controller, we could control the speed of the motor and simulate the load condition of the two-stage gearbox. The measuring system, as shown in the right picture of Figure 5, consisted of four free-field 4189-A-021 model microphones from Brüel & Kjær (Copenhagen, Denmark) and a data acquisition instrument from HEAD Acoustics (Herzogenrath, Germany). In our study, the four free-field microphones were arranged to provide a four-channel microphone array, where they were arranged symmetrically with a hemispherical enveloping surface and the coordinates were set according to the ISO 3745:2003 standard to collect all the gears' acoustic signals. Then, the microphone array and data acquisition instrument were connected using a Bayonet Nut Connector (BNC) interface for data transmission. In addition, the data was recoded using Artemis 6.0 software, which is shown in bottom-left of Figure 5.

In our experiments, we chose the low-speed shaft of the two-stage gearbox as the object for detecting the gears' fault patterns under different working conditions. The gears' fault patterns, as shown in Figure 6 [34], consisted of a tooth fracture, pitting, and wear. We set the motor at three speeds—900 rev/min, 1800 rev/min, and 2700 rev/min—by controlling the frequency and adjusting the magnetic brake using two load conditions—0 Nm and 13.5 Nm—via tension control to simulate different working conditions. Regarding those conditions, we believe that the acoustic signal that we obtained can be viewed as only containing the gears due to the general assumption that the interference of other parts of the gearbox, such as the bearing and shaft via vibration, was minor. All the acoustic signals of the gears were recorded as an audio file that was 60 s long for further analysis.

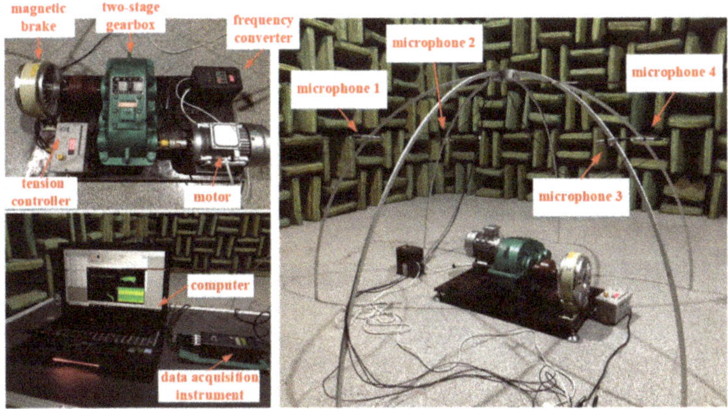

Figure 5. Experimental system in a semi-anechoic chamber.

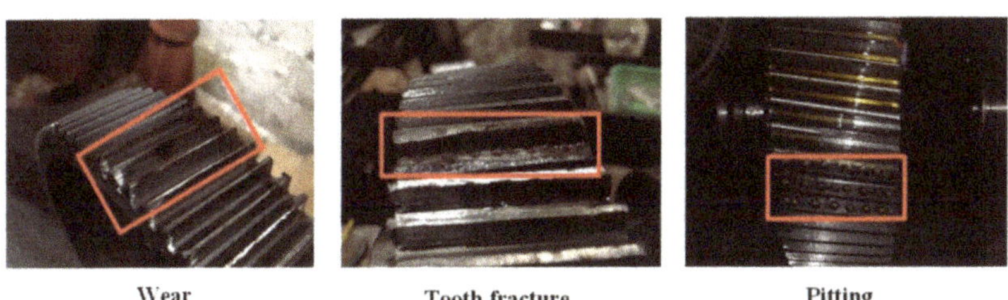

Figure 6. Fault pattern of the gears.

3.2. Dataset

In order to verify that the method we proposed is effective and feasible under different working conditions, we built two different datasets, A and B, that represented the two load conditions of 0 Nm and 13.5 Nm, respectively. Each dataset contained an audio file of four types of gears (one normal type and three fault types) at three different speeds, and each type was recorded to produce a four-channel audio file that was 60 s long. Each file was divided into 1-s samples with no overlap because 1-s samples are an optimal size for analysis based on empirical experiments in audio processing tasks. Then, each dataset contained 21,600 samples with 18,000 used as training samples and 3600 used as testing samples.

3.3. Implementation Detail

We used the cross-entropy as a loss function to train the attention-based multi-scale CNN model for multi-fault type classification under different working conditions. We applied the Adam algorithm in the training step to optimize the model, where learning rate was set to 0.003. In addition, we used rectified linear units as activation functions for each layer. When training, the dropout layer, which was followed by the fully connected layer, was employed to prevent overfitting with a 0.5 dropout rate. Finally, the early stopping approach and the no-improvement-in-10-epochs strategy was adopted to identify the number of epochs via the testing set.

4. Experimental Result and Analysis

4.1. Time and Frequency Analysis in Different Working Conditions

The time and frequency domain information of the acoustic signals that we obtained from four types of gears at three different speeds under two load conditions are shown in Figure 7. The subplots (a), (b), and (c) represent the time and frequency domain signal of gears under two load conditions at 900 rev/min, 1800 rev/min, and 2700 rev/min, respectively. The left panel of each subplot shows the four types of gears in a no-load condition and the right panel shows the same type in a 13.5-Nm-load condition.

Comparing the time and frequency domain signals from the subplots, we can see that the signal amplitudes of the gears were different in the time and frequency domains under variable speed conditions. To be specific, with the increasing of the operation speed, the maximum amplitude of the majority types of gears in the time domain also increased under the same load condition. Meanwhile, from the frequency domain signal, we found that the magnitude of the frequency amplitude for the same types of fault signals under variable speed conditions was different, but the distribution of the amplitude was not affected by the varying speed. For example, in the frequency domain, the normal gears had a higher amplitude at 2700 rev/min speed condition than the same type of gears at 900 rev/min and 1800 rev/min under the no-load condition, but the distribution of the amplitude, which was concentrated in the range of 0–500 Hz and 1000–1200 Hz, was consistent in the three different speed conditions. Based on the description above, we could infer that the variable speed caused the amplitude modulation phenomenon of the acoustic signal, but did not influence the frequency modulation.

Furthermore, comparing the time and frequency domain signals under the no-load condition and load condition at 900 rev/min, 1800 rev/min, and 2700 rev/min, we observed that the acoustic signals for four types of gears seemed to be obviously different in the two load conditions. From the time-domain signal, we saw that the gears with a wear type and the gears with pitting under the no-load conditions had a higher amplitude range than the same fault type under the 13.5-Nm-load condition at 900 rev/min, but it was the opposite at 1800 rev/min. Moreover, the waveform of each type of gear in the time domain were different, which means the temporal structure and energy modulation patterns of the signal under the two load conditions were diverse. Meanwhile, according to the magnitude and distribution of the frequency amplitude in the ranges of 0–500 Hz and 1000–1200 Hz, we found that the frequency signal of the four types of gears under the two loads were also different, especially for the normal and pitting types of gears. As for the normal gears at 900 rev/min, the frequency component under the no-load condition was around 0–500 Hz and 1000–1200 Hz, while the frequency component under the load condition was around 1000–1200 Hz. Equally, the gears with the pitting type had the same phenomenon at 1800 rev/min. This indicates that the acoustic signals of gears under different load conditions were not only affected by the amplitude modulation, but also by the frequency modulation. In addition, another interesting phenomenon we observed was that the type of gear signal that we obtained under the no-load condition at one speed may follow the same magnitude and distribution of frequency amplitude under the load condition at another speed. For example, the normal gears under the no-load condition at 900 rev/min had a similar amplitude and distribution of frequency to the same gear type under the 13.5-Nm-load condition at 1800 rev/min. This means that the gear fault diagnosis under variable load conditions is more complex and difficult than that of a variable speed condition. Therefore, we proposed an attention-based multi-scale CNN model for gear fault diagnosis under variable load conditions.

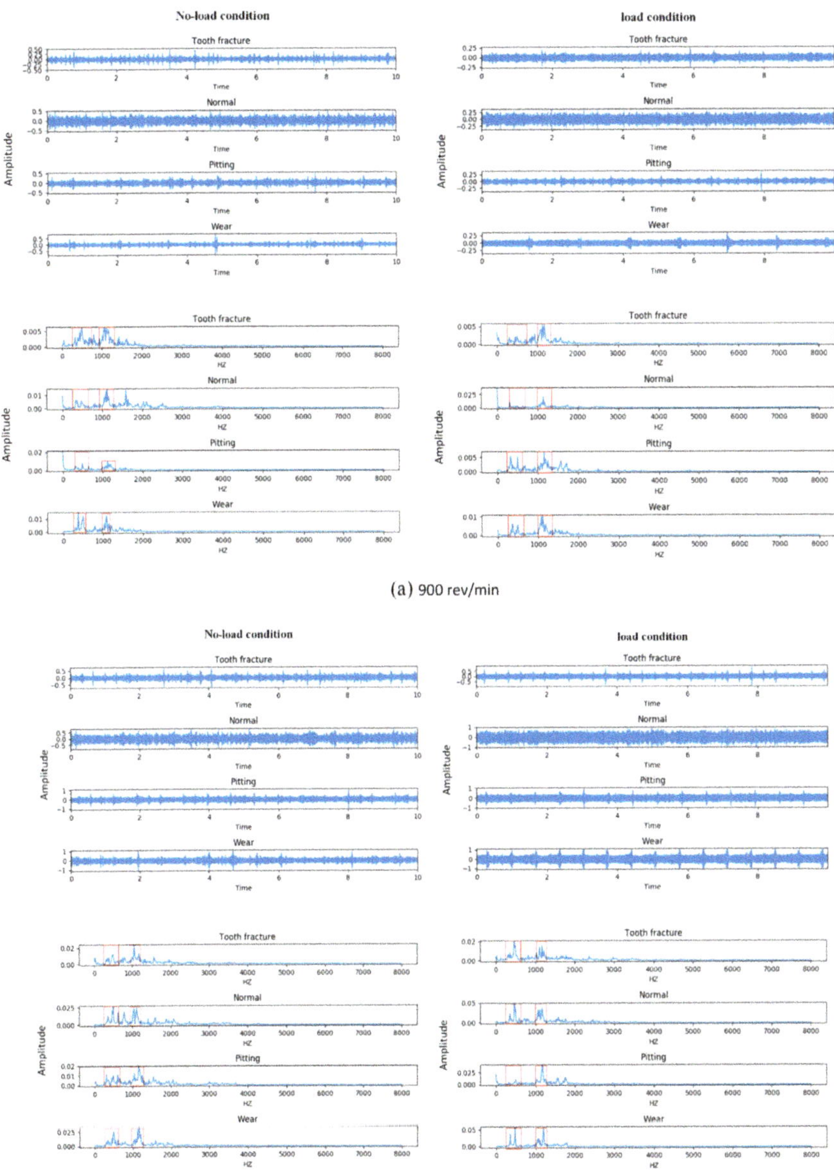

(a) 900 rev/min

(b) 1800 rev/min

Figure 7. *Cont.*

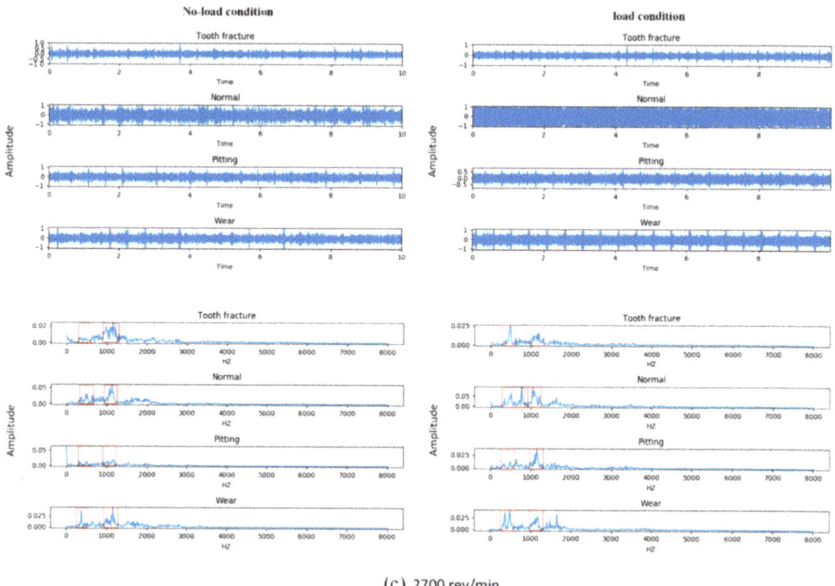

(c) 2700 rev/min

Figure 7. Four different types of gear signals in the time and frequency domains under two load conditions: (**a**) signal at 900 rev/min, (**b**) signal at 1800 rev/min, and (**c**) signal at 2700 rev/min.

4.2. Effectiveness of the Multi-Scale Convolution Operation

In order to verify the hypothesis that the multi-scale convolutional learning structure is superior to the single-scale convolutional structure in for gear fault diagnosis tasks, we first compared the performance of our multi-scale convolutional neural network with that of a low-scale convolutional neural network, mid-scale convolutional neural network, and high-scale convolutional neural network with no attention mechanism. These three models remained at only one scale each, which is shown in Figure 2. The input and the rest of the network structure is the same as in Figure 4 for fair comparison. The two evaluation methods were designed to evaluate the performance of models under different load conditions:

In evaluation A, we used the training samples of dataset A to train these CNN models and test it on the test samples of dataset B.

In evaluation B, we used the training samples of dataset B to train these CNN models and test it on the test samples of dataset A.

The classification accuracy was used as the evaluation criterion. The equation is defined as:

$$ACC = \frac{N_1}{N_2},\tag{10}$$

where N_1 represents the number of test samples that were predicted properly and N_2 represents the total number of test samples.

The test results of the multi-scale network and single-scale network in the two evaluation methods are shown in Table 2. From the result, we saw that the recognition accuracy of the multi-convolutional neural network reached 81.1% and 71.0% for the two evaluation methods, respectively. This was an improvement of 2.5%, 2.3%, and 1.6% compared with the low-scale, mid-scale, and high-scale networks for evaluation A, respectively. Also, the multi-scale convolutional neural network achieved the best accuracy for evaluation B. The improvement from the single-scale model to the multi-scale model proved that the model, which contained different scales, was capable of learning more discriminative features from the waveforms.

Table 2. Prediction accuracy of multi-scale CNN and single CNN model.

CNN Model without Attention	Accuracy (%)	
	Evaluation A Train on A Test on B	Evaluation B Train on B Test on A
Multi-scale	81.1	71.0
Low-scale	78.6	70.8
Mid-scale	78.8	64.4
High-scale	79.5	62.3

To further understand how the multi-scale convolution operations help to improve the performance of fault pattern recognition, we visualized the frequency magnitude of the response of the multi-scale feature maps Conv1 of the model in Figure 8. As indicated in this figure, the 32 filters were viewed as band-pass filters to learn a particular frequency area, and each filter was sorted based on their center frequencies. From the left picture of Figure 8, we observed that the curve of the center frequency almost matched the sound feature of the human auditory system. This means that the low-scale structure was able to extract features from all frequency areas. Conversely, the high-scale structure, which is shown in the right of Figure 8, was located in the low-frequency area with fine-grained filters. It shows that the high-scale structure tended to concentrate on low-frequency components and ignore high-frequency information. Moreover, the mid-scale performed between the low-scale and high-scale networks. In general, a model with a narrow kernel size can cover all frequency areas but obtains a low-frequency resolution, and a model with a wide kernel size does not have sufficient filters in the high-frequency range but gives good frequency resolution. It indicates that learning structures of different scales can extract discrepant features based on what they can efficiently represent. This may explain the result that we present in Table 2 where the multi-scale models obtain a better performance than the single-scale models.

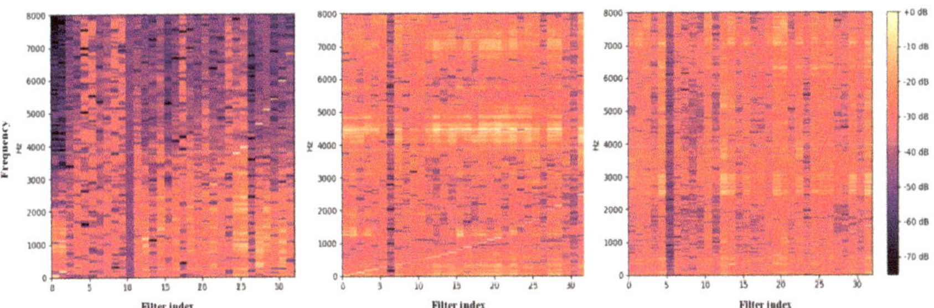

Figure 8. Frequency magnitude response of the multi-scale convolutional filters of the first layer. The filters are sorted by their center frequency. Left shows the frequency response of the low-scale network, middle shows the frequency response of the mid-scale network, and right shows the frequency response of the high-scale network.

4.3. Comparison of a Standard CNN Model and Attention Models

To verify the effectiveness of the attention mechanism, we applied the attention mechanism to a multi-scale convolutional neural network and each single-scale network for a fair comparison. The performances of the models are summarized in Table 3.

Table 3. Prediction accuracy when applying an attention mechanism.

CNN Model With Attention	Accuracy (%)	
	Evaluation A Train on A Test on B	Evaluation B Train on B Test on A
Multi-scale	**93.3**	**82.8**
Low-scale	86.7	76.2
Mid-scale	87.6	75.1
High-scale	86.3	76.4

In Table 3, it is shown that the performances of each model presented a significant improvement when using an attention mechanism compared with the standard model, which did not use an attention mechanism, especially in the case of the multi-scale convolutional model. The attention-based multi-scale CNN model achieved a 93.3% accuracy in evaluation A and 82.8% in evaluation B. This was 12.2% and 11.8% higher than the accuracy of the standard multi-scale CNN model. Meanwhile, by using an attention mechanism, the improvement range of recognition performance of the single-scale model was from 6.8% to 8.8% for evaluation A and from 5.4% to 14.1% for evaluation B. The test results indicate that the attention mechanism was effective at gear fault diagnosis under different load conditions.

To provide a better understanding how a temporal attention mechanism helped to improve the performance of the multi-scale CNN model in the gear fault diagnosis task, we visualized the results of randomly selected filters from the multi-scale pool layer and temporal attention output for four different types of gear input signals under the two load conditions. Figure 9 shows the visualization result of the attention-based multi-scale CNN model for evaluation A. The first and third rows represent the waveform under the no-load condition and 13.5-Nm-load conditions, respectively. The second and fourth rows represent the attention output corresponding to the waveform of the two load conditions, respectively. From this figure, we found that the temporal attention mechanism, which was based on the multi-scale CNN model, was able to adaptively focus on the relevant temporal information from the different waveforms of the two load conditions while reducing the impact of the data distribution variation. Furthermore, from the attention output, we found that the attention weights of the four gear types were different for different time stamps. For example, the tooth fracture condition in gears had three high-weighted areas at time stamps in the ranges of #30–#50, #100–#200, and #430–#500 frames, while the high-weighted areas of the gears with pitting were in the ranges of #180–#230, #300–#320, and #450–#480. This may indicate that the learned temporal attention was able to detect the essential feature information required to distinguish these four types of gears under the two load conditions.

Figure 10 shows the visualization results of the attention based multi-scale CNN model for evaluation B. From this figure, we found two phenomena. First, the learned temporal attention was also able to locate the meaningful temporal parts from the different waveforms of the two load conditions for evaluation B. Meanwhile, the four types of gears could be easily distinguished according to the distribution of the time stamps. This visualization result was similar to that obtained for evaluation A. Second, the high-attention-weights area of the four gear types for evaluation B were not consistent with the high-attention-weights area for evaluation A. This may be due to the fact that the temporal attention mechanism, which was trained on different datasets, could generate different attention weights to focus on different temporal parts.

The above visualization results may explain why the temporal attention mechanism was effective at gear fault diagnosis under different load conditions.

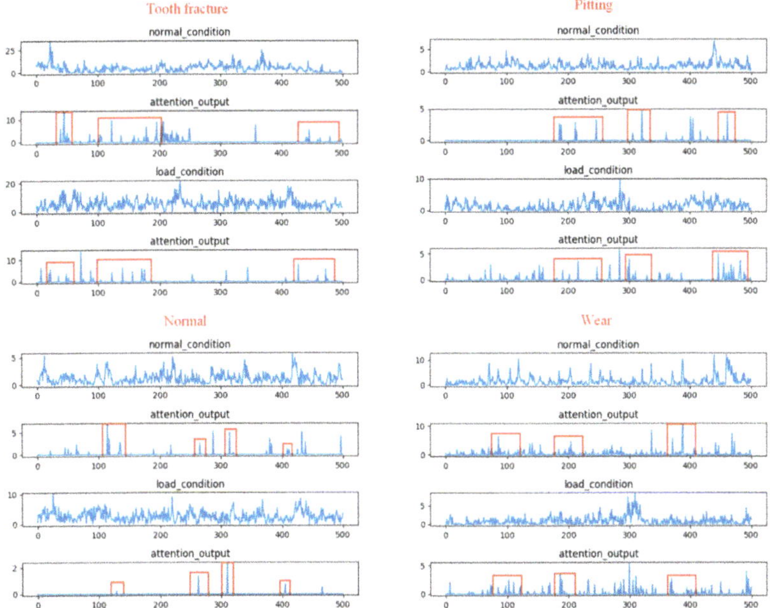

Figure 9. Visualization results of randomly selected filters from the multi-scale pool layer and temporal attention output for four different types of gear input signals under the two load conditions corresponding to the attention based multi-scale CNN model for evaluation A.

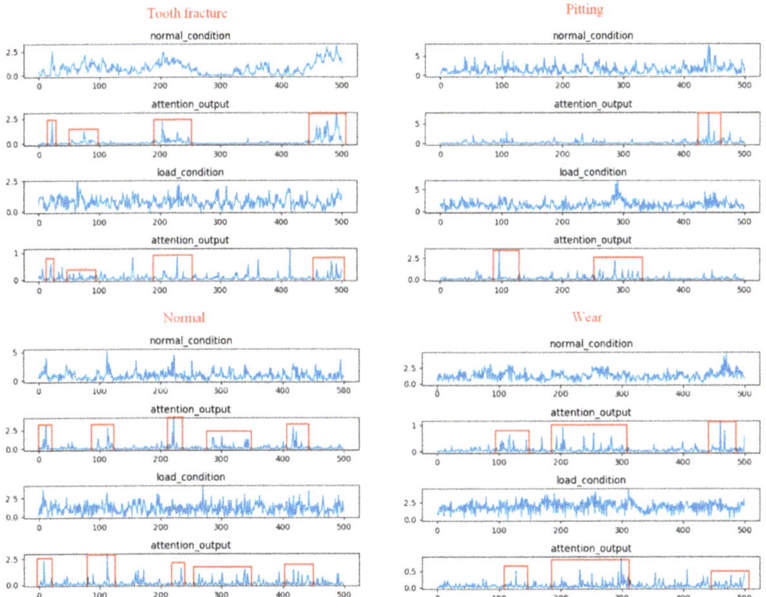

Figure 10. Visualization results of randomly selected filters from the multi-scale pool layer and temporal attention output for four different types of raw gear signals under the two load conditions corresponding to the attention based multi-scale CNN model for evaluation B.

4.4. Comparison between the Attention-Based Multi-Scale CNN Model and Other Methods

In this section, we compare our model with traditional methods that combine one of the most powerful acoustic features, namely MFCC (Mel-frequency cepstral coefficients), with a convolutional neural network and several classic fault diagnosis methods, which have been successfully used in fault pattern recognition tasks based on time, frequency, manual features, and machine learning algorithms, to analyze the performance of our model. Meanwhile, the end-to-end model, which is based on time and frequency domain signals that we proposed in a previous work, was also used for further comparison.

We adopted the commonly used parameter [35,36] to construct the MFCC feature, the first derivative of the MFCC feature (MFCC-delta) and the second derivative of the MFCC feature (MFCC-delta-delta) as matrix features. Then, we fed it into the convolutional neural network, whose structure is the same as module (b) of the multi-scale CNN model in Figure 3 to provide a fair comparison. The procedure of the classic fault diagnosis methods can be divided into two parts: manual feature extraction and fault identification. The manual features that we extracted include the root-mean-square error (RMS) [37], spectral centroid [38], and Mel spectrogram in the log domain [39], which represent the popularity acoustic features in the time, frequency, and time–frequency domains, respectively. We fused these manual features at the feature level for improving the representation of the fault information. Then, we fed it into several classic machine learning classifiers [40–45] that are widely used in fault diagnosis tasks for comparison. The end-to-end model that we proposed in a previous work [34] used time and frequency signals as raw input signals to detect the gear fault patterns. The test results of those methods on two datasets are shown in Table 4.

From the results, we found that the accuracy of the MFCC-delta CNN and the MFCC-delta-delta CNN had similar performances, which were better than the MFCC CNN, but the accuracy of the best traditional method, namely the MFCC-delta CNN, was 10% lower than our attention-based multi-scale CNN for evaluation A. Moreover, the prediction accuracy of all methods declined for evaluation B, but those traditional methods gave worse performances compared with our model. The above conclusions indicate that our attention-based multi-scale CNN structure could learn more discriminative features than traditional manual features by combining attention-based multi-scale information.

Furthermore, the recognition accuracy of our attention multi-scale CNN model reached 93.3% for evaluation A and 82.8% for evaluation B. This was 5.4 % and 6.0% higher than the accuracy of the best classic diagnosis method, namely manual features + k-nearest neighbor (KNN).

In addition, our attention multi-scale CNN model showed an improvement of 3.6% when compared with the end-to-end CNN model for evaluation A. Furthermore, our model achieved at least a 1.7% improvement over the end-to-end CNN model for evaluation B. The improvement proved that the multi-scale CNN model based on the attention mechanism was able to adaptively learn more efficient frequency representations using attention based multi-scale band-pass filters from raw waveforms without frequency input signals.

In Figure 11, we provide the confusion matrix in order to further analyze the performance of our proposed method regarding the two evaluation methods. From the confusion matrix for evaluation A, we observed that most fault patterns under different working conditions could obtain a high classification accuracy, except for gears with pitting types at 1800 rev/min and 2700 rev/min. The two categories appeared to be easily misclassified due to the signal of pitting types at 1800 rev/min under the no load condition being similar to the same type at 2700 rev/min under the 13.5-Nm-load condition, as caused by the amplitude modulation and frequency modulation, the phenomena of which is discussed in Section 4.1. Furthermore, from the confusion matrix for evaluation B, we noticed that the recognition accuracy declined for some classes, especially for gears with pitting types at 1800 rev/min. According to Figure 10 in Section 4.3, we found that the temporal parts, where the attention mechanism was located, were totally different for pitting-type gears, which means that the discriminative feature of pitting-type gears that the model learned were different at 1800 rev/min under the two load conditions. This may explain why the pitting-type gears displayed a lower performance for evaluation B. As for the rest of the classes, we suspect that the easy misclassification was due to some extra information between the acoustic signals and fault types existing in the load working condition but not in the no-load working condition such that the training the model in the load condition and testing it in no-load condition could be viewed as diagnosis fault patterns with noise, which led to some degree of overfitting in our model. In general, the attention-based multi-scale CNN model that we propose still had better generalization capabilities under variable load conditions compared with the other methods.

Table 4. Prediction accuracy comparison of our attention-based multi-scale CNN model and other methods. GBDT: Gradient Boosting Decision Tree, KNN: k-nearest neighbor, SVM: support vector machine.

Method	Feature	Recognition Model	Accuracy (%)	
			Evaluation A Train on A Test on B	**Evaluation B** Train on B Test on A
Attention-based multi-scale CNN	**Time signal**	**Multi-scale CNN**	**93.3**	**82.8**
MFCC CNN	MFCC	(b) module	78.7	59.4
MFCC-delta CNN	MFCC-delta	(b) module	83.3	58.8
MFCC-delta-delta CNN	delta-Deltas	(b) module	82.6	57.4
End-to-end stacked CNN	Time–frequency signal	–	89.7	81.1
Multiple feature + KNN	Multiple feature	KNN	87.9	76.8
Multiple feature + SVM	Multiple feature	SVM	83.2	66.7
Multiple feature + GBDT	Multiple feature	GBDT	71.5	48.4

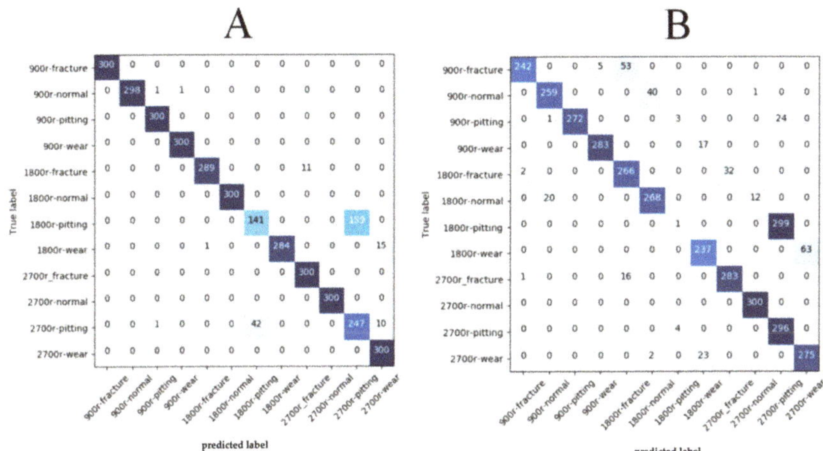

Figure 11. Confusion matrixes for the proposed attention-based multi-scale CNN model. The left matrix shows the statistics for evaluation (**A**), while the right matrix shows the statistics for evaluation (**B**).

Finally, to better show the performance of the attention-based multi-scale CNN model for the two evaluation methods, we visualized the prediction result of the model by using a t-SNE (t-distributed stochastic neighbor embedding) algorithm (Figure 12). The t-SNE algorithm was operated on the output matrix of the last fully connected layer to reduce the dimensionality to conveniently show the classification result in three-dimensional space. From these visual results of three-dimensional space, we observed that most features clustered successfully around the two evaluation methods, which also proved that our attention-based multi-scale model achieved a better performance at acoustic signal-based gear fault diagnosis under different working conditions.

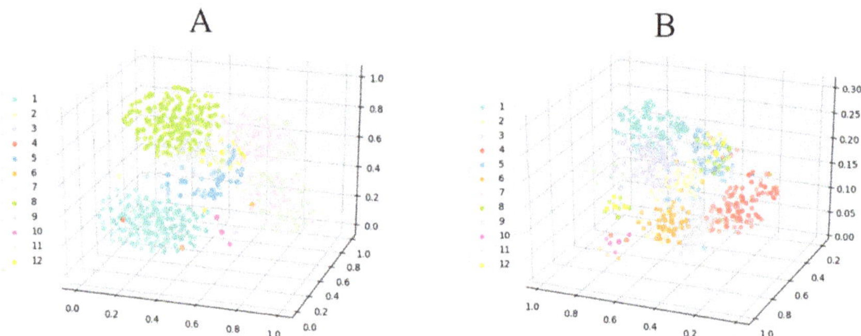

Figure 12. Feature visualization via t-SNE (t-distributed stochastic neighbor embedding). Left shows the feature representations for the last fully connected layer of attention-based multi-scale CNN model for evaluation (**A**). Right shows the feature representations for evaluation (**B**).

5. Conclusions

In this paper, a novel ABD method was proposed for gear fault diagnosis under different working conditions based on a multi-scale convolutional learning structure and attention mechanism. By using a multi-scale convolutional learning structure, our model was able to automatically mine more efficient feature representations from raw acoustic signals. It achieved better performance than the single-scale models. Based on the multi-scale convolutional learning structure, a novel attention mechanism, which operated on the convolutional layer, was applied to adaptively extract relevant fault information and reduce the data distribution variation under different working conditions. The experimental result for the two evaluation methods showed that the accuracy of our model reached 93.3% and 82.8%, respectively. All the performance metrics were higher than those of the standard CNN model, end-to-end CNN model based on time and frequency domain signals, and other traditional fault diagnosis methods with manual features. This indicates that our model was more effective at acoustic-based gear fault diagnosis under different working conditions. Furthermore, we analyzed the discrimination of different scale convolutional learning structures using feature representations and visualized the attention output to provide insight into the reason for the performance improvement in the gear fault diagnosis task. In future, the attention-based mechanism method can be further developed in the ABD field. We will continue to explore the effectiveness of the attention-based method for bearing fault diagnosis, the multi-fault diagnosis of gears, and the coupling fault diagnosis of gearboxes. Meanwhile, we will extend our attention-based method into other fault diagnosis fields that are out of the controlled environment, such as fault diagnosis in normal environmental conditions and strong environmental noise conditions.

Author Contributions: Y.Y. contributed to formal analysis, methodology, software and writing-original draft. S.Z. contributed to validation and writing-review&editing; S.Y. contributed to funding acquisition, supervision and writing—review&editing; G.G. contributed to data curation and formal analysis. All authors have read and agreed to the published version of the manuscript.

Funding: We gratefully acknowledge the financial support of National Natural Science Foundation of China (NSFC) under grant no. 51275325.

Acknowledgments: We gratefully acknowledge the support of NVIDIA Corporation with the donation of the Titan X Pascal GPU used for this research and the support of National Institute of Measurement and Testing Technology with the provision of the semi-anechoic laboratory.

Conflicts of Interest: The authors declare no conflict of interest.

References

1. Zhang, X.; Wang, L.; Miao, Q. Fault diagnosis techniques for planetary gearboxes under variable conditions: A review. In Proceedings of the 2016 Prognostics and System Health Management Conference (PHM-Chengdu), Chengdu, China, 19–21 October 2016.
2. Lei, Y. Research Advances of Fault Diagnosis Technique for Planetary Gearboxes. *J. Mech. Eng.* **2011**, *19*, 59. [CrossRef]
3. Feng, Z.; Zuo, M.J. Fault diagnosis of planetary gearboxes via torsional vibration signal analysis. *Mech. Syst. Signal Process.* **2013**, *36*, 401–421. [CrossRef]
4. Zou, J.; Shen, Y. Review of Gearbox Fault Diagnosis Techniques under Variable Conditions. *J. Mech. Transm.* **2012**, *8*, 124–127.
5. Yang, Y.; Wang, H.; Cheng, J.; Zhang, K. A fault diagnosis approach for roller bearing based on VPMCD under variable speed condition. *Measurement* **2013**, *46*, 2306–2312. [CrossRef]
6. Wang, Z.; Jia, L.; Kou, L.; Qin, Y. Spectral Kurtosis Entropy and Weighted SaE-ELM for Bogie Fault Diagnosis under Variable Conditions. *Sensors* **2018**, *18*, 1705. [CrossRef]
7. Lei, Y.; Lin, J.; Zuo, M.J.; He, Z. Condition monitoring and fault diagnosis of planetary gearboxes: A review. *Measurement* **2014**, *48*, 292–305. [CrossRef]
8. Liu, H.; Zhang, J.; Cheng, Y.; Lu, C. Fault diagnosis of gearbox using empirical mode decomposition and multi-fractal detrended cross-correlation analysis. *J. Sound Vib.* **2016**, *385*, 350–371. [CrossRef]
9. Chen, J.; Zhou, D.; Lyu, C.; Lu, C. An integrated method based on CEEMD-SampEn and the correlation analysis algorithm for the fault diagnosis of a gearbox under different working conditions. *Mech. Syst. Signal Process.* **2017**, *113*, 102–111. [CrossRef]
10. Xing, Z.; Qu, J.; Chai, Y.; Tang, Q.; Zhou, Y. Gear fault diagnosis under variable conditions with intrinsic time-scale decomposition-singular value decomposition and support vector machine. *J. Mech. Sci. Technol.* **2016**, *31*, 545–553. [CrossRef]
11. Zhang, X.; Jiang, H. LCD denoise and the vector mutual information method in the application of the gear fault diagnosis under different working conditions. In Proceedings of the IOP Conference Series: Materials Science and Engineering, Kuala Lumper, Malaysia, 15–16 December 2018.
12. Chen, X.; Feng, Z. Time-Frequency Analysis of Torsional Vibration Signals in Resonance Region for Planetary Gearbox Fault Diagnosis Under Variable Speed Conditions. *IEEE Access* **2017**, *5*, 21918–21926. [CrossRef]
13. Wang, X.; Makis, V.; Yang, M. A wavelet approach to fault diagnosis of a gearbox under varying load conditions. *J. Sound Vib.* **2010**, *329*, 1570–1585. [CrossRef]
14. Lu, W.; Jiang, W. Diagnosing Rolling Bearing Faults Using Spatial Distribution Features of Sound Field. *J. Mech. Eng.* **2012**, *48*, 68–72. [CrossRef]
15. Hou, J.J.; Jiang, W.K.; Lu, W.B. The application of NAH-based fault diagnosis method based on blocking feature extraction in coherent fault conditions. *J. Vib. Eng.* **2011**, *24*, 555–561.
16. Hou, J.; Jiang, W.; Lu, W. Application of a near-field acoustic holography-based diagnosis technique in gearbox fault diagnosis. *J. Vib. Control* **2013**, *19*, 3–13. [CrossRef]
17. Lu, W.; Jiang, W.; Yuan, G.; Yan, L. Gearbox fault diagnosis based on spatial distribution features of sound field. *J. Sound Vib.* **2013**, *32*, 2593–2610. [CrossRef]
18. Glowacz, A. Fault Detection of Electric Impact Drills and Coffee Grinders Using Acoustic Signals. *Sensors* **2019**, *19*, 269. [CrossRef]
19. Glowacz, A. Acoustic fault analysis of three commutator motors. *Mech. Syst. Signal Process.* **2019**, *133*, 106226. [CrossRef]
20. Zhang, H.; Zhang, S.; He, Q.; Kong, F. The Doppler Effect based acoustic source separation for a wayside train bearing monitoring system. *J. Sound Vib.* **2016**, *361*, 307–329. [CrossRef]
21. Zhang, D.; Entezami, M.; Stewart, E.; Roberts, C.; Yu, D. Adaptive fault feature extraction from wayside acoustic signals from train bearings. *J. Sound Vib.* **2018**, *425*, 221–238. [CrossRef]
22. Zhang, B.; Li, W.; Hao, J.; Li, X.L.; Zhang, M. Adversarial adaptive 1-D convolutional neural networks for bearing fault diagnosis under varying working condition. *arXiv* **2018**. Available online: https://arxiv.org/abs/1805.00778 (accessed on 9 May 2018).
23. Sankaran, B.; Mi, H.; Al-Onaizan, Y.; Ittycheriah, A. Temporal Attention Model for Neural Machine Translation. *arXiv* **2016**. Available online: https://arxiv.org/abs/1608.02927 (accessed on 9 August 2016).

24. Pappas, N.; Popescu-Belis, A. Multilingual Hierarchical Attention Networks for Document Classification. *arXiv* **2017**. Available online: https://arxiv.org/abs/1707.00896 (accessed on 15 September 2017).

25. Chorowski, J.K.; Bahdanau, D.; Serdyuk, D.; Cho, K.; Bengio, Y. Attention-Based Models for Speech Recognition. *Comput. Sci.* **2015**, *10*, 429–439.

26. Chan, W.; Jaitly, N.; Le, Q.; Vinyals, O. Listen, attend and spell: A neural network for large vocabulary conversational speech recognition. In Proceedings of the 2016 IEEE International Conference on Acoustics, Speech and Signal Processing (ICASSP), Shanghai, China, 20–25 March 2016.

27. Chiu, C.C.; Sainath, T.N.; Wu, Y.; Prabhavalkar, R.; Nguyen, P.; Chen, Z.; Kannan, A.; Weiss, R.J.; Rao, K.; Gonina, E.; et al. State-of-the-art Speech Recognition with Sequence-to-Sequence Models. In Proceedings of the 2018 IEEE International Conference on Acoustics, Speech and Signal Processing (ICASSP), Calgary, AB, Canada, 15–20 April 2018; pp. 4774–4778.

28. Zhang, Z.; Xu, S.; Zhang, S.; Qiao, T.; Cao, S. Learning Attentive Representations for Environmental Sound Classification. *IEEE Access* **2019**, *7*, 130327–130339. [CrossRef]

29. Li, X.; Chebiyyam, V.; Kirchhoff, K. Multi-stream Network With Temporal Attention For Environmental Sound Classification. *arXiv*. 2019. Available online: https://arxiv.org/abs/1901.08608 (accessed on 24 January 2019).

30. Zhu, B.; Wang, C.; Liu, F.; Lei, J.; Huang, Z.; Peng, Y.; Li, F. Learning Environmental Sounds with Multi-scale Convolutional Neural Network. In Proceedings of the 2018 International Joint Conference on Neural Networks (IJCNN), Rio de Janeiro, Brazil, 8–13 July 2018.

31. Dai, W.; Dai, C.; Qu, S.; Li, J.; Das, S. Very Deep Convolutional Neural Networks for Raw Waveforms. In Proceedings of the IEEE International Conference on Acoustics, Speech and Signal Processing (ICASSP), New Orleans, LA, USA, 5–9 March 2017; pp. 421–425.

32. Szegedy, C.; Liu, W.; Jia, Y.; Sermanet, P.; Reed, S.; Anguelov, D.; Erhan, D.; Vanhoucke, V.; Rabinovich, A. Going deeper with convolutions. In Proceedings of the IEEE Conference on Computer Vision and Pattern Recognition (CVPR), Boston, MA, USA, 7–12 June 2015.

33. Zhu, Z.; Engel, J.H.; Hannun, A. Learning Multiscale Features Directly from Waveforms. *arXiv* **2016**. Available online: https://arxiv.org/abs/1603.09509 (accessed on 5 April 2016).

34. Yao, Y.; Wang, H.; Li, S.; Liu, Z.; Gui, G.; Dan, Y.; Hu, J. End-To-End Convolutional Neural Network Model for Gear Fault Diagnosis Based on Sound Signals. *Appl. Sci.* **2018**, *8*, 1584. [CrossRef]

35. Cotton, C.V.; Ellis, D.P. Spectral vs. spectro-temporal features for acoustic event detection. In Proceedings of the IEEE Workshop on Applications of Signal Processing to Audio and Acoustics (WASPAA), New Paltz, NY, USA, 16–19 October 2011.

36. Salamon, J.; Jacoby, C.; Bello, J.P. A Dataset and Taxonomy for Urban Sound Research. In Proceedings of the 22nd ACM International Conference on Multimedia, Orlando, FL, USA, 3–7 November 2014.

37. Ganchev, T.; Mporas, I.; Fakotakis, N. Audio features selection for automatic height estimation from speech. In Proceedings of the Hellenic Conference on Artificial Intelligence, Athens, Greece, 4–7 May 2010; Springer: Berlin, Germany; pp. 81–90.

38. Paliwal, K.K. Spectral subband centroid features for speech recognition. In Proceedings of the IEEE International Conference on Acoustics, Speech and Signal Processing, ICASSP'98 (Cat. No. 98CH36181), Seattle, WA, USA, 15 May 1998.

39. Tjandra, A.; Sakti, S.; Neubig, G.; Toda, T.; Adriani, M.; Nakamura, S. Combination of two-dimensional cochleogram and spectrogram features for deep learning-based ASR. In Proceedings of the 2015 IEEE International Conference on Acoustics, Speech and Signal Processing (ICASSP), Brisbane, QLD, Australia, 19–24 April 2015.

40. Liu, Z.; Zuo, M.J.; Xu, H. Feature ranking for support vector machine classification and its application to machinery fault diagnosis. *J. Mech. Eng. Sci.* **2013**, *227*, 2077–2089. [CrossRef]

41. Li, C.; Sanchez, R.V.; Zurita, G.; Cerrada, M.; Cabrera, D.; Vásquez, R.E. Multimodal deep support vector classification with homologous features and its application to gearbox fault diagnosis. *Neurocomputing* **2015**, *168*, 119–127. [CrossRef]

42. Praveenkumar, T.; Saimurugan, M.; Ramachandran, K.I. Comparison of vibration, sound and motor current signature analysis for detection of gear box faults. *Int. J. Prognostics Health Manag.* **2017**, *8*, 1–10.

43. Liu, L.; Liang, X.; Zuo, M.J. A dependence-based feature vector and its application on planetary gearbox fault classification. *J. Sound Vib.* **2018**, *431*, 192–211. [CrossRef]

44. Vanraj Dhami, S.S.; Pabla, B.S. Hybrid data fusion approach for fault diagnosis of fixed-axis gearbox. *Struct. Health Monit.* **2018**, *17*, 936–945. [CrossRef]
45. Kou, L.; Qin, Y.; Zhao, X.; Fu, Y. Integrating synthetic minority oversampling and gradient boosting decision tree for bogie fault diagnosis in rail vehicles. *J. Rail Rapid Transit* **2019**, *233*, 312–325. [CrossRef]

MDPI
St. Alban-Anlage 66
4052 Basel
Switzerland
Tel. +41 61 683 77 34
Fax +41 61 302 89 18
www.mdpi.com

Sensors Editorial Office
E-mail: sensors@mdpi.com
www.mdpi.com/journal/sensors